D0872380

THE CULTURAL LIVES OF WHALES AND DOLPHINS

THE CULTURAL LIVES OF **WHALES AND DOLPHINS**

HAL WHITEHEAD AND LUKE RENDELL

THE UNIVERSITY OF CHICAGO PRESS
Chicago and London

Hal Whitehead is a University Research Professor in the Department of Biology at Dalhousie University in Halifax, Nova Scotia. Supported by the Marine Alliance for Science and Technology, **Luke Rendell** is a lecturer in biology at the Sea Mammal Research Unit and the Centre for Social Learning and Cognitive Evolution of the University of St. Andrews, Scotland.

The University of Chicago Press, Chicago 60637
The University of Chicago Press, Ltd., London

Printed in the United States of America
24 23 22 21 20 19 18 17 16 15 1 2 3 4 5
ISBN-13: 978-0-226-89531-4 (cloth)
ISBN-13: 978-0-226-18742-6 (e-book)
DOI: 10.7208/chicago/ 9780226187426.001.0001

Library of Congress Cataloging-in-Publication Data
Whitehead, Hal, author.
The cultural lives of whales and dolphins / Hal Whitehead and Luke Rendell.
pages ; cm
Includes bibliographical references and index.
ISBN 978-0-226-89531-4 (cloth : alk. paper) —
ISBN 0-226-89531-9 (cloth : alk. paper) —
ISBN 978-0-226-18742-6 (e-book)
1. Whales. 2. Dolphins. 3. Social behavior in animals.
4. Animal communication. I. Rendell, Luke, 1973- author. II. Title.
QL737.C4W47 2015
599.5—dc23
2014020610

♾ This paper meets the requirements of ANSI/NISO Z39.48–1992 (Permanence of Paper).

Dedicated to the memory of

CHRIS RENDELL *and* FRANKIE WHITEHEAD

CONTENTS

1 | CULTURE IN THE OCEAN?

Ocean of Song

We love wilderness, the parts of the earth where humans have little impact. So much of the planet is eroded, polluted, and dominated by people. Well, not people directly. It is rarely the mere physical presence of large numbers of humans that degrades—it is, rather, what we do, as well as our products, our methods of exploiting the land, the plants, and the animals, the effluents of our industries, and the things that we build. All these are the results of human culture, the body of knowledge, skills, customs, and materials that each generation inherits and builds on and that surround us every moment of our lives. We are born with the genetic template of *Homo sapiens*, but we cannot become fully human without what we learn from each other. Human culture accumulates, so the good can become very, very good—like the routine treatment of medical conditions lethal not a century ago—and the bad, such as our pollution of the earth and its atmosphere, can get worse. This feature of our societies is a large part of what makes humans unique. The effects of our cultures are very nearly omnipresent, affecting the entire earth. The one major part of our planet's surface where humans and our cultures are least apparent is the deep ocean.

So we love to sail the deep ocean. Unless crossing a shipping lane, a fishing ground, or a garbage-strewn central-ocean gyre, we see few signs of humans outside our twelve-meter sailing boat. Out here, it would be easy to believe we have managed to escape the mess humanity has made of the earth. In reality, we have not. There are far fewer turtles and sharks and whales than even a hundred years ago, before humans learned such effective ways of killing them. The deep-ocean waters are more polluted and acidic than they used to be. But it feels like wilderness. We do not directly see the lack of ocean wildlife—or the pollutants.

Far out in the ocean, we have escaped the vast dominance of human cultural impact, although to make this escape we have to use the seafaring knowledge and technology that humans have built up over many generations. This accumulation began before 5,000 B.C., when the earliest known depictions of sailing boats appeared (plate 1).[1] Fishers in developing countries use simple sailing boats, basically logs with some piece of material for the wind to catch, which have not changed much for millennia. But during the late Middle Ages sailing ships became some of the most technologically advanced elements of human culture, and human mobility took a great stride forward. The yachts we sail for our research, with their fiberglass hulls, stainless-steel fittings, and Dacron sails, are technological descendants of those ships (plate 2). They are products of a system of cumulative cultural evolution that allow humans to cross oceans reasonably reliably, a remarkable achievement for a terrestrial mammal.

As we sail, every half hour we listen to the ocean through a hydrophone, an underwater microphone towed behind our boat on a hundred-meter cable. We hear waves, and sometimes dolphins. Quite often there is the deep rumble of ships. We can hear the ships farther than we can see them, and their rumble signifies that this is not the wilderness that it appears.

Despite this, on recent voyages through the Sargasso Sea in the western North Atlantic, we heard another type of sound more often than the whistles of dolphins or the throb of ships. Not one sound, but an extraordinary range of sounds, high sweeping squeals, low swoops, barking, and ratchets. All are part of the song of the humpback whale. In February 2008 we heard humpbacks at 45 percent of our half-hourly hydrophone listening stations over two thousand kilometers of ocean between Bermuda and Antigua. As we will explain later, we think that humpback song is a form of nonhuman culture. A humpback whale learns the song from other humpback whales and passes it on. Some liken it to human music, others to the songs of birds; it has elements of both. Within the frequencies that we can hear on our hydrophone and over thousands of kilometers of ocean, the culture of the humpback whale dominates the acoustic environment of the ocean, as it has for millions of years. Human cultural supremacy over the surface of the earth is recent and not quite complete. If we could have listened at lower frequencies, below the limits of the human ear, we would have heard rumbles and groans of other whales—the finback and the blue—their songs competing in the lowest frequency bands with the recent sound of ships. Could these be other nonhuman cultures?

This book is about the culture of the whales and dolphins, known collectively as the cetaceans. What is it? Does it even exist? If it does, why? What might it mean? It is also about our evolving understanding of nonhuman societies and, through them, what it means to be human. We are carried by rafts of insights hard won from the oceans by scientists all over the world.

"Culture Changes Everything"

To biologists like us, culture is a flow of information moving from animal to animal.[2] The movement of information is the basis of biology. Life happens and creatures evolve because information is transferred. Every new piece of life is built from templates of other life. Most of these templates are genes, and we have learned an immense amount about the living world from biologists' focus on genes. But there are other ways of moving information around. The great evolutionary biologist John Maynard Smith identified cultural inheritance, this process of learning from others, as the most recent major evolutionary transition in the history of life on earth. He labeled it "much the most important modification" of genetically based evolutionary theory.[3] So an animal may eat a certain food because of preferences largely coded in its genes or because it learned from others that the food is good. An animal may also develop preferences through individual learning, for instance, working out that something is good to eat through its own experimentation. In fact, virtually all the information that moves around through cultural processes originates in this way. However, individual learning on its own does not involve information transfer between organisms and so cannot transform biology in the manner that cultural transmission does.

These processes can interact in different ways. A bird may have the genetically driven instinct to migrate but learn the route from others. Some behavior can be acquired either way. For instance the calls of cuckoos (and many other birds) can mostly develop without social inputs, whereas canaries, finches, and other birds in the oscine suborder learn at least some aspects of their song from others, so their song is a form of culture.[4] Genetic determination and social learning are, however, fundamentally different processes. Tellingly, the cultural songs of the oscine birds are generally more complex, sometimes much more so, and more diverse than the genetically driven nonoscine calls.

We use the phrase "genetic determination" with respect to behavior here and will do so again. However, we do this as shorthand. What we really mean is a large genetically inherited causal component. Genes do not code for behavior—they code for proteins and control the production of those proteins. How genes come to affect behavior is a complex process, intertwined with other factors such as development, maternal effects, and environmental experiences, a system that we still do not fully understand. Biologists have left behind the nature/nurture debate, for good reason, and we have no desire at all to resurrect it here.[5] Unfortunately, discussing the various ways an animal comes to behave the way it does quickly becomes tedious in the extreme without using shorthand in this way. Nearly all behavior that has been well studied is found to require some form of experience to develop properly. It is also true that there are species-typical behaviors that develop even among animals raised in isolation and that vary across populations in ways that are completely consistent with a relatively large genetically inherited causal component. This is what we mean by "genetic determination." It can be contrasted with behavior that requires a significant social input to develop fully. It does not mean we should expect to find a gene "for" that behavior. Contrary to what you might read in the popular press, things are just more complicated than that.

Human language is another example of these complex interactions. While still arguing about the details, most who have studied its evolution conclude that we are born with a genetic template that allows us to learn *a* language effortlessly between the ages of one and four, but *the* language we learn is completely determined by social input during this period—we learn it from others.[6] It is part of our culture.

Animals, including humans, acquire their culture in fundamentally different ways from their genes. During sexual reproduction the genes from two parents shape the new offspring, and in asexual reproduction there is just one parent. These genes are present at the beginning of a life and stay more or less unchanged until death or before being passed on. In contrast, culture may be acquired at any age, from a wide range of models, including parents, siblings, peers, teachers, role models, and, in our material-based culture, media like books and web servers. In many cases recipients actively choose the culture giver. Cultural information from various sources may be combined and altered and then passed on with different content or in a different form. A mother teaches her daughter a recipe for a cake. From watching TV chefs and talking to her friends the daughter adds new ingredients. One day she accidentally cooks the cake at a higher temperature. It tastes better. The improved cake recipe is passed on to her son. In contrast the genes the son received from his mother that, together with those from his father, determine his eye color are virtually identical to those she received from her parents. As another example, we trace many of the methods that we use to study sperm whales at sea, such as identifying individuals using photographs and tracking groups using directional hydrophones (underwater microphones), to innovations made our colleague Jonathan Gordon during the 1980s. With Jonathan aboard, the sailing was a backdrop to tinkering, as he worked on methods that could allow us to begin to understand the whales. Wires ran hither and yon, devices were attached here and there. Jonathan would climb the mast to take photographs of the whales lying parallel to the horizon and thus measure them—that worked—and attach underwater cameras to the boat to watch them underwater—that didn't. Many of Jonathan's techniques were inspired by field methods introduced by the American scientists Roger and Katherine Payne for right and humpback whales a decade earlier, and these in turn had roots in the work of terrestrial scientists like Jane Goodall. Both the photo-identification and tracking of sperm whales have increased in efficiency since the 1980s with experience and technical developments such as digital cameras and on-board processing of sounds, as well as the incorporation of new research methods such as the collection of sloughed skin for genetic analysis. These are the techniques we show to our students, who will develop them further in the decades to come. Cake baking, sperm whale science, and most other human behavior develops from a complex blend of cultural inputs.

So the cultural transfer of information is, potentially at least, much more flexible than genetic reproduction. The products of genes change only on intergenerational timescales. In some cases, especially when a culture is conformist and learned largely from parents, it can be as stable as the products of genes—elements of Judaism, for example, have changed little over thousands of years, but few would argue that the religion could exist without cultural transmission. At the other end of the scale, when culture is learned primarily from peers, it can be highly ephemeral, spreading fast and dying faster—think pop music or fashion. For a short time such cultures—whether a boy band or a "seasonal look"—can have immense influence on behavior, and then they are gone. The abilities to meld cultures and modify them before passing them on allows for the rapid evolution of extraordinary cultural products: jumbo jets and the Internet, hip-hop and nouveau cuisine. Even when culture does not accumulate it can be very useful, basically because other individuals are a rich vein of information about what works and what does not.[7]

So, when culture takes hold of a species, everything changes. Extraordinary new ideas are developed from old ideas and passed on. Things are produced. The things can be technology or art or language or political systems. Interactions with the environment change. New ways of exploiting, polluting, or caring for the earth arise. Nations and ethnicities are formed. And it all feeds back into genetic evolution as those able to deal effectively with all this information and its consequences do better, living longer and having more offspring survive.[8] But there is more. In the words of Peter Richerson and Robert Boyd: humans' "extreme reliance on culture fundamentally transforms many aspects of the evolutionary process. The evolutionary potential of culture makes possible unprecedented adaptations like our

1.1. Ken Norris, scientist, naturalist and original thinker, was the first to contemplate deeply the cultures of whales and dolphins. Here he celebrates his birthday with characteristic creativity. Photograph courtesy of Flip Nicklin/Minden Pictures.

modern complex societies based on cooperation with unrelated people, and some almost equally spectacular maladaptation, such as the collapse of fertility in these same modern societies."[9] The result is the extraordinary evolutionary and ecological trajectory of modern *Homo sapiens* over the past ten to twenty thousand years. Culture is the principal reason why humans are so different from other species. But, in terms of the significance of culture, are we so different from *all* other species?

The Idea of Whale Culture

In the 1960s people started to study whales and dolphins in the wild, spending significant time observing their behavior.[10] Over the next twenty years, they grew to know cetaceans a little, but that little was enough for the beginning of speculation on the role of culture in their lives. Most prominent among these pioneers was Ken Norris (fig. 1.1). The American zoologist inspired numerous friends, colleagues, and students—the next generation of whale and dolphin scientists, many of whom he supervised—as well as interested members of the public. He gave us a new view of cetaceans as intensely social animals.[11] A superb naturalist, fine scientist, and generous teacher, Norris spent a good part of his life with whales and dolphins, both wild and captive. He observed carefully and devised new ways of looking at the animals. He talked and wrote in a folksy way, but he was clear and careful in what he said. In 1980 Norris felt that dolphin "learning capabilities provide for a high level of flexibility in nature and this is translated into local variations in group behavior that we might call culture."[12] By 1988 he went further, concluding that some of the social patterns that he observed were "clearly cultural."[13]

That cetacean social patterns are cultural is a radical concept, but, despite Norris's insight and eloquence, the idea of cetacean culture stayed mostly beneath the surface of cetology, the study of whales and dolphins, during the remainder of the twentieth century. However, over the last decade or so, the idea of whale culture has taken a stake in the awareness of scientists and reached the general public.[14] The idea may be out there, but there is little clarity as to its extent or what it means. In 2001 we published a scientific review titled "Culture in Whales and Dolphins."[15] It received thirty-nine commentaries from academics in a wide range of disciplines with perspectives ranging from "a whale of a tale: calling it culture doesn't help" to "culture among cetaceans has important philosophical implications . . . [they] should now be included with us in an extended moral community."[16] These reactions emphasize both the difficulties some scientists had with our interpretation of the evidence, as well the potentially profound implications if it were to be accepted.

The scientific study of cetacean behavior has continued apace since Norris's observations, comparing and contrasting the evolution of behavior in the sea with that on land. The idea of dolphin intelligence, for example, is so fascinating partly because it evolved and operates in an environment completely different from that which produced human intelligence.[17] The parallels and contrasts with human intelligence tell us much about dolphins, as well as about humans. Likewise, contrasting the social behavior of apes and dolphins can help us understand the forces behind the evolution of our own societies as well as theirs.[18] In the same way, we will argue, some whales and dolphins have a form of culture that evolved and operates in a radically different environment, with some remarkable similarities to human culture but also some profound differences. This culture, we believe, is a major part of what the whales are. Understanding it properly will tell us not just about them, but about ourselves as well.

What This Book Is About

In this book we consider the case for whale and dolphin culture. Chapter 2 considers "culture." How is it used, defined, and studied? In chapter 3, we think about whales and dolphins as mammalian colonizers of the deep ocean. What did they bring with them into the ocean, and what did they evolve there? How did these adaptations lead to culture? Then in chapters 4–7 comes the evidence, clear in some cases, indicative in others. Inspired by the critics of animal culture, we then submit this evidence to scrutiny in chapter 8. From our perspective the evidence mostly holds up, but we try to lay out where different perspectives might lead to different conclusions. Chapters 9–11 address the implications of whale and dolphin culture for the evolution of the capacity for culture, for the coevolution of genes and culture more generally, as well as for the ecology of the ocean and for conservation. The final chapter considers what, if anything, whale culture implies for how humans see and treat whales and dolphins.

What This Book Is Not About

This book is not an ode to whales. The study of whales and dolphins has had at times an unfortunate history, with unsupported claims of higher intelligences and pseudoscientific attempts at "cross-species communication." Today, tension persists because these animals have come to play a strong symbolic role in modern human cultures, and they captivate and fascinate us in ways that few other animals can. Scientists like us are drawn to the whales and dolphins because of many of the same attributes. We think there are good reasons for this attraction, but these reasons are based on scientific evidence—data and observations, hard-won and painstakingly documented by a community of scientists who share our fascination. This community has worked hard to rescue the study of cetacean behavior from misguided mysticism and this book is one part of that effort, insofar as we attempt to place the study of cetacean culture on a firm scientific footing. Part of our aim is to introduce the evidence to a broad range of both scientists and nonscientists, some of whom may have previously regarded these ideas with justified suspicion. We would be naïve, however, not to realize that pairing whales and dolphins with evocative words like "culture" risks us being seen alongside those who claimed that dolphins were "more intelligent than any man or woman."[19] We have therefore tried to separate as clearly as possible the evidence—the scientific documentation—from our views about what it means. You, the reader, will therefore be able reject our views without doing the same to the evidence (although this will be trickier in the few cases where it was actually us who collected the data!). Some perceptive scientists have already done this, as we shall describe. We are obviously going to present our view, but our overriding desire is to engage you via the evidence in a debate that is very much ongoing across several research communities, rather than simply convince you that we are right.

This book is also not about culture in nonhuman animals other than cetaceans. We will touch on other groups of animals, especially in chapter 12, discussing, for example, behavior in primate communities and the cultural transmission of birdsong, but they are not our focus here. There are fascinating and important cultural processes operating in these animals, and we will point to where you can find out more about them. A full discussion of nonhuman culture would necessarily incorporate this knowledge, but that is not our aim here, and for us there is more than enough to talk about in the world of the whales and dolphins.

2 | CULTURE?

Before we cast off and head to sea to explore the cultures of whales and dolphins, we need in our seabag a concept of what we mean by the word "culture." Try to answer the question for yourself: What is culture? Your answer will probably be eye-openingly different to some or maybe all of the ideas that follow. It was once considered so self-evident that culture is uniquely human that many early definitions actually included the words "human" or "man." As anthropologists began to engage with the theory of evolution by natural selection, and the evolutionary continuities it implied between humans and other animals, it became clear to some that this would not do. The debate that then started is still very active. Nonetheless, we need to pin down a concept of culture, or we will be adrift, and we need one that doesn't tie us to the immovable quay of anthropocentrism.

What Is "Culture"?

Definitions of culture abound. They range from the very simple "transfer of information by behavioral means" to unwieldy aggregations such as "the mass of learned and transmitted motor reactions, habits, techniques, ideas and values—and the behavior they induce" or Edward Tylor's widely cited concept of culture as "that complex whole which includes knowledge, belief, art, morals, law, custom, and any other capabilities and habits acquired by man as a member of society."[1] In colloquial speech "culture" is used in two principal ways: a very general "the way we do things" and a more pretentious "intellectual and artistic activity."[2] The former includes quite simple attributes such as how we greet each other and eat, as well as the side of the road that we drive on and our language. The "high" concept of culture—sometimes it is called "high culture"—is more restrictive to religion, poetry, opera, and the like. This second type of culture is usually considered a sign of sophistication. From our perspective this high culture is a subset of culture in general. Arguably, it is more elaborate than the methods a hunter-gatherer uses to hunt and gather, but from an ecological and evolutionary perspective, it's less important. The basic idea of culture seems to be about learning from each other, but there may be additional requirements depending on one's concept of culture. Is a recipe for granola culture? What about a way of doing question 4 on physics homework, or the route for driving downtown to avoid traffic jams that your father-in-law discovered?

Some academics are preoccupied by definitions of culture—it is an occupational hazard. Scholars from many disciplines study culture, and definitions are important because they delineate what is and is not to be included in that study. Culture is central to anthropology, art history, and some areas of psychology and archaeology. But it is also studied by sociologists, biologists, economists, and historians. A scholar's definition of culture often highlights what he or she studies. So, experimental psychologists often insist that specific psychological mechanisms for transferring knowledge, namely, imitation and teaching, are a necessary condition for culture.[3] They study imitation and teaching but pay little attention to how often these mechanisms are employed in everyday real-world cultural transmission. Cultural anthropologists aim to show "how variation in the beliefs and behaviors of members of different human groups is shaped by sets of learned behaviors and ideas that human beings acquire as members of society" and, consequently, may define culture as "sets of learned behaviors and ideas that humans acquire as members of society."[4] They then study behavioral variation between human groups. This implies that the study of culture is fundamentally the realm of cultural anthropologists and that humans are its only bearers. These distinctions sometimes seem to reflect an academic turf war because, in interdisciplinary exchanges, scholars often appear horrified that anyone would think of not using "their" definition.

To be useful to us, a concept of culture must be concerned with things that can be observed, without the necessity of knowing about internal states or constructs (for example, beliefs or values). It must at least allow for the possibility of nonhuman culture and should not exclude things that we commonly consider cultural in humans. Rather than aiming culture at one's favorite species or academic interest, another approach is then to look for "the most useful way to define culture," in the words of Kevin Laland and his colleagues.[5] They, together with other biologists, some sociologists, a few anthropologists, and those who study culture using mathematical models, use a definition such as: "Behavior patterns shared by members of a community that rely on socially learned and transmitted information."[6]

You will notice that this is a very broad definition. Many scholars of human culture would dismiss it as absurdly inclusive. A broad definition like this is useful, however, primarily because it helps us understand the evolutionary roots, and spread, of culture in humans and other species. Many types and grades of culture are included within the definition. These types and grades are important and interesting, but more important and interesting is the whole phenomenon of culture.[7] What this definition captures elegantly is that which makes culture an evolutionary transition—the flow of information independent of genetic inheritance. Another definition that we like is from the cultural theorists Peter Richerson and Robert Boyd: "Information capable of affecting individuals' behavior that they acquire from other members of their species through teaching, imitation, and other forms of social transmission."[8]

The major difference between these definitions is whether culture is the information that affects the behavior, or the behavior resulting from the information. An understanding of this distinction between knowledge and behavior, between what we know and what we do, is still being worked out by scholars of cultural evolution.[9] While human culture can be stored in books or bytes or buildings, most cultural information, and that important part that actually affects how we behave, resides in brains.[10] We think that culture can be either the information—for instance, scientific knowledge—or the behavior—for instance, a dance step. Therefore,

> culture is information or behavior—shared within a community—which is acquired from conspecifics through some form of social learning.[11]

As you can see, this definition is really very similar to, in fact virtually interchangeable with, those of Laland, Richerson, and Boyd. In this book, as in normal speech, "culture" will sometimes refer to behavior ("playing soccer is part of our culture" or "song is part of whale culture") and sometimes to the information ("my taste in architecture comes from my culture" or "the culture of whales includes rules for the evolution of the song").

Our preferred definitions correspond roughly with the nonscientific concept of general culture as "the way we do things."[12] Culture, then, is behavior or information with two primary attributes: it is socially learned and it is shared within a social community. While two of the definitions we have given specify that culture can only be transmitted within a species, fascinating experiments on small forest birds have shown that nesting preferences can be transmitted between species, so in the future this aspect may be revisited.[13] Fundamentally, culture contrasts with genetic inheritance as a way that information moves from animal to animal.[14]

In the remainder of this chapter we will consider the phenomenon of culture: first its roots in social learning, then its setting, the community. We will discuss the diversity of culture; how it changes; how we—Hal and Luke—think about culture and study it; how and why it went wild in one species, *Homo sapiens*; and finally the contentious issue of culture in nonhumans.

What Is Social Learning?

Modern humans live in an "information age." We don't just mean the modern Internet era though. Ever since its origin in the mists of prehistory, and especially since we became able to record it, culture has been the core of our ability to adapt to every terrestrial habitat on Earth.[15] The information that an individual accumulates is key to his or her success, whether success is defined financially, reproductively, or in summed happiness. We are born with some information, mainly encoded in our genes, but some also in the form of "maternal effects"—a healthy or alcoholic mother, for instance—and other inheritances, perhaps financial and social, that come with our kinship lineage.[16] These inheritances, however, do not make a person. We go on to acquire vast swathes of other important information throughout our lives. Some is individually learned but, at least in the case of humans, much more is learned socially. We imitate—try to copy exactly what others do—we emulate others by trying to achieve their goals, and we are taught. We make important discoveries just by being with others, discoveries that we would not make if alone. All these are forms of social learning, the bedrock of culture. The human propensity for social learning allows us our culture and all that it brings, and, it is sometimes argued, the lack of sophisticated social learning abilities, especially imitation and teaching, in other animals condemns them to live without "complex cultural features."[17] Social learning did not, however, originate in humans—it is a good trick that has been identified in species across the entire animal kingdom and takes a range of different forms.

The neatest definition of social learning we know is "learning that is influenced by observation of, or interaction with, another animal or its products."[18] There is nothing particularly special about the learning part of social learning—it is the influence of others, or their products, that makes the difference. Social learning results in the transfer of behavior or information from one individual, whom we can call the demonstrator, to another, the observer. More often than not it results in the behavior of the observer becoming more similar to that of the demonstrator, but it doesn't necessarily have to. Such a description might bring words like "copying" or "mimicry" to mind, but, remember, social learning can also work through the products of others. For example, black rats that live in Israeli pine forests live almost entirely on the seeds that they extract from pinecones. As the seeds are protected by the tough scaled outer casing of the cone, the rats use a complex technique for getting the seeds: they systematically strip the protective scales following a spiral around the cone until only the bare shaft that contains the seeds is left.[19] If you take some of these rats into the lab, breed them, then present the young with whole pinecones in isolation, they never learn the skill. However, if you let the adults partly gnaw some cones, then give them, half-opened, to the young, then many do learn the extraction process, even without seeing an adult stripping a cone.[20] This is still social learning.

So we can think, and psychologists have thought a lot, about what the observer and demonstrator are doing, behaviorally and psychologically, to make the observer's brain come to contain some of the same information as the demonstrator's. First, let's consider the observer. There are several ways that the observer's behavior can change following social interaction with a demonstrator, or the products of the demonstrator's behavior. We have listed the processes that could be involved in table 2.1. There are a lot, but don't take the entries as a definitive and complete categorization of social learning. These categories, for the most part, are based on little more than conceptual reasoning and some experiments that explored, in a few limited situations, what is required for social learning to occur. We still don't really understand how learning works at the level of the neuron, so the categories of social learning are still up for grabs. Processes like imitation and emulation end up at the complex end of the list because they are presumed to rely on relatively sophisticated mental models of the world that can incorporate the concepts of self and others, as well as the notion that others may have different experiences and motivations. Given the level of detail in the list in table 2.1 and the assertions that have been made about which processes might support human culture, you might think that we have a good idea about how people use these learning processes as they acquire their culture, but this just isn't the case.[21]

Through all these mechanisms, then, an observer may learn. But what about the demonstrator? The demonstrator can ignore the other animal, behaving the same as if it was not present, but it can also change its behavior to enhance the learning process or even to inhibit it. If the demonstrator changes its behavior to enhance learning, then this is teaching, at least by some definitions.[22] Thus the different social learning mechanisms of the observer can be enhanced by the teaching behavior of the demonstrator (we will discuss teaching in more detail in chap. 7). For instance, a mother is teaching through local enhancement if she takes her calf to a potentially important place that she would not have gone to otherwise; she is teaching through imitation or emulation if she repeats an activity more than she normally would to make it more likely that the calf imitates her behavior or emulates her goal.

Social learning can be a complex process—just think of how you learned to read or to behave appropriately in a particular social setting, say, when being given a gift by an elderly relative. Learning each of these tasks probably involved several of the learning processes that we have listed, as well as interactions between them, sometimes enhanced through teaching by parents or others. Efficient social learning can be tough. So we use a range of shortcuts that help us to get it generally right, including considering the context.[23] We may actively engage in social learning when we are not happy with our behavioral repertoire in the current circumstances. We may choose behavior to copy depending on how frequently it is performed in our community, perhaps "doing as the Romans do" when in Rome. However, in a few cases, for instance, when nearly everyone is starving, copying rare behavior, say, that of eating an unusual root, may make sense. We may also choose particular demonstrators from whom we copy, preferentially choosing individuals like ourselves, our kin, prestigious individuals, successful individuals, or dominant individuals.

Social learning may be tough, but it is often much less tough than figuring stuff out on your own. That is why we do it. And sometimes it is not so tough; we find ourselves acquiring information effortlessly, as when we hum tunes that have become subconsciously fixed into our brains.

Social learning can have downsides though. When you figure something out yourself using trial and error, you can often be confident that you got it right, or nearly right. However, what you get from others offers no such assurance, and you can end up learning something useless or, even worse, harmful. We can get caught in what have been called "information cascades," where incorrect information propagates through a community because individuals, for whatever reason, prefer to copy others rather that trust their own judgment.[24] When this happens in communities of financial traders, it can bring entire economies to their knees. Thus, making good use of the full potential of social learning presents a considerable challenge for the learners themselves, whether they be human or whale, and a considerable challenge for those trying to understand it.[25] But it is a challenge we need to face up to if we are to truly understand culture in any species.

What Is a Community?

Our definition of culture includes the word "community." By "community" we mean a collection of individuals that is largely behaviorally self-contained and within which most individuals interact, or have the potential to interact, with most others.[26] Thus a "community" is defined by the

Table 2.1. Social Learning Processes, in Order of Complexity or Cognitive Demand

Name of Process	Definition	Example of How It Works
Stimulus enhancement	The demonstrator's behavior draws attention to a particular stimulus.	A dolphin watches another dolphin probing indentations in the sandy bottom of their habitat without any obvious result. The observer dolphin starts paying attention to these features, notices a flat fish emerging from one, and then figures out a way of scaring the fish from the indentations. In this case, the observer does not learn the behavior directly from the demonstrator, but being with the demonstrator stimulates new behavior in the observer.
Local enhancement	The demonstrator attracts the observer's attention to a particular location, which leads to the observer learning about the location and its attributes.	Whale calves who follow their mothers to a particular summer feeding ground will likely learn about that ground and its resources. Once there, if they do this learning independently of other animals, then only local enhancement is operating, but they may also use some of the other methods of social learning from the other whales that are using the ground to enhance their behavior.
Observational conditioning	The observer learns the relationship between a stimulus and behavior by observing other animals respond to the stimulus in this way.	Adult sperm whales quickly form a tight defensive group when killer whale calls are heard. Young sperms notice this and do the same when they hear the calls.
Social enhancement of food preferences	The observer picks up on cues that a demonstrator eats certain food.	A particularly obvious form of this might be when false killer whales pass around food that they have caught (Norris and Schilt 1988).
Response facilitation	An individual is more likely to do something after its performance by another. Then the observer learns about the contexts in which the act is appropriate and its consequences.	A young male might be more likely to initiate courtship behavior after observing it in an older male. He then learns about the consequences of performing courtship in different circumstances. The basic courtship behavior itself could be instinctive, individually learned, or socially learned by other mechanisms.

Table 2.1. Continued

Name of Process	Definition	Example of How It Works
Social facilitation	If the mere presence of other individuals makes certain behavior more likely, then this can influence the observer's learning processes.	A young dolphin may learn about "drafting" (Weihs 2004) by swimming beside other dolphins and adjusting its position and motions for minimum effort while remaining beside the other dolphin.
Observational response-reinforcer learning	By observing the demonstrator, the observer becomes aware of a relationship between a behavior and its consequences.	The dolphin might learn that probing the sandy bottom with its beak is beneficial when the sand has a slight indentation after watching others do this because sometimes this behavior exposes a hiding fish. Note that in this case the dolphin associates the behavior of probing with a reward.
Contextual imitation	If an animal knows how to do something, then by watching others it can learn the specific contexts when this behavior works.	In this case, the young male dolphin sees when females are responsive to sexual advances by other males and makes his own advances in conditions when responses are more likely to be favorable.
Production imitation	An animal observes an action, or a sequence of actions, that it has never performed before itself and then starts repeating what it has observed.	The learning of the song of the humpback whale is essentially production imitation.
Emulation	Emulation is about copying the goal of a behavior rather than the specific form.	Whereas a dolphin might imitate another placing a sponge on its beak so that it could rout around in sharp cavities flushing out prey without getting scraped, an emulator might observe this behavior and then achieve a similar result by holding a stick in its mouth.

Note. In each case we give an example of how this process might operate among whales and dolphins. These examples are realistic but not necessarily proven. This list follows a recent classification by biologists William Hoppitt and Kevin Laland (2008). There are other, rather different, classifications of social learning (e.g., Whiten et al. 2004), which indicates that the divisions between different kinds of social learning are not always clear.

social relationships that knit it together. For any social human, whale, insect, or rat, the great majority of its social interactions and social learning experiences are with other members of their community. As a consequence, socially learned information can stay partially or totally contained within a community. This containment within communities can lead to individuals within them behaving more like each other and less like those in other communities. So communities may develop idiosyncratic cultures. As we will discuss later, conformity may reinforce differences between the behavior of different communities, and they may develop cultural ethnic markers that accentuate community boundaries.

An individual can be a member of two or more communities defined on the basis of different types of social interaction. These communities can be nested within one another, such as urban communities (Pittsburgh) within national identities (the USA), in the case of humans, or pods within clans for killer whales, but they need not be. Communities may overlap in different ways. We are mostly members of several unnested communities, each of which can be based on genealogy, geography, profession, interest, or whim. So you might be a Canadian Catholic teacher who builds model airplanes and supports the Toronto Maple Leafs ice hockey team. None of these communities map onto each other, although the number of non-Canadian Maple Leaf supporters is limited.

Sometimes a population cannot be divided into communities. Maybe social life is very fluid. Or perhaps individuals have static home ranges and interact with those whose ranges overlap with theirs. If there are no major barriers in the habitat, then there is nowhere to draw lines between communities. Can there then be culture? Yes, but in these cases the entire population becomes the community. We do not expect sharp cultural divisions within the population, but much of their behavior may be cultural. There may be clines, with more of this behavior over here and more of that over there. These uniform cultures can be hard to identify as culture. It is obvious that which side of the road we drive on is cultural because, although all members of a particular community consistently drive on one side, the side varies between communities, and transplanted individuals (say, from left-driving UK to right-driving USA) make the switch very quickly. Neither genetic causation nor individual independent learning can produce such a pattern. But suppose all humans drove on the left throughout the world. Then this pattern could be cultural, but it could also be the result of a human instinct to pass other humans on their right sides. It is hard to prize out the underlying mechanism when we all do the same thing.

Human communities have changed shape and dynamics over time.

Hunter-gatherer communities were usually well-defined and fairly small. With agriculture, urbanization, and industrialization, communities became larger and more complex. We went from tribal villages to city states to nation states, with many variants along the way. Now the Internet and social media are changing the shapes and natures of our communities extremely rapidly.[27] The development of multicultural societies means that cultural communities need not map so neatly onto geography. Some cultural communities, such as those embracing graffiti art, can spread around the world with only a few enthusiasts in any one geographic community, while if we move to a new country we can find ourselves living next door to people with a quite different cultural heritage to our own. Cultural preferences can thus actually create new communities.

Whale communities are also extraordinarily variable. Baleen whale communities seem to be large, loose, and geographically based, whereas the communities of some large toothed whales are tight, multilayered, and arranged around matrilines. Dolphin communities vary considerably, sometimes even within a study area.[28] It is very unlikely that these communities have changed shape as fast as those of modern humans, but heavy whaling on highly social species such as sperm whales may have had profound effects on their social and community structures. At the start of each of chapters 4–6 we summarize the kinds of communities formed by baleen whales, dolphins, and the large toothed whales, respectively.

Potentially, a community could contain more than one species. Dolphins frequently form mixed-species schools.[29] However, we suspect that these schools are usually transient and that behavioral interactions between individuals of the different species are not often repeated or individualized. Thus these interspecies members of the same school are not of the same community, and cultural transmission between the species is likely not very important. As ever, there are exceptions, and these include the human-dolphin fishing cooperatives that we discuss in chapter 5, where information appears to move both within and between species.

The Diverse Forms of Culture

Culture comes in many forms. These forms are shaped by the transmission process. Culture, as we conceive it, can be assimilated through any of the forms of social learning we have discussed, sometimes in combination. As we mentioned, there are scientists who disagree, asserting that only "complex" processes like imitation and emulation can support culture.[30] However, these scientists don't all agree on which specific processes are the important ones or how to distinguish the processes. Furthermore, the evidence that human culture is supported solely by these complex processes is, at best, dubious.

The relationship between learners and demonstrators is also important: "vertical" cultures are learned from parents, "oblique" transmission is from other members of an older generation in a form of information transfer that we have institutionalized in schools, while peers of similar age exchange "horizontal" cultures. Vertical cultures can be very stable. Religious rituals and languages, for example, can be traced back thousands of years. In the case of language, the vertical transmission can be so strong that it comes to mirror the transmission of genes from parent to child and results in linkages between genetic and linguistic diversity.[31] In contrast, horizontal cultures—fashion and popular music styles, for instance—are faster moving and subject to very rapid change and are also generally much more transitory.[32] In the extreme, fads can arise, spread through, and then disappear from a community in a fraction of a human generation.

Culture can also be classified according to the kinds of behavior that are affected by it—whether it affects your vocal behavior—for example, your language or accent—your foraging behavior—the kind of food you eat—or your social behavior—the appropriate way to greet someone you are meeting for the first time. Culture affects just about every kind of behavior we humans exhibit, from walking to whistling to philosophy. This is part of what makes human culture so vital. Potentially, culture could be important for just about any behavior of any other animal. To assess the significance of nonhuman culture, we need to understand where it operates.

These different attributes of culture have important consequences for its possessors. We will present evidence in later chapters that whales and dolphins possess various types of culture affecting a broad range of behavior. These include stable vertical cultures, transmitted faithfully between generations, and horizontally transmitted fads, which arise and die out in much less than a generation. We will describe how we think culture affects communication, tool use, social behavior, and foraging in these animals.

Cultural Evolution

Cultures evolve. The culture associated with a community changes over time, moving this way and that, pushed by several forces.[33] Take a tune—a form of culture. A tune can be invented, modified, have mistakes introduced, and ultimately may be lost from the community that hummed it. This change in content over time is known as cultural evolution.[34] One of the most important recent insights in the study of this change is the realization that cultural change is loosely analogous to genetic evolution and, thus, can, with appropriate tailoring, be potentially studied and modeled with the same tools that we use to study the way genes evolve. This has helped us understand both how cultural evolution is similar to genetic evolution and how it is different. For example, you can have only two genetic parents, who bestow their genetic inheritance once, at your conception. You can, however, have multiple cultural "parents," from whom you can inherit cultural information at any point in your life. This alone can significantly change the evolutionary dynamics of the system.[35] We will illustrate these evolutionary forces using two examples, human religion and whale song.

Variation is the raw material of both cultural and genetic evolution. For instance a febrile prophet has a strange vision that becomes part of religious dogma or a whale accidentally sings a new version of one phrase that appeals to the other whales and becomes incorporated in the animals' song. Mistakes can be made when information passes between individuals. Human scribes can introduce mistakes into religious texts—in the King James Bible, for example, how many stalls does King Solomon have for his horses? Forty thousand (1 Kings 4:26) or four thousand (2 Chron. 9:25)? The apparent contradiction is most likely due to a copying error in the number, equivalent to dropping a zero.[36] Similarly, whales can lose track of how many times they have repeated a given phrase of song. Such changes lead to cultural evolution: religious rituals and song content gradually change as the effects of the random mutations build.

Evolutionary forces can be more or less random if new variants are neither favored nor at a disadvantage under the prevailing selection pressures. In this case, the fate of new variants depends mostly on the effects of random sampling—if a whale sings a new song, but nobody happens to be listening, it won't go far. When the genetic makeup of populations changes in this way it is called genetic drift, and there are areas of culture that appear to behave in similar ways. For example, the "in" color of the fashionistas seems to drift randomly, at least to us nonfashionistas. However, cultural change need not be random. Some cultural variants may be preferentially adopted and spread by several means. The most obvious is what Robert Boyd and Peter Richerson call "guided variation."[37] Change can be channeled by preference for content, especially if the preference is for cultural variants that seem successful. For instance, if females seem to prefer rhyming in whale songs sung by males, variants that have better rhymes may be preferentially sung by males.[38] Similarly, modern religions that legitimize material wealth seem to be attractive, spreading fast at the expense of more austere variants, while religions that demand human sacrifice have not really stood the test of time.

Instead of a simple preference of one cultural variant over another, there may be biases in the transmission of different variants, leading to another set of evolutionary forces. The content of a cultural variant may make it more likely to be transmitted. For instance louder sections of whale song may catch on more quickly than quieter alternatives, and evangelical rather than contemplative aspects of religion may spread faster: "Proclaim the lord!" goes farther and faster than "Meditate on the lord." These are examples of content-based transmission bias.

Other factors may bias transmission. In model-based bias, we pay attention to who is demonstrating the behavior. Whales might prefer a song not because it is loud or common but because it is being sung by another male who appears to be having great success in attracting mates. We might be more inclined to take up a new religion not because it has sensible truths or because its hymns have lovely tunes or because it is the state religion but because a movie star we like proclaims his devotion.

In frequency-dependent bias, common variants are disproportionately likely to spread. Humpback whales seem driven to sing broadly the same song that everyone else is singing. Practicing the state religion, whatever it happens to be, is simply a must in some societies. Conformity is when individuals disproportionately prefer the behavior that is most frequent in their community. We conform for two primary reasons. We may believe that the majority are usually right and, thus, conform as a shortcut to finding the best way to behave. This is called conformist transmission by the cultural theoreticians.[39] Alternatively, in normative conformity, we may conform because fundamentally we want to do what everyone else is doing, right or wrong. The conformity is more important than the actual behavior. In some cases the behavior to which the normative conformity applies acts as a symbolic marker of a community, and this gives culture particular potency. The communities marked by languages, clothes, rituals, and other symbols become more distinct socially and so even more distinct culturally. The cultural communities, whether they be human states, religions, or football teams, become active competitors first for space in the brains of the members of the population and then for the population's resources. Conformity in general is thought to stabilize cultural evolution, although it can also promote rapid change should a new variant cross some threshold of popularity for other reasons.[40]

A final force for cultural evolution is good old natural selection. Those individuals using a particular cultural variant survive better and thus are better able to act as models. A religion that promotes fecundity will likely increase the number of its practitioners. And a cultural trait of avoiding poisonous food will spread because the proportion of the population with this trait will tend to increase, as those without this behavior are less likely to survive and reproduce. For example, anthropologists Joe and Nathalie Henrich have shown how cultural food taboos for pregnant and lactating women in Fijian villages target fish species known to pose a poisoning risk and estimate that these taboos reduce the chances a pregnant or lactating woman will suffer fish poisoning by up to 60 percent.[41]

Natural selection, drift, and at least some forms of biased transmission have parallels in genetic evolution but guided variation does not.[42] The ability to select, filter, alter, and combine available cultural variants makes culture potentially very powerful, but making use of this flexibility and thus gaining the power can be hard. Think of how many different and conflicting cultural influences were out there when you selected a career and how many you have or will have been exposed to just today. Making good use of all that advice and information is not easy. As we discuss later, a brain that is effective in this regard may become a premium asset in environments rich in social information.

These are all forces affecting cultural inheritance and cultural evolution. But now bring in genes, and there are two streams of information running through cultural populations. The streams can interact, and when they do, evolution becomes more complex. These interactions go by the name of "gene-culture coevolution." Natural selection of genetically determined traits affects gene frequencies and natural selection of culturally determined traits affects the distribution of cultural variants, but genes may also influence cultural evolution, and culture can feed back into selection that affects genetic evolution. The renowned biologist and thinker E. O. Wilson calls the study of gene-culture coevolution "one of the great unexplored domains of science."[43] We'll briefly look at two well-studied examples in human history to illustrate these points.

The first concerns lactose intolerance, a condition that makes most adult humans unable to digest lactose (table 2.2). The exceptions are generally members of populations with a cultural history of dairy farming. To digest lactose, young mammals produce the enzyme lactase. Adult lactose intolerance is actually the norm in mammals—once weaned, there is no need for lactase as adult mammals do not usually consume milk products, so the gene to produce it is switched off, and it's not a problem because there is no lactose in their diet. However, some prehistoric human populations domesticated cows and developed a culture of pastoral farming, which made

Table 2.2. Evolutionary Interactions between Genes and Culture: The Case of Lactose Tolerance

	Culture	
Genes	No Dairy Farming	Dairy Farming
Lactase production switched off in adults	Ancestral state—almost all mammals switch off lactase production after weaning	Gene-culture coevolution: lactose intolerant individuals at selective disadvantage as unable to use dairy products
Mutations cause persistent lactase production in adults	Natural selection acts against lactase persistence to prevent wasteful production of enzyme	Gene-culture coevolution: adults able to digest lactose have nutritional advantage over intolerant individuals

a lactose-rich source of nutrition readily available to everyone. It turned out that very small changes in the genes controlling the production of lactase can have the effect of keeping the lactase gene switched on into adulthood.[44] Analysis of these genetic switches in both European and African cultures with a dairy-farming history indicates individuals carrying an activated lactase gene into adulthood were strongly favored by natural selection, producing anywhere from 1 to 20 percent more surviving children than lactose-intolerant adults. In evolutionary terms, this is very strong positive selection, and it has resulted in these genetic changes becoming very common in pastoralist populations within a few thousand years.[45] This example is really the type specimen of gene-culture coevolution in human societies.

In our second example, culture actually leads to demographic disaster.[46] The Eastern Highlands of Papua New Guinea are home to the Fore people. The South Fore population comprised around seven thousand individuals in the early 1960s, before they were hit with an epidemic of a disease called kuru, which killed twenty-five hundred people between 1957 and 1977, almost all of them adult women. Understandably, this had devastating consequences on the population. Kuru is a neurological disease, thought to be closely related to Creutzfeldt-Jakob disease, the human form of mad cow. Sometime before the outbreak, the Fore people had, for reasons that remain unclear, adopted from their neighbors the cultural practice of consuming the bodies of their relatives. For similarly unclear reasons, adult men did not take part in this. However, it does seem clear that by consuming the brain tissue of their relatives who had themselves died of kuru, the Fore had spread the disease through their female population. Thus, a culturally transmitted practice led to catastrophe, both for the Fore and for their genes.

Coevolution of genes and culture can happen in all kinds of ways but is particularly strongly affected by the way the culture is transmitted. Genes are only inherited in one way, but culture has many paths. You can get your culture from your parents. This vertical culture propagates in parallel with genes, and so gene-culture coevolution in this case can strongly resemble gene linkage, where different genetic elements are inherited together because of their proximity on the chromosome or because of the interactions of their products. In chapter 10, we will discuss some potential examples of this process in whales and dolphins. When the genes and culture are transmitted by different routes, however, such as a teacher whose students are not her children ("oblique" transmission), then the coevolutionary system can become particularly complex. By concentrating time and effort on teaching, her ideas may spread well, but that time and effort cannot then be used to raise children, so her genes may not spread well at all. Complexity is also added when cultural variants have attributes that improve their transmission to the detriment of genes. Celibate priests may have more time to preach the gospel, and giving up sex for the Lord may make them particularly appealing teachers (it is, after all, a very honest signal of devotion, given how much most people like sex), but their genes have little future. Conversely, genes may have attributes that promote their transmission but reduce the holder's value as a cultural model, for instance, when men buy into the mythology that demands blonds be perceived as both above-average mates because they are attractive and also rather dim-witted. Both perceptions are unjustified. However, if men act on those perceptions, the blond genes spread but not the ideas of the blonds.[47]

Such interactions can lead evolution down strange paths. For instance, cultural conformity naturally reduces the variability of behavior inside a group and, thus, makes differences between groups more stark.[48] Then the ears of evolutionary biologists perk up because the groups become potential units of "group selection," the controversial theory that groups, rather than individuals or genes, can be the units of selection for evolution through natural selection. In essence, groups of individuals whose members behave "for the good of the group" rather than in their own selfish interests prosper and so the altruistic characteristics spread. One problem for genetic group selection is cheating. An individual who lives in the "good group" and receives its benefits but does not do any helping himself will do even better than his altruistic groupmates, so the cheating gene spreads, eventually destroying the good group. Hence, although genetic group selection has recently made a bit of a theoretical comeback in special circumstances, it has largely been rejected by evolutionary biologists since the 1970s.[49]

Theoretically, cultural group selection is considerably more feasible. Cheating is prevented by several factors including conformity, punishment, and a desire to imitate particularly prestigious individuals.[50] Conformity also ensures that migration of individuals between groups does not break down the differences between groups, as individuals can change their cultural behavior in their own lifetimes, whereas in the genetic case, even very low rates of migration can rapidly erode genetic differences between groups. Cultural group selection explains large-scale cooperation among unrelated humans who sacrifice some of their own interests for the good of the group, leading to structures like political parties and armies, which don't make much sense in the light of standard genetic evolutionary theory. Peter Richerson and Robert Boyd conclude that: "Group selection on cultural variation has been an important force in human evolution."[51]

Other Perspectives on Culture

Because culture is so vital to humans, we have whole academic disciplines arranged around its study: history, sociology, anthropology, art history, political science, media studies, and so on. All have developed characteristic perspectives on what culture is and how it should be studied. Historians generally study written records, archaeologists physical artifacts, and cultural anthropologists human societies. There are huge debates between and within these disciplines as to how to proceed. A major subfield of anthropology is often called cultural anthropology, yet there are some postmodern anthropologists who question the very value of the term "culture."[52] Given all these potential ways of looking at the culture of humans, what approach should we take toward nonhuman culture?

Should we look to anthropology, which sometimes sees itself as the "science of culture"?[53] Kim Hill, one of the most evolutionarily oriented anthropologists, and one of the few actually willing to discuss animal culture, notes: "Redefinition of the word culture by non-anthropologists, absent significant exchange with cultural anthropologists, seems intellectually inappropriate, just as it would if cultural anthropologists were to redefine the term gene without seriously engaging molecular biologists about the issue."[54] The cultural anthropologists, when they even admit to any utility to what we students of nonhuman culture do, would have us examine animal "traditions" from the perspective of human "culture." This might have some validity when considering chimpanzees or bonobos, humans' closest relatives, although it rather relegates these creatures to being mere precursors to humans, rather than complex and unique species in their own right, with a distinctive and ongoing evolutionary history.[55] For whales and dolphins, living in an utterly different habitat, at the end of a very long, effectively independent evolutionary trajectory, taking what humans do as an ideal seems profoundly wrong. Humans are in some cases useful models, for instance, in helping us compile the lists of social learning mechanisms or forces for cultural evolution that we discussed in previous sections of this chapter. We will also compare human and cetacean cultures. However, our focus is on what cetaceans do for themselves.

Another argument against using anthropology as the foundation of studies of culture in other animals is the turmoil within anthropology itself, where some are arguing that the concept of culture is no longer useful to the field, which appears from outside to rest entirely on it.[56] Cultural anthropologists themselves have considerable difficulties with culture. As one of their number, Ralph Grillo, notes, "Anthropology's internal debates" seem "trapped in ever-decreasing circles of argument and counterargument, with 'culture' well and truly problematized out of existence."[57] Nonetheless, our definition of culture is not so different from those to be found in introductory cultural anthropology textbooks.[58]

Cultural transmission is a process that has similarities to, and differences from, genetic transmission. It is undoubtedly affected by genetics and, in turn, itself affects genetics. Cultural transmission, like genetics, is a population phenomenon, in that culture can be characteristic of a population and can vary within a population. In contrast to the division, and sometimes rancor, in parts of the social sciences over the last decades, the biological sciences have been quite harmonious and extremely productive.[59] Biologists, whose research subjects range from the biochemistry of genes to the complexities of ocean ecology, generally agree on the "modern synthesis" as the foundation of their work. The modern synthesis, developed in the 1930s and 1940s, is a body of theory and evidence that fused Charles Darwin's natural selection with Gregor Mendel's genetics and has been the bedrock for the extraordinary advances in biology over the last seventy years.[60]

The modern synthesis is continually being reviewed and refined. Nongenetic biological inheritance, such as epigenetic effects, is being added to that synthesis.[61] So is culture. The idea that culture can be viewed as a form of inheritance goes back to Charles Darwin, who sometimes did not clearly distinguish between what we now recognize as genetic and cultural inheritances.[62] As Darwin knew nothing of genetics, this was understandable. In the 1970s Richard Dawkins catalyzed the study of culture using evolutionary methods with his coining of "memes," the cultural analog of genes.[63] Although memes can help us think about cultural evolution, we find the meme concept to be overly restrictive, as culture does not usually come in discrete gene-like packages.[64] It is more complex and messy. Robert Boyd and Peter Richerson capture this messiness in their explorations of cultural evolution using the methods and thinking of population biology.[65] They envisage a wide range of cultural transmission processes governing a wide range of cultural variants that interact with each other and with genetic evolution. So culture "is part of biology."[66] This approach, which we have partly described already, is called dual inheritance theory or gene-culture coevolution.[67] It tries to bridge the perspective of most social scientists, in which only culture determines human behavior, and the biologically oriented sociobiologists and evolutionary psychologists, who use genetic evolution to explain what humans do. Although it has other proponents, this way of looking at culture is particularly well explained by Boyd and Richerson in their groundbreaking book *Culture and the Evolutionary Process* and their more recent and less technical *Not by Genes Alone*.[68]

E. O. Wilson once claimed that "genes keep culture on a leash."[69] Of course, on one level this is true: without the genes to build a brain capable of social learning, you can forget about gene-culture coevolution. Once you have them though, the picture changes. We can visualize this by imagining a dog owner called Gene walking his dog Rembrandt. While the two are clearly inextricably linked, what this picture does not tell us is whether Gene is an adult bodybuilder or a six-year-old child, or whether Rembrandt is a Chihuahua or Great Dane. This, as many dog owners can testify, is really important to know if we want to understand the dynamics of their walk in the evolutionary park. Gene-culture coevolution theory gives us the tools to quantify the relationship between Gene and his dog and to begin to figure out who is walking who, when. As with many dog walks, the direction of the pull can go either way.

Biologists, especially those who study populations, mostly find themselves nodding in agreement when reading Boyd and Richerson's books. Many of the odd elements of human behavior that don't make much sense from the perspective of genetic evolution through natural selection, such as armies and celibate priests, become less mysterious when cultural evolution is added to the mix. In contrast, unfortunately, few social scientists seem to have much interest in adding this evolutionary perspective to their research.

We love Boyd and Richerson's work. Their books contain precise theory illuminated with ideas and case studies from psychology, anthropology, history, linguistics, agriculture, religion, and many other disciplines. However, they see their work as being about humans and suggest that "all other animals, including our closest ape relatives, have rudimentary capacities for culture compared with ourselves."[70] We are not so sure of the truth of this assertion, so we look at the behavior of whales, we measure it, and try to estimate the relative importance of genes and culture in the variability that we see. We think about the evolution of their cultures and, more fundamentally, the evolution of their capacity for culture. We consider how genetic and cultural evolution may interact. And we ask ourselves whether, in some species of whale, culture has become so dominant that it has taken over large parts of their genetic evolution, their ecological interactions with the rest of the world, and what they really are.

Studying Culture

For those of us who study culture, definitions structure our research. Nature, however, doesn't care about our love for categories, and any definition should be seen simply as a tool to help focus our understanding of what is happening in the natural world. This leads to the perspective of Kevin Laland and his colleagues that we should use a definition or concept of culture that facilitates and stimulates the broadest range of research. Narrow definitions direct attention to sterile arguments about whether this or that behavior is culture, whether this or that species has culture. Our broad definition (information or behavior—shared by community—which is acquired from conspecifics through some form of social learning) draws us to more interesting and important questions.[71] Why did culture evolve? What are the different kinds of culture? Why do species differ in how they use culture? How does culture affect ecology? How does culture influence genetic evolution?

The broad definition also directs attention toward a particular style of research. Quite a number of scientists who study nonhuman cultures structure their work as an experimental or statistical test of the null hypothesis that "species *X* does not possess culture [under my definition]" or "behavior *Y* is not culture [under my definition]" against alternatives that the species has culture or the behavior is culture. This traditional scientific approach of testing null hypotheses, based on the work of philosopher Karl Popper, is becoming recognized as sterile and flawed, even dangerous.[72] Instead, scientists are increasingly using techniques aimed directly at uncovering the structure of nature, whatever it is, without null hypotheses or alternatives. For instance, information theoretic methods developed by Japanese statistician Hirotsugu Akaike and others aim to find out which of a set of models most closely approximates the reality of the process that is being observed.[73] Bayesian methods take what we know about a process and then use new information to update our knowledge. We believe that such techniques are the best way to look at the evolution, variety, extent, and effects of culture, defined broadly, for animals in the wild. But, we must admit, we have not got very far yet.

With whale culture, broad definitions and broad approaches to its study are even more desirable. We cannot keep the animals in controlled, captive environments, and only very rarely has anyone managed informative experiments in the wild. We watch, collect data, and try to piece the picture together. There is so much that we don't know that any null/alternative hypotheses setup will almost certainly miss the richness and probably the essence of what the whales are doing. We whale scientists only stumbled into whale culture in the first place, and our discoveries are still largely the result of long-term, large-scale observation together with happenstance—a lot of happenstance.

This book is sprinkled with anecdotes: observations of whales and dolphins, or sometimes of a single whale or dolphin, doing something. In behavioral science, "anecdote" is a loaded term. It generally refers to a one-off observation, and these have long been held in suspicion as a basis for scientific inference. However, anecdotes come in different flavors. There are anecdotes that refer to an event seen by someone not experienced in observing behavior, often retold, sometimes many times. There are also anecdotes that come from records of behavioral observations by scientists experienced with a given species, perhaps including a video or acoustic recording of the event. These two kinds of anecdote are not the same. We are not alone in believing that the latter are in fact extremely valuable sources of information.[74] Whale culture is not usually amenable to controlled experiments, and the statistical analysis of largish data sets only works with some kinds of behavior, for instance, vocal dialects and movement patterns. For strange kinds of play behavior, tool use, and hints of morality we are usually left with anecdotes.[75] Some forms of culture, such as morality, are just occasionally outwardly exhibited in behavior, while others, such as physical greeting ceremonies, are hard to view and recognize when performed by cetaceans. So we see the behavioral manifestations of these cultures rarely and are left with the anecdote. The anecdotes in this book are one-offs but not, we hope, unreliable. We only use anecdotes with reliable provenances—usually information recorded at the time by observers experienced with the species in question. So we know that the behavior happened at least once. Such anecdotes tell us that at least some whales and dolphins can do such things, thus setting a minimum standard for the capabilities of the animals.

If there are several anecdotes of somewhat similar behavior then this indicates that the general behavior may be fairly widespread and the capacity for such behavior nearly universal.

The most satisfactory scientific evidence for culture in whales or dolphins comes from quantitative analyses of behavioral variation with space, time, and social structure. Whales and dolphins are highly vocal animals, and it is usually much easier to record sounds than to observe the visual behavior of aquatic creatures, hence much of the behavioral data we have is acoustic. Such studies of humpback whale songs, killer whale calls, and sperm whale codas are the bedrock of what we know of cetacean culture.[76] There are also good quantitative data on movements for quite a number of cetacean species. However, movements are not generally considered as cultural features to the same degree as dialects and tools, presumably because less "advanced" types of social learning, such as local enhancement rather than imitation, are assumed to be involved.[77] We suspect this view is wrong and, since data sets on whale movements have not been fully exploited, that there are in fact insights waiting to be found on the cultural transmission of movement patterns in the ocean. There are some careful quantitative studies of other "cultural" phenomena, such as the recent work on sponging behavior in the Shark Bay dolphins.[78] However, when we go beyond sounds and dialects, we are mostly drawing conclusions from the results of studies not specifically aimed at identifying culture.

The Key, and Maybe Key, Attributes of Human Culture

Human culture is extraordinary. Walk through a modern city, or even a village of hunter-gatherers, and we are surrounded by the products of human culture: the things humans have made from animate or inanimate objects; the sounds of humans, whether speech, song, musical instruments, or tools; the way these things and sounds are organized in space and time; how the humans themselves behave; and how they walk and look and greet one another. All these are culture. Our culture includes languages, cuisines, religions, technologies, political systems, literature, art, social conventions, and so much more.

This seems a world away from a songbird learning a part of his courtship song from a neighbor. Why has human culture "gone viral"? What are the key elements of human culture? Scholars have been drawn to this question from many standpoints. For some it stems from a fascination with human culture itself. Others are intrigued by the origins of modern humans. What were the roles of changes in the general capacity to use culture, or specific cultural elements, such as language or the use of fire, in the emergence of modern humans from the ancestors we share with chimpanzees? Still others are fascinated by the differences, as well as similarities, between the culture of humans and that of other animals. In the writings of such scholars we have sought the "key attributes" of human culture, the underlying factors that make human culture so powerful and diverse, and have hence given us an ecological role and understanding of our world unlike that of any other species.

One of the most useful sources in this search was the book *The Question of Animal Culture* edited by Kevin Laland and Jeff Galef and published in 2009.[79] Laland and Galef asked a number of scholars who had examined nonhuman cultures from a range of perspectives to write a chapter. They instructed the authors, who included Hal, to address five key questions, including: "If you feel a definition of culture is helpful, what is yours?" "Which animals, if any, exhibit culture?" And, "In what ways are animal and human culture similar and different?" The most pro-animal-culture chapters are at the start of the book and are largely written by zoologists. For an animal culture enthusiast the tone gets darker as one proceeds through the book, with the chapters becoming less and less favorable to the idea. We nonetheless found these last chapters, mostly written by anthropologists, particularly useful in identifying the key attributes of human culture. While the definition of culture used in all chapters included transmission by social learning and an element of sharing, as in our definition, the anthropologist authors of the later chapters added additional requirements. These additions were what they thought made human culture special and why they thought animals did not have culture.

The key attributes that emerged from this search were: technology, cumulative culture that builds up over time so that individuals are doing things that no single individual could reasonably invent in their own lifetime, morality, culturally transmitted and symbolic ethnic markers, and the ways that culture affects biological fitness. "Fitness" here means the rate at which individuals, and perhaps groups, reproduce themselves and can be a function of their genes as well as their culture.[80] Note, however, that the term as used in an evolutionary context confers none of the positive connotations associated with our everyday use of the term to describe someone's physical condition. Individuals can become more fit, in the biological sense, by means that we would consider morally heinous, such as rape or murdering competitors, as well as by more acceptable behavior, such as collaborative hunting. We will say a little about each of these key attributes here and, toward the end of the book, consider how cetacean culture matches up to each these attributes.

Language almost made the list of key attributes. We put language in, took it out, put it in, took it out . . . Part of the reason for our dilemma is ongoing debate about how much of language is itself a product of human culture as opposed to a component of it. Among those studying the evolution of language, there is a school of thought, led by Noam Chomsky, that language is strongly determined, up to and including the level of grammatical structure, by our genetic inheritance—that we possess a "universal grammar" and that the only cultural input is the form of words we plug into that grammar as children.[81] From this perspective, it is easy to see how language could be seen as a prerequisite for culture. This view has been challenged in recent years, though, by linguists like Simon Kirby who, while accepting that language evolution is a complex interplay of biological and cultural factors, emphasize the role of cultural transmission itself in the evolution of language, arguing that full-blown language requires a significant history of cultural transmission, and what he calls "iterated learning," to already have occurred.[82] If true, then it is cultural transmission that is the prerequisite for full-blown language, not the other way round. We're not sure which side of this debate will ultimately prove to be right, but while the debate exists, it seems problematic to assert that language is a key component of culture. Certainly, attempting an account of human culture without language would be futile, but whether that justifies its inclusion on a list of essential attributes is less clear, at least to us. So we're going to fudge it—we'll make language "maybe key" and mention it later when we match the whales and dolphins against the key attributes of human culture.

Michael Tomasello, a psychologist who has done influential research delineating just what makes human culture special, has listed his own "key characteristics of human culture": *universality*, by which he means that some cultural traditions are practiced by virtually everyone in a community, such as a language or a religion; *uniformity*, individuals within communities performing the cultural behavior in the same way; and *history*, the pattern of cumulative change in behavior over time.[83] "History," as Tomasello uses the word in this context to mean "cumulative culture," is already on our list. Universality and uniformity seem less fundamentally important. Some elements of human culture, such as language and religion, are often nearly universal and uniform within a community (although they clearly are not in many modern societies). But other important parts of our culture, for instance, medical techniques and weaponry, are not. So we have left universality and uniformity off our list.

As we look around a modern city, or even that hunter-gatherer village, technology is the predominant manner in which culture is expressed.[84] Much of what we see, touch, hear, and smell is human technology or its products. Technology has changed the environment of humans and the ecology of Earth. It has radically changed geographical distributions, population sizes, life histories, and gene flows not only for humans but also for many other species. Without technology humans would not be able to live in the Arctic or Antarctic or to cross oceans, would almost never live beyond eighty years, would never marry those born thousands, or even hundreds, of kilometers away. Technology comprises the tools, techniques, and crafts that we learn. Technology allows us to construct our own ecological niches and, thus, to change our environment.[85] Technology does not have to be cultural. Individuals could each learn the technology by themselves. Galápagos woodpecker finches use twigs or cactus spines to pry bugs from trees, but they seem to learn this individually rather than from each other.[86] Almost all human technology, however, propagates by social learning and thus forms part of our culture. A potter will usually learn from watching or perhaps being taught by another potter, and the construction of a bus involves masses of information being passed through many routes among many individuals by teaching, reading, imitating, and other forms of social learning.

Human technology can be so powerful and effective because it accumulates. Cultural traditions are learned socially, built on by individual learning and innovation or by combining different socially learned information, and then passed on, ratcheting the culture toward more complexity and, usually, more effectiveness in some respect. So we have submarines, symphony orchestras, and French cuisine, and we build libraries to hold this vast store of accumulated culture that has long outstripped our poor brain's ability to keep track of it all. While this accumulation is most obvious in the construction of elaborate technologies, it is necessary for the development of many other parts of human culture as well, from stories to horticultural methods to political systems. Without this cumulative property, cultural complexity is constrained to what one individual can discover for his- or herself without access to the knowledge of others. The effects of noncumulative cultures on other areas of biology will be relatively small. Michael Tomasello, who was the first to point out the significance of accumulation when considering nonhuman cultures, believes there has been "no convincing demonstration of the ratchet effect or any other form of cumulative cultural evolution for chimpanzees or any other nonhuman animal."[87] How we came to accumulate culture is still an open question. Tomasello believes the key is that humans learn from each other in unique ways—involving joint attention—that mean our culture gets passed on with a particularly high-fidelity form of imitation; teaching, as well, plays a role in how we pass on culture.[88] Without this fidelity, cumulative culture is impossible. Support for Tomasello's position can be found by demonstrating how humans compare to other primates in interacting with puzzles that have cumulative solutions and in some mathematical models of how cumulative culture might work.[89] These are powerful arguments then, but they are not the only game in town. It may come down simply to having enough people around or the relative strengths of conservatism and conformity in those people's learning strategies.[90] Entire books have been and will continue to be written on this fascinating question. We move on by emphasizing that there is no doubt that cumulative culture is a key characteristic of modern human culture.

Morality can be viewed as having a sense that some things are "right" and others "wrong" and then behaving consequently.[91] Morality makes cooperative and prosocial behavior possible on large scales. Biologists expect animals to be kind to relatives because of what we call "kin selection"—our relatives share our genes, so a gene that leads to its bearers assisting their relatives helps itself and spreads.[92] In small societies, reciprocal arrangements can develop: I help you and so you help me. It is, however, a widely (though not universally) held view that, in the absence of strange genetic effects like haplodiploidy,[93] neither kin selection nor direct reciprocity can lead to the large-scale cooperation in large societies that is, most of the time, a human characteristic.[94] Something else must be going on.

Punishment is one possibility: many unrelated individuals can form a stable and effective army or navy if each knows that he will be shot or hung from the yardarm if he deserts.[95] But the army or navy is usually much more effective if its members believe that the fight is right than if it is held together by fear. At the beginning of the War of 1812, the tiny fledgling U.S. Navy had some remarkable victories over the huge and previously almost unbeatable British Royal Navy.[96] Part of the reason for the Americans' unexpected success was that their ships were manned by volunteers who were righteously indignant over several incidents in which the Royal Navy had, prior to the war, seized sailors from American merchant ships. In contrast, the majority of the British sailors were victims of the press-gang, a legalized form of kidnap. A sense of morality can induce extraordinarily large-scale, and sometimes personally dangerous, cooperative behavior, like navies, armies, and missionary expeditions. Punishment can enforce, and so reinforce, morality. The punishment may be more efficiently given, and more responsively received, if it is a consequence of "doing wrong" rather than just going against the will of the powers that be. Morality, by promoting cooperative behavior, can change the whole population biology of a species, and it seems to be an important product of gene-culture coevolution in human societies.[97]

Another way to achieve effective large-scale cooperation is through ethnicity. Rather than "right" and "wrong" governing behavior, it is "us" and "them." Individuals cooperate with those belonging to "their group" whom they perceive as like themselves and may be antagonistic toward those they perceive as being from a different group.[98] Kin selection might give a natural "us" (relatives) and "them" (nonrelatives) on the scale of families, but not with hundreds or thousands of unrelated animals, except, as noted above, with out-of-the-ordinary genetic systems. It breaks down in human societies organized around bands as small as thirty.[99] In humans, ethnicity is achieved through culturally transmitted and symbolic markers: crucifixes, berets, college colors, accents in speech, uniforms, and so on.[100] As a result, "human societies are a spectacular anomaly in the animal world. They are based on the cooperation of large, symbolically marked in-groups."[101] Ethnicity and morality can of course combine, giving the sense that "we" are "good" and "they" are "bad."

For our evolutionary approach to whale culture, a crucial question is whether cultural behavior affects fitness. Do individuals with some cultural variants do better in the evolutionary race than individuals with alternative variants, thus spreading their genes, as well as their ideas? Different religious groups may reproduce at different rates even though they live in the same area.[102] In the United States during parts of the nineteenth and twentieth centuries, newly arriving Catholic immigrants had higher birth rates than the more established Protestant Yankees, thereby proportionately increasing the incidence of Catholicism in America, as well as, incidentally, any genes characteristic of the members of the religion. And it is not just religion. All kinds of cultural traits affect human fitness, both now and in prehistory. Weaponry and medical techniques are obvious examples. Thus, if culture affects fitness it can drive evolution along pathways it would not otherwise have taken.

So human culture has some extraordinary attributes. Do other animals share any of them? There is considerable skepticism for each of these attributes. In terms of technology the gap seems obvious, although William McGrew caused quite a stir in 1987 when he suggested that there was little difference between the tool kits of Tasmanian aboriginals, perhaps the technologically simplest of human groups, and Tanzanian chimpanzees.[103]

The lack of accumulation in nonhuman cultures has been a frequent argument in the case for the overwhelming superiority of human culture. For instance, in 1988 anthropologists Joe Henrich and Robert Boyd found "no evidence that nonhuman traditions change cumulatively over time or allow the development of behaviors that individuals could not learn on their own."[104] Animal morality, too, is often dismissed. In 2009, the anthropologist Kim Hill, whose thoughts about defining culture came up earlier, could not find evidence in any nonhuman for "socially learned conventions, ethics, rituals, religion, or morality."[105] This is not, however, a universal view. Donald Broom, for example, sees morality rather differently, and from his biologist's perspective finds evidence for it in a wide range of animal species, as do ethologist Mark Bekoff and philosopher Jessica Pierce, who argue that our human-centered view of morality is far too narrow.[106] The primatologist Frans de Waal also argues that empathy, one of the emotional roots of morality, has evolutionary roots far deeper than our own species.[107]

Ethnic markers have been carefully considered by perhaps the most open-minded of all anthropologists to think about nonhuman culture, Susan Perry, who studies capuchin monkeys in Costa Rica. The extraordinary behavior patterns of the animals that she studies, for instance, poking fingers up companions' noses, easily constitute culture by our definition.[108] She writes: "Although it seems likely to me that some animal species will eventually be proved to exhibit some limited between-group variation in social norms and a strong sense of group identity, I am far more skeptical that there will ever be evidence for linkages between socially learned traits or symbols and group identity."[109]

Kim Hill dismisses our final key attribute—that culture affects biological fitness—in nonhumans, believing that "group selection on cultural traits" does not "take place at any appreciable rate in animals."[110] There can be little doubt that cultural transmission interacts significantly with genetic evolution in animal species. Biologists Peter and Rosemary Grant have spent decades studying Darwin's finches on the Galápagos Islands and have shown how the cultural transmission of song and song preferences through the paternal line plays a crucial role in evolutionary change and even speciation among these icons of evolutionary theory.[111] There is little evidence, though, that this has led to the kind of group-level selection that is one of the theories advanced to explain some unique features of human societies.

Finally, language, our maybe key trait, is generally perceived as a major difference between humans and all other life forms. In the words of Peter Richerson and Robert Boyd: "Language is often given pride of place as the watershed between humans and other animals."[112] While we are still exploring the frontiers of just how complex animal communication can be, there is simply no evidence that any nonhuman possesses the kind of open-ended and recursive communication system we recognize as language.

In chapter 12 we will line up what we know of the cultures of whales and dolphins against these key attributes of human culture. But, in the meantime, we need to bear in mind that the habitats and evolutionary histories of humans and whales are entirely different, so, as the philosopher Thomas White explains, evaluating whales or dolphins according to human features is profoundly one-sided.[113] We should also think about how humans might line up against the key attributes of cetacean culture. There is a problem here. As humans, we may find it very difficult to identify or articulate those attributes that are key to the whales but not shared by us.

The Animal Cultures Debate

The idea that culture is important for nonhumans, including whales, has a history of controversy. In the 1930s–40s, biology was given a strong theoretical basis in the form of evolution through natural selection—natural selection as first suggested by Charles Darwin and Alfred Russel Wallace and then formalized with genes as the units of selection in the modern synthesis. The modern synthesis was not particularly about behavior, but behavioral theorists in and around the 1970s realized it could be applied to behavior as well as morphological, physiological, or anatomical features. This new field was called behavioral ecology or, largely in the United States, sociobiology. Advocated comprehensively in E. O. Wilson's book *Sociobiology* and summarized eloquently by Richard Dawkins in *The Selfish Gene*, behavioral ecology made a fine job of explaining why animals do what they do.[114] Its application to human behavior was, and is, controversial.[115] For the study of nonhuman behavior, however, behavioral ecology became a hugely successful scientific paradigm. From the 1980s onward, scientific papers describing the behavior of animals invariably started and ended with how the research was situated within the theory of behavioral ecology. We, and most of our scientific colleagues, found the theory very appealing and felt it well explained the behavior of animals. In the field of animal behavior, behavioral ecology became "normal science," in the terminology of the philosopher of science, Thomas Kuhn.[116] Suggesting that culture could be a major driver of the behavior of nonhumans challenges this paradigm—making it "revolutionary science," according to Kuhn—and, as with other challenges, was resisted. However, in contrast to the opposition facing most other scientific revolutions, the attacks are not coming from the stalwarts of "normal science." Since the inception of their theory, behavioral ecologists and sociobiologists have largely accepted the possibility that culture might have an important role in determining behavior, along with genes. E. O. Wilson, for instance, cowrote *Genes, Minds and Culture: The Coevolutionary Process* and Richard Dawkins coined the term "meme," the cultural analog of the gene.[117]

This is not to say that proposing culture as an explanation for animal behavior does not meet resistance from our closest colleagues, but it does so only from the angle of questioning what the evidence is that a particular behavior results from some kind of social information. There are now, however, enough solidly demonstrated examples that this does in fact happen in many species for the study of social learning to be accepted as a valid and growing field within mainstream animal behavior science.[118] No, while behavioral ecologists may question the evidence and suggest alternative explanations, they are generally not appalled by the very notion of chimpanzee or whale culture.[119] The fiercest critics come mostly from anthropology and psychology. Here, it is the very concept of animal culture that is anathema, not the nature of the evidence. It is part of the paradigm in most of the social sciences, insofar as the social sciences have paradigms, that humans are unique in having culture or, at least, in being overwhelmingly cultured. Culture in other species, if it exists, is an epiphenomenon, not terribly important. It is the challenge to this paradigm that is being resisted.

Critics of all stripes argue against the evidence put forward. They pick away at the (necessarily) spotty evidence from field studies, suggesting that this or that pattern of behavior could have arisen genetically or through environmental correlations. In their laboratory experiments, chimpanzees don't imitate and rats don't teach—thus: no culture. On the other side of the debate, field scientists are convinced that culture is a major part of the lives of the animals that they study, but how can they show it?

People have thought about nonhumans having culture for a long time.[120] Aristotle noted that at least some birdsong was learned. Darwin thought many animals possessed "inherited habits," and although he did not know how inheritance worked, his conception of these inherited habits was very similar to what we now think of as culture.[121] Following Darwin, many of those who studied the behavior of animals in the late nineteenth and early twentieth centuries believed that socially learned traditions were important in shaping the behavior of at least birds and mammals. However, once genes had become central to biology in the modern synthesis, thoughts that culture might have a role in animals other than humans faded.[122] For a while it was nearly all about genes.

After the end of the Second World War, ethologists began to study the behavior of animals quite broadly and rigorously, and the trend against nonhuman culture reversed. These studies at first indicated and then showed that in two very different kinds of animals, social learning—and "culture" for those who wished to use that term—is an important determinant of what animals really do. The clearest case was birdsong. Birds were the model organism for many, perhaps most, of the early ethologists, people like Konrad Lorenz and Niko Tinbergen. Many aspects of bird behavior are quite amenable to experimental study, and most especially their songs. It soon became clear that elements of the songs of many birds are socially learned, and social learning seemed the most plausible explanation for the spread of a technique by which blue tits opened the tops of British milk bottles.[123] Birds seemed to have culture.

In the 1950s and 1960s, another group of animals also began to receive the culture label, the primates. This development occurred first in Japan, where, both in society generally and among scientists, the dichotomy between humans and other primates is much less strict than in the European Christian tradition. Japanese scientists noted socially learned traditions in groups of Japanese macaques, most famously the spread of sweet potato washing among the monkeys on Koshima Island, where they dunked the tubers in the ocean to remove sand before eating them.[124] The Japanese and other scientists were usually cautious in discussing these patterns, often referring to them as "precultural" behavior, "protoculture," or "traditions" rather than unqualified "culture."

The study of primate culture moved to another level in 1978 when William McGrew and Dorothy Tutin described variation in the grooming handclasp between groups of chimpanzees.[125] Unlike the Japanese monkey traditions this was "arbitrary" behavior not involved in resource extraction, and McGrew and Tutin addressed the question of culture head on. They concluded that the evidence on grooming handclasps satisfied most but not all of eight conditions they listed for culture.[126] Over the next few years primate behavior was increasingly described as "culture," without qualifiers. A highlight of this period was the publication of McGrew's book *Chimpanzee Material Culture* in 1992.[127] In this book and in his papers, McGrew showed the value of making systematic comparisons both across social groups and for a range of types of behavior, what he called an "ethnographic" approach. To virtually all field biologists studying chimpanzees, as well as orangutans, capuchin monkeys, and some other species, this approach made sense. It was clear that the animals that they had spent so much time with learned from each other, that they had culture, and that comparing what happened in different groups or at different times was a good way to look at this. However, not everyone was happy with McGrew's ethnographic approach.

The promotion of culture as an important influence on behavior in chimpanzees and other nonhumans was attacked by two prominent psychologists. Jeff Galef found little evidence that nonhumans either teach or imitate, and, given that he felt that culture propagated only through these processes, which he argued were different in humans compared to any other animal, he concluded that it is misleading to think of the evolution of culture in animals.[128] Galef also emphasized that culture in humans and what others call culture in animals were analogous—that is, evolved independently—rather than homologous—that is, similar through common descent. While the homology-analogy contrast is clearly a potentially important issue for those interested in chimpanzee culture, because of the recent common ancestors of humans and chimpanzees, from our cetological perspective this is an unnecessary controversy. The common ancestor of humans and whales was a small, probably fairly solitary, mammal, likely without much, if any, culture. The cultures of whales and those of humans, or chimpanzees, are analogous not homologous, and the fact that they evolved independently makes their similarities and differences particularly interesting. The second major attack on nonhuman culture came from Michael Tomasello. He reinforced Galef's arguments, adding the potentially important point that of all the social learning processes only imitation and teaching can lead to cumulative cultures, which, as we have noted, are one of the most key attributes of human culture. Galef and Tomasello come from a null-hypothesis-testing, experimental psychology background. The null hypothesis is something like "chimpanzees do not possess culture," with culture being defined by something like "traditional behavior transmitted by imitation or teaching." They could not show in their own or others' experimental studies that captive chimpanzees could imitate or teach, so did not reject the null hypothesis. No culture.

The "chimpanzee culture wars" were on.[129] The field scientists responded, most prominently in a remarkable paper published in the journal *Nature* in 1999. The study, headed by psychologist Andrew Whiten, charted the incidence of thirty-nine chimpanzee cultural behaviors at seven study sites and concluded that "the combined repertoire of these behavior patterns in each chimpanzee community is itself highly distinctive, a phenomenon characteristic of human cultures."[130] Their basis for saying that these behaviors were cultural was that they did not seem obviously linked to some variation in ecology (that might guide individual learning along different paths in different places) nor could it be explained by obvious genetic differences, an argument termed the "method of exclusion." Similar studies of orangutans and capuchin monkeys followed.[131]

Unlike some anthropologists, the psychologists Galef and Tomasello had been specific in their criticism of animal culture. They set out arguments, hypotheses really, that could be falsified. For instance, their claim that chimpanzees did not imitate could be refuted. It subsequently was refuted—chimpanzees do imitate—in several studies, including some by Tomasello and his colleagues.[132] However, Galef and Tomasello believe that other criticisms stand. For instance, Tomasello still maintains that there has been no "convincing demonstration of the ratchet effect or any other form of cumulative cultural evolution for chimpanzees or any other nonhuman animal."[133]

While this culture war was ostensibly about nonhuman cultures in general, it was, and is, highly focused on chimpanzees and, to a lesser extent, on other primates. All the evidence on the culture of birds, especially the detailed studies of birdsong, was largely and rather unfairly put aside, perhaps on the basis that birds are perceived as "one-trick ponies" with just one cultural behavior—song—both by the proponents and opponents of nonhuman culture.[134] Whales were also initially outside the discussion.

In 2001, though, we wrote a review titled "Culture in Whales and Dolphins," and cetaceans joined the fray.[135] Our paper attracted commentaries from those on both sides of the chimpanzee culture wars. The proponents of chimpanzee culture were generally positive about whale culture. For instance, Christophe Boesch wrote: "The sacrilegious proposition of the existence of cultures in whales and dolphins should open the discussion of cultures in other animals, allowing us to find what is unique in human cultures."[136] And Andrew Whiten, who was the primary author of the famous 1999 chimpanzee culture paper, believed that we actually underestimated the imitative ability of at least some cetaceans.[137]

In contrast, those who did not like chimpanzee culture were far from convinced by our arguments for whale culture. The anthropologist, Tim Ingold wrote: "It is sad to see such rich empirical material, about such wonderful creatures, harnessed to such an impoverished theoretical agenda," by which he specifically referred to our use of the word "ethnography" but, more broadly, seemed to dislike the dual inheritance perspective that we embraced.[138] Jeff Galef, following his critiques of chimpanzee culture, went on about analogs and homologues of human culture, as well as a lack of conclusive evidence for imitation or teaching.[139] We pushed back, titling our response to the commentaries, with tongues somewhat in cheeks, "Cetacean Culture: Still Afloat after the First Naval Engagement of the Culture Wars."[140]

An important criticism made by culture skeptics concerns how difficult it is to really know what causes behavior to develop in wild animals.[141] Cultural transmission can cause variation in behavior within a species. It is not enough, though, simply to demonstrate that animals in one place, or one group, behave differently to those in other places or groups. Environmental variation can also produce differences in behavior between communities in several ways. For instance, if animals can learn a behavior with no social input, then in places with a particular food or tool present these may get used, whereas their use will be absent from communities where the necessary food or tool does not occur. This behavioral variation is not culture; it results from individuals learning in different ways on their own. Genetic variation can also lead to differences in behavior. For an example we'll take gibbon romance.[142] Gibbon couples across Southeast Asia sing duets together in the mornings. The songs of these apes vary systematically between populations. However, this is not the result of cultural transmission. There are strong correlations between genetic and vocal differences between populations. Most crucially, when gibbons of different species mate, the songs their offspring produce are intermediate hybrids of their parents' songs. Thus there is little role for cultural transmission in their development. All of this makes the method of exclusion a tricky beast, which for intellectual health and safety must always be labeled "Handle with care!"

There have been several attempts at peacekeeping in the culture wars.[143] Kevin Laland and his colleagues have probably been most effective.[144] They have criticized as well as commended those on both sides of the issue. Laland was the primary editor of the 2009 book *The Question of Animal Culture*, which brought the views of the pro- and anti-animal culture scientists together (although birds were almost completely ignored). They recommend a broad and simple definition of culture—basically the one that we have adopted—as being most useful. Such a definition shifts attention from whether a particular behavior type fulfills all the conditions for "culture" to the types of cultural behavior that seem to be present. However, Laland and his colleagues also criticized some of the inferences that pro-culture scientists have made from their data, suggesting alternative noncultural explanations. For instance they propose that some of the variation in the thirty-nine "cultural" chimpanzee behavior patterns might be, at least partially, ecologically or genetically determined, through processes such as we've just discussed.[145] As an alternative, they argue strongly that we must develop better methods of studying culture in both wild and captive animals. We must, they rightly caution, be wary of slipping into simplistic thinking about potential causes for a given behavior. If we force ourselves to decide that it's either ecology, or genes, or culture, then we are doomed to fail in understanding behavioral development.[146] For example, one of the most powerful benefits we get from our culture is an ability to adapt flexibly to the prevailing ecological conditions, be it in the Arctic or the jungle. We should expect in some cases a relationship between behavior and ecology even if the behavior is culturally transmitted. We therefore need to get away from the "either culture or genes or environment" trichotomy and toward asking, "How important are culture, genes, and the environment" in the development of behavior? This is, however, rather challenging if you are talking about large, wild animals that you can only partially observe part of the time, such as most cetaceans. Laland and his colleagues have themselves worked to develop methods to overcome this challenge, as we have ourselves, and we'll meet some of these techniques later on.[147]

We have got ourselves an idea of what we mean by "culture." We have seen how controversial the notion that nonhumans might have culture is in some quarters. We have also seen how quickly things are changing in the way we understand these issues. We have a way to go before we have a good methodology for identifying the importance of culture to wild animals, but over the past decade the picture has changed in favor of nonhumans of some species possessing culture. Without the information they learn from each other, their behavior would be very different. We hope that scholars with as much insight as Robert Boyd and Joe Henrich would not write today, "Unlike other animal species, much of the variation among human groups is cultural," as they did in 1998.[148]

There has been a lot to chew over in this chapter. Let's hope, though, we've now built ourselves a not-too-leaky vessel of ideas to carry us out to where the fun really starts—at sea, with the dolphins and whales.

3 | MAMMALS OF THE OCEAN

Whales, Seals and Mermaids

Mammals evolved on land, so it is hardly surprising that not many have a presence in the ocean. The deep ocean is an environment that contrasts radically with the land in just about every important way. Even the coastal waters, while more land-like in some respects, are quite different from neighboring shores. Mammals occasionally reach the ocean incidentally, such as rats on ships, or accidentally, such as wallabies on logs, but in neither case are the mammals subsisting on marine resources. A few terrestrial mammals get sustenance in the ocean, such as the fishing pigs of Tokelau, but these are celebrated as oddities.[1] About 130 of the approximately 5,500 mammalian species consistently make their real living in the ocean. These 130 species are from four orders: the Primates, Carnivora, Sirenia, and Cetartiodactyla.

Today the dominant marine mammal is a primate, *Homo sapiens*. We are dominant not because we evolved genes that coded for useful adaptations to an oceanic life but because of our culture. Kayaks and ocean-going ships are technological culture. The know-how to navigate a kayak or an ocean-going ship is part of our informational culture, as is the means to fish from them. Until very recently human impact on the oceans was restricted to waters just off the coasts, but now we, and our effects, are everywhere.[2]

A second marine order is the Carnivora, a diverse group of mammals that includes cats, dogs, bears, weasels, otters, and giant pandas.[3] Most of these mammals are bound to the land, but there are important exceptions. Polar bears and sea otters make a living in the ocean, although they usually stay close to ice or land. The primary oceanic members of the Carnivora are thirty-three species of seal, sea lion, and walrus, collectively known as the pinnipeds. The pinnipeds can use the deep ocean, but they are dependent on land or ice to breed and so have a limited presence far from land or the icy poles. Nevertheless, in several ocean areas, particularly the colder ones, pinnipeds are important animals. They are big, some weighing several metric tons, and their populations can number in the millions.

Pinnipeds entered the ocean much later than the two other principal marine mammal groups, the sirenians and cetaceans. About twenty-three million years ago they evolved from bearlike ancestors. The pinnipeds retain terrestrial features that other marine mammals have lost, such as giving birth out of the water, being covered with hair, having hind limbs, a fairly regular set of mammalian teeth, and keeping their nostrils on their snouts, rather than moving them to the tops of their heads like whales. Are they only partway through an evolution toward becoming truly marine mammals? In a few million years will the descendants of today's seals look and behave more like whales? Or has evolution found them semiaquatic niches where they get some of the benefits of being terrestrial? They give birth to pups with very limited mobility in a habitat with few or no natural predators but make good use of the bounty of the oceans. Evolution tends toward specialization, at least for the big issues such as whether to be terrestrial or aquatic, so the pinnipeds are rather a mystery with their successful dual lifestyle, although they share it with sea turtles and seabirds.[4]

Unlike the pinnipeds, the sirenians live their whole lives in the water. This ancient order, descended from elephant-like creatures, is unique among the marine mammals in that its members are herbivores, eating algae and sea grasses. There are currently four species of sirenian: three manatee species in the Atlantic and Caribbean and one species of dugong in the Indian and Pacific oceans. All are about three meters long, stout animals weighing up to fifteen hundred kilograms, not built for speed but very efficient consumers of sea grasses. Ancient seafarers, long starved of the company of women, considered them to be mermaids, although there is a considerable mismatch between their stout, homely bodies and Disney's Ariel. A fifth sirenian, Steller's sea cow, was much larger, eight meters long, but no prettier. The sea cows lived in the northern North Pacific and ate kelp. Humans exterminated sea cows in 1768.

The fourth group of marine mammals, the cetaceans—whales, dolphins, and porpoises—are fully marine. Unlike the pinnipeds they almost never come out of the water onto land or ice except for brief lunges by dolphins or killer whales onto beaches to catch seals or fish that they have driven ashore; becoming stranded out of water typically results in death. Unlike the herbivorous sirenians, they are not tied to shallow coastal waters. The cetaceans are members of the order Cetartiodactyla, which also includes deer, pigs, camels, and cattle. Their closest "terrestrial" relative may well be the hippopotamus. The abundance of transitional forms in the cetacean fossil record make it an archetypal exemplar of evolution at both macro and micro levels. If you have creationist friends (or are one yourself!), then take note.[5] They started their oceanic journey fifty million years ago, dipping into the warm Tethys Sea, which lay roughly where the Middle East is today. Their limbs became webbed, and other adaptations to marine life followed. Like the pinnipeds they are highly streamlined and flexible, allowing them to swim fast. Like the sirenians they lost their hind limbs and almost all their hair. Their tails became flukes for efficient propulsion, their forelimbs stiffened into steering flippers, and their nostrils moved to the tops of their heads so they could breathe more easily while swimming fast. So they became sleek, streamlined predators of the high seas that bore a superficial resemblance to large fish, apart from the horizontal tail.

As these cetardiodactyls made their way into the Tethys Sea, they radiated into all kinds of strange forms—protocetids, basilosaurs, dorudons, and others—with a wide range of ways of making a living in the ocean.[6] Most went extinct, but a few of these archaeocetes persisted. They in turn radiated, so that the living cetaceans are an extremely diverse group of mammals (see fig. 3.1). They are diverse in size, ranging from Hector's dolphin at about 1.4 meters long to the blue whale, which can reach to thirty-two meters. The cetaceans have made their way into just about every marine habitat, except the greatest ocean depths. Narwhals, belugas, and bowhead whales swim among the arctic ice, while humpback dolphins live in the southern Red Sea, which contains some of the hottest and saltiest ocean waters on the planet. River dolphins are found in land-locked Nepal and Bolivia, thousands of kilometers from the sea.

Of all the contrasts within the cetaceans, most fundamental is the division between toothed whales and baleen whales. The toothed whales, technically called odontocetes, include about seventy-three species of porpoises, dolphins, and whales, the biggest being the sperm whale.[7] There are only about fourteen species of baleen whale, formally the mysticetes. These are generally larger animals, with the smallest adult being a six-meter-long pygmy right whale.

The odontocete-mysticete split occurred about thirty-five million years ago. While both groups are streamlined, largely hairless, possess blowholes on top of their heads, have flukes and flippers but lack hind limbs, they each evolved characteristic adaptations to aquatic life. For odontocetes, the key attribute is echolocation, or sonar.[8] Like many bats they make clicks and use the returning echoes to sense their world. The echolocation clicks of sperm whales are very powerful, allowing detection of prey and other things at longer ranges than vision or other senses can manage underwater.[9] Sonar can also be very precise. Dolphins and porpoises with their higher frequency echolocation can build detailed pictures of their surroundings. Sonar users can sense the bottom and sides of the ocean—islands, seamounts, and coasts—as well as each other, getting detailed pictures of the shapes and inner structures (as in ultrasound scans of human

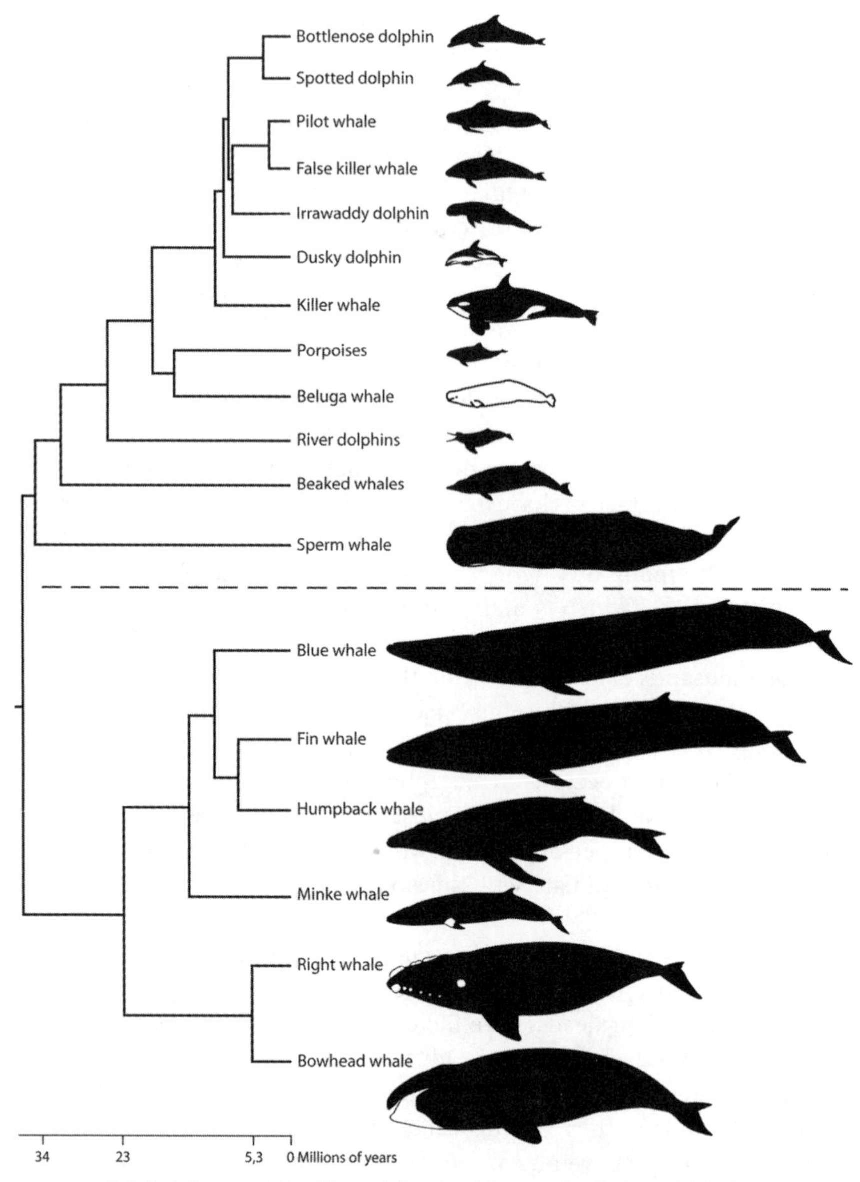

3.1. Evolutionary relationships and diversity of Cetacea, the whales and dolphins. Of the roughly eighty-seven living cetacean species, this diagram shows eighteen species (or groups of species), those that we principally discuss in this book. The dashed line divides the odontocetes, or toothed whales (*above*), from the mysticetes, or baleen whales (*below*). (The phylogeny is from McGowen, Spaulding, and Gatesy [2009].) Copyright Emese Kazár.

pregnancies—ultrasound is sonar) of potential prey, social partners, and competitors. Sonar can also alert them to the presence of predators, but primarily the sonar of the toothed whales is about finding food.

The major adaptation of the baleen whales is quite different in form but has the same consequence: better foraging. The mysticetes modified their mouths, losing teeth and developing plates of fibrous baleen that hang in rows from each side of the upper jaw.[10] The baleen plates can be quite short—a half meter in the gray whale—or very long—four meters in the bowhead whale—but their basic function is the same. On the inner sides they fray into separate fibers, and the fibers on the different plates form a mesh, so the array of baleen is somewhat like a comb with the hair stuck to it.[11] The baleen whales use these combs to filter food from the water. There are two general ways in which the mysticetes use their baleen filters. The skimmers—bowhead, sei, and right whales—swim steadily along with their mouths slightly open. Water enters the mouth through a gap in the baleen at the front and leaves through the baleen. Food is caught on the inside of the baleen and periodically shoveled down the animal's throat by its tongue. The gulpers—including blue, fin, humpback, and minke whales—open their jaws wide, engulfing whole schools, or large parts of schools, of prey. They then use their tongues to press the water out through the baleen, whose filter keeps the food inside the mouth. Using either mechanism, baleen whales can strain huge quantities of schooling marine animals from the waters in just a short time. Such major mouthfuls have allowed the baleen whales to support the largest animal bodies ever to have existed and to get their annual nutritional requirements in just a few months.

These highly developed foraging structures, the echolocation of the odontocetes and baleen of the mysticetes, make the cetaceans special and have allowed them to become huge. But there is much more to these extraordinary animals. They are major players in many marine ecosystems, are socially complex, are cognitively advanced, and have the largest brains on Earth.

Mammals: Dominant Land Animals

We are mammals. So are most of the animal species that we have domesticated to provide us with food, to clothe us, to transport us or our products, to guard us or our belongings, or simply to share our homes. Dogs, cats, cows, pigs, sheep, goats, llamas, and horses are among the most prominent, but there are others. Most of the animals that we hunt in the wild are also mammals, as are some of our most annoying pests—think raccoons or rats—as well as many of the wildlife species that we find most interesting, such as chimpanzees and elephants. To humans, the default "animal" is a mammal. But mammals are really rather odd animals. They are quite different from most other creatures on the earth or in the sea.

About 400 million years ago, during the Devonian period, some fishes—relatives of modern coelacanths and lungfish—started to develop characteristics allowing them to leave the water and make use of the resources of the land (fig. 3.2). Their pectoral and pelvic fins evolved into legs, giving rise to tetrapods ("four legs"), and they took their oxygen from the air by breathing rather than from the water using gills. These early tetrapods were amphibians, animals that still needed water as a medium in which to lay their shell-less eggs. About sixty million years later, a major evolutionary development took place in a group of tetrapods. These small lizard-like animals, which we now call amniotes, produced eggs with shells. The eggs could survive on land, and this allowed the amniotes to move into drier areas.

Roughly twenty-five million years later, or about 315 million years ago, there was a profound evolutionary split in the amniotic line. The sauropsids led to all existing reptiles, as well as birds and dinosaurs, whereas the synapsids became mammals. The first synapsids were very reptile-like, but mammalian characteristics gradually evolved. Their descendants, the therapsids, developed efficient jaws and specialized teeth, the ability to stabilize internal body temperature, hair, a more efficient upright walking gait, and lactation.[12] These and other adaptations took the therapsids and then their progeny, the mammals, which appeared about 200 million years ago, into some new ecological niches. They could run fast, make their living in cold weather, and give excellent care to a small number of offspring rather than hoping for the survival of a few among many eggs.

About 165 million years ago there was a major cleavage in the mammalian line itself. One branch led to the monotremes, strange mammals that lay eggs and can be venomous. Few monotreme species survive today, the duck-billed platypus being the most famous. The other branch evolved placentas and live birth. The line of these placental mammals itself divided 148 million years ago, giving rise to the marsupial mammals and the eutherian mammals. The marsupials—kangaroos and their relatives—have babies that are born small and live for a long while in their mothers' pouches, surviving and developing through suckling. In contrast, the eutherian mammals put more effort into gestation. They gave birth to larger, more developed offspring, able to survive outside of a pouch from birth.

The early premammal therapsids did well, becoming diverse in size, shape, and ecological role, some ending up looking quite similar to large

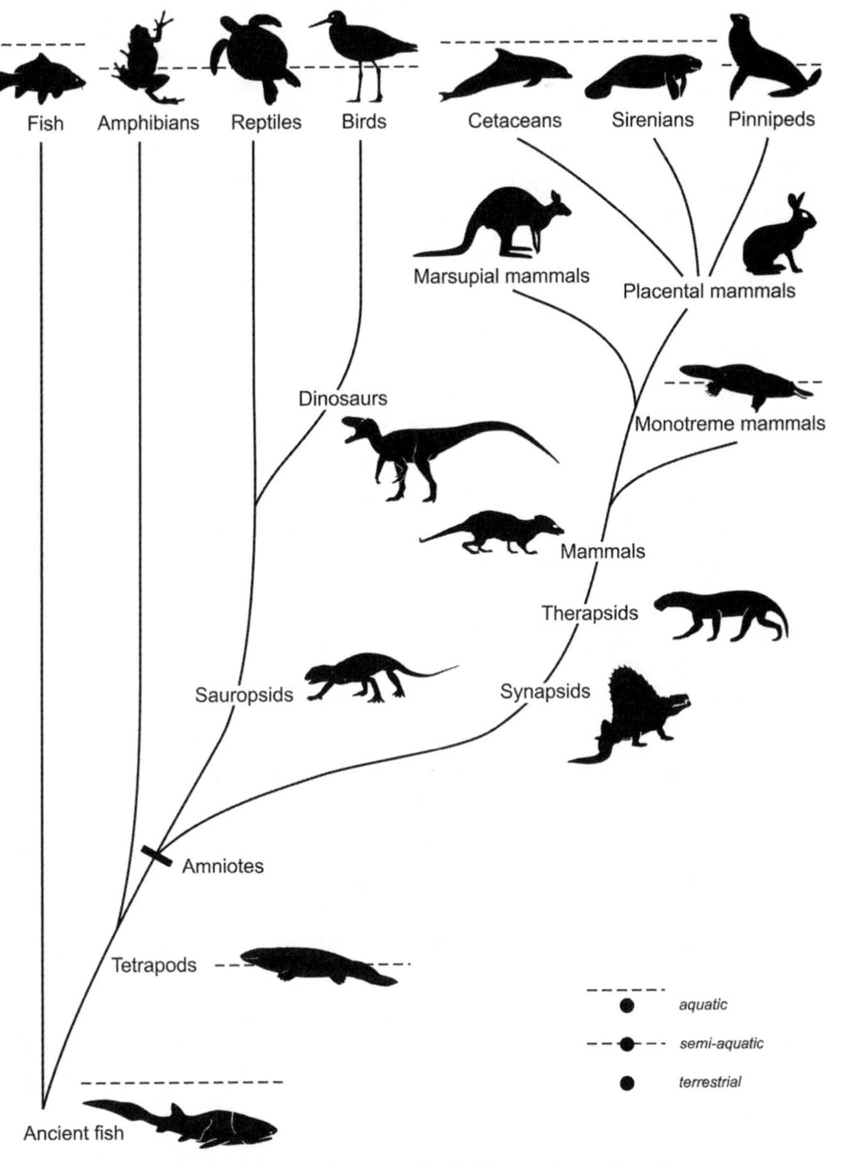

3.2. Evolution of the marine mammals, showing aquatic, semiaquatic, and largely terrestrial groups of species. Copyright Emese Kazár.

modern terrestrial mammals. They were the dominant land animals 275 million years ago. But in the next fifty million years the therapsids were eclipsed by species with more reptilian form from the sauropsid line, especially the dinosaurs. Eventually all the therapsids, except the mammals, became extinct. These early mammals were small creatures, and they stayed that way for the next 140 million years of their evolutionary history, apparently unable to maintain large body sizes in a dinosaur-dominated world. The little mammals evolved a number of very useful attributes, including high-frequency hearing and specialized teeth.[13] The largest mammals over this huge time period reached about fifteen kilograms, but most were much smaller.

But then, sixty-five million years ago, came the Cretaceous-Tertiary mass extinction. The cause of this cataclysmic event is hotly debated. Maybe it was the impact of one or more asteroids, maybe the eruption of many volcanoes, perhaps a sea-level rise. Whatever the cause, many forms of life were badly affected: plants, plankton, and fish lost species as well as biomass. Other organisms, including insects, turtles, and snakes, survived pretty well. The most dramatic result of the Cretaceous-Tertiary extinction was the end of nearly all of the dinosaurs. The only survivors are their avian descendants, the birds.

Without dinosaurs the mammals blossomed. They returned to the large-body-size niche that their therapsid ancestors had once held, and then some. From the small creatures scuttling around out of the way of the dinosaurs, they moved into all kinds of habitats and niches, sometimes becoming very large. There were huge herbivores, as well as top-level predators. The largest terrestrial mammals, weighing about seventeen metric tons, appeared about thirty-two million years ago, very different from the dinosaur-suppressed small creatures that mammals had been for most of their evolutionary history.[14]

Today, terrestrial mammals live on the earth, under the earth, in forest canopies, and in the air, as well as in lakes and rivers. They are found in rain forests and deserts, on mountain tops, and patrolling over the arctic ice and have even made it to space. They eat vegetation, insects, and almost all other forms of life, including other mammals. They change the nature of the land and its ecosystems with their predation, grazing, burrowing, and trampling. Terrestrial mammals may be ecosystem-controlling "keystone species," like elephants, or ecosystem engineers, like beavers. Well before humans, mammals dominated much of the land. They achieved this because of their anatomical and physiological adaptations. Like other terrestrial vertebrates they breathe air and possess backbones, but they also have hair, are warm-blooded, give birth to live young, nurse their infants with milk, and have powerful jaws. Mammals additionally possess what might be called second-order characteristics: they can be adaptable, opportunistic, intensely social, and intelligent.

Of all the habitats colonized by the mammals none was as extreme as the ocean. It was a huge challenge for very land-oriented animals to achieve success in the ocean. But they did.

The Ocean

The ocean covers the majority of planet Earth, contains most of the living world, and is a primary driver of Earth's biosphere. Life evolved in the ocean and continues to evolve there. The ocean is a really important place, literally vital—for its own inhabitants, obviously, but also for terrestrial life.

The most fundamental characteristic of the ocean is water. Although chemically simple—two hydrogen atoms with each atom of oxygen—water has unusual and important properties. These properties are fundamental to life being possible in the ocean and give a framework for how life exists, in the ocean and beyond.

Water is dense, about 840 times as dense as air—roughly as dense as most life forms, as they are primarily made of water. This means that marine organisms fight no battle with gravity and possess none of the structures that we need on land to combat it. In the ocean, there are no tree trunks. The closest analogy would be the stipes of kelp, allowing kelp to form "forests." But these stipes do not hold the kelp up—they just hold it in place. At low tide the kelp collapses. Likewise, marine animals can have flexible skeletons or no skeleton at all. This makes it much easier to become large. The major problem with the density of the ocean comes in the depths, where the weight of all that seawater bears down, creating enormous pressures.

Related to density is viscosity—or, basically, friction. It is about sixty times easier to move through air than water. The importance of this friction depends on how big one is and how fast one is moving. It is much more significant for the little creatures. Thus, while a killer whale can cruise at about ten kilometers per hour, krill, weighing about 0.2 grams, can achieve only about 0.2 kilometers per hour.[15] Therefore, in the viscous ocean the little animals move slowly, giving the big ones a big advantage. There are no darting ocean insects. Some medium-sized animals, like flying fish and leaping dolphins, leave the water when they want to move really fast.

Water has the greatest heat capacity of all common liquids. That is why we usually use it to cool our car engines, as well as to move heat around our homes. It is why a hot bath is so splendid on a freezing day and falling into an icy river so painful. Maintaining body temperature is hard work in water, especially when the temperatures of the body and the water are very different. Because temperature is generated throughout the body, but lost and gained through its surface, small animals, which naturally have more surface area per unit volume, suffer much more from heat loss than do larger ones. The only ocean animals that attempt to maintain a particular body temperature are some of the larger fish, like swordfish, and the marine mammals, none of which are small.

Water dissolves other substances better than any other common liquid. This allows it to function as the medium for chemical communication using hormones within animal bodies. Seawater includes all kinds of dissolved substances, including, of course, salt. Many of these substances are important for marine life, but none has the significance of oxygen, which all animals need to power their bodies. Seawater is about 0.5–0.9 percent oxygen at the surface. Most marine animals use gills to get this oxygen into their bodies. A few, including the marine mammals, come to the surface and breathe the air. Coming to the surface may have costs in time, energy, or vulnerability to predators, but it has benefits, too, primarily that air is 21 percent oxygen.

Few animals just sit, or blunder along, waiting for good things, like food or mates, or dangerous things, like predators, to come their way. They sense their environment and change their physiology or behavior, and they communicate with each other. One can sense and communicate through a variety of channels, primarily what we call the five senses: touch, taste, smell, hearing, and sight. Moving from the air to water changes the relative benefit of each of the senses. Chemical signals are not dispersed as widely or predictably under water as in air, so taste and smell have less value for marine animals. Some penguins seem to be able to smell areas of high productivity and swim toward them over large ranges, but, tellingly, they do so by smelling the air that they breathe, not by tasting the water that they swim through, even though the productivity they are aiming for is in the water.[16] Sight is also degraded in the ocean because light is absorbed by water. At depths of a few hundred meters there is virtually no light, even in the middle of the day, and even just beneath the surface one can rarely see more than about twenty meters, less than the length of a blue whale.

One sense that does do better underwater is sound. It travels about four times faster than in air. More important, and in contrast to light, sound is less attenuated by water than air. While there are a few sounds of terrestrial mammals that travel over kilometers—the roars of lions, the rumbles of elephants, and the howls of wolves—most sounds of most terrestrial mammals are lost at much shorter ranges. In contrast, many underwater sounds of marine mammals can be heard in quiet conditions at a kilometer and some travel very much farther.[17] Thus sound is generally the best way to sense and communicate in the ocean, especially over ranges of more than a few meters.[18] This simple fact has profoundly affected the evolution of whales and dolphins.

The most apparent contrast between the land and the ocean as habitats for animals is in their physical structures. The earth's surface is solid and is textured by hills and valleys and rocks and trees. All are pretty permanent. Together they form a boundary between solid substrate and air. On this two-dimensional surface most terrestrial animals live their lives. Some fly a bit, others burrow, but most of the action is in two dimensions. In contrast the ocean is three dimensional and, above its bottom, has virtually no solid features. Even the boundary between ocean and air is often in motion. The coastal waters are intermediate with more structure, less fluidity, and less of a third dimension than the deep ocean, but they still show much less permanence than the land.

Relationships between life forms depend on the habitat in which they live. A fluid, three-dimensional habitat stifles control. Many terrestrial mammals effectively maintain exclusive territories within which they allow no other members of their species, except dependent offspring and perhaps the occasional mate. In the ocean, there are no boundaries to mark, and, even if there were, they would be huge. Think about the difference between patrolling a circle or the surface of a whole bowl. No animals of the open ocean defend geographically defined territories. Even defending objects, such as a fish school or a potential mate, is usually prohibitively difficult in three dimensions. This means that in the ocean competition is usually a scramble—who can get the most per unit time—rather than a contest in which only one competitor gets the spoils. In scramble competitions, the emphasis changes from the competitors themselves to the resources. One expects less antagonism between members of the same species in the fluid, three-dimensional ocean.

The third dimension of the ocean works quite differently from the other two. One has usually to travel a long way north, south, east, or west before the ocean changes in any predictable way, but a diver leaving the surface within the first few tens of meters passes through a range of depth bands that have strong effects on life. The pressures increase, and the temperatures generally drop, especially as one passes through the thermocline at anywhere between fifty and several hundred meters. Going down, light levels decrease so that at depths of more than a few hundred meters there is no photosynthesis, and, with a few remarkable exceptions, no primary productivity.[19] Life in the dark ocean is largely dependent on what comes down from the surface waters. A few hundred meters is also the depth of the oxygen minimum layer, where dissolved oxygen is so scarce that animals with gills—that is most marine animals—are in a bind. Some, like the vampire squid, have evolved physiologies to deal with low oxygen levels, but all must use energy sparingly.

These structural contrasts are the most basic, and most obvious, differences between terrestrial and marine habitats. Other differences are not as apparent but are also profound. The environment of any creature varies in space and time over all kinds of scales. A deer can move a few meters, perhaps to the shade of a tree. The light level and temperature may go down, the humidity may rise. If it traveled a few kilometers, the changes might be a little greater but probably not much. The amount of change might be similar over timescales of minutes and hours. In fact, on land the amount of change in the environment is quite similar over a wide range of scales of both time and space, once regular cycles like day-night and season have been accounted for. This pattern, in which variability is similar over a range of scales, is sometimes called white noise, as the color white is an equal mixture of short and long wavelength electromagnetic radiation. In the ocean, conditions change little over hours or meters but can be radically different if one waits a few weeks or moves a few hundred kilometers. Lower frequencies and longer wave lengths predominate—as they do in red light—hence the ocean is said to have a "redder" structure.[20] This fundamental contrast has consequences all the way up the food chain. On land, resources tend to be quite evenly distributed in space and time. In the ocean, especially the deep ocean, it is much more boom and bust.[21] We have sailed for days, seeing almost nothing in the way of marine life, but then come across a patch of ocean teeming with leaping fish and dolphins and diving birds. As in the case of physical structure, the shallow coastal waters tend to be intermediate between the land and the deep ocean in their pattern of variation. Using the color metaphor, we might consider them somewhat pink.

So ocean animals have had to adapt to a fluid, highly variable, three-dimensional world, a world where gravity is countered by the density of water, where pressures build quickly with depth, where everything moves all the time, where heat is hard to conserve and sound travels much better than light. There are advantages to being big in a water world: speed and ease of motion, heat retention, fewer potential predators, and the ability to store energy during the lean times between the oceanic bonanzas. A major disadvantage of size on land—the fight with gravity—is gone. As a result, compared with terrestrial ecosystems, a relatively much greater proportion of ocean biomass is in big animals.[22] These large creatures not only form a large part of the oceanic biome, they often structure it.[23] Many of the bigger oceanic creatures are mammals. Before we consider why the marine mammals have been so remarkably successful in the ocean, we will start with a more fundamental question. Having evolved initially on land, how do they survive in the radically different oceanic world?

How Mammals Survive in the Ocean

It seems a shame. It took millions of years of evolution for the amphibians, amniotes, synapsids, therapsids, and mammals to develop the range of extraordinary features that made them so effective on land, only for the marine mammal lineages to ditch many of these same adaptations. The losses were most complete for the cetaceans, intermediate for the sirenians, and only partial for the pinnipeds. Mammals had developed intricately engineered limbs that allowed some species to run very fast, others to dig, and still others to swing through the trees or even to fly. In the cetaceans, the hind limbs disappeared completely, while the forelimbs became stiff paddle-like flippers.[24] A sense of smell, external ears, and external testicles were all lost, along with almost all hair. The diverse mammalian dental toolbox which could rip, grind, and pierce, regressed into rows of nearly identical conical teeth or, in some species, no teeth at all.

Some of these losses made the animals more streamlined, a major benefit in the much more viscous ocean. Other attributes that had little utility in the ocean, like external ears, simply disappeared. There were also modifications that evolved to deal with a viscid fluid environment. Cetaceans and sirenians turned their tails into efficient propulsive flukes, so they moved and looked rather like large fish, with the important difference that their tails beat up and down rather than side-to-side. Their largely hairless skin sheds epithelial cells at an extraordinary rate, acting as "antifouling," inhibiting marine invertebrates like barnacles from settling on them.

Despite all these changes, the marine mammals retain some features characteristic of terrestrial mammals that might on first consideration appear to be disadvantages in an oceanic environment. Most obvious of these are the maintenance of a warm body temperature and the breathing of air.

To maintain body temperature, marine mammals kept large body sizes, at least thirty kilograms for an adult, and developed insulating blubber layers.[25] This blubber had the additional benefit of being able to store energy, so the animal can live for long periods with little or no food. The thermal balance problem is particularly severe for the newborn, which is much smaller than its parents, and worse for cetaceans, which, unlike pinnipeds, give birth in the ocean. All marine mammals give birth to only one relatively large offspring at a time. Newborns can be up to 20 percent of their mother's mass in the smaller cetacean species. This is extreme. Few, if any, large terrestrial mammals give birth to young that are more than 15 percent of their own weight.[26] The minimum size for a newborn marine mammal seems to be about 0.6 meters long and five kilograms.[27]

Even with large size and good insulation, high body temperatures are energetically expensive. A resting bird or mammal burns five to ten times the energy of a similar-sized lizard or cold-blooded fish.[28] Marine mammals use, and thus need, lots of energy. They have developed a very diverse range of foraging methods that help them get that energy. For the cetaceans, there are two major adaptations. As we described at the beginning of this chapter, the odontocetes developed echolocation and the mysticetes filter feeding.

Breathing air means coming to the surface regularly. The surface may be a long way from where food is, as it is for elephant seals and sperm whales, which can feed over a kilometer below the surface. It is also a potentially dangerous place. With easy access to oxygen, either by breathing in the case of other marine mammals—such as killer whales—or by using the high levels of dissolved oxygen in the water—in the case of sharks—surface predators can be energetic and fast.

The marine mammals adapted to minimize the drawbacks of breathing air. Physiological changes allowed them to use oxygen more efficiently and dive for longer.[29] Their myoglobin, which holds oxygen in the muscles, evolved to become more electrically charged and therefore better at holding onto oxygen, so their muscles became huge oxygen stores.[30] These changes are quite remarkable in the case of the deepest divers, like Cuvier's beaked whales, which can go for over an hour between breaths.[31] There were other adaptations that minimized the drawbacks of breathing air. For instance, the cetacean blowhole moved to the top of the head, so whales and dolphins can breathe easily while swimming fast.

But marine mammals still must come to the dangerous surface waters to breathe. Terrestrial mammals use structures like trees and burrows for protection, sometimes as safe dens for infants. But there is nothing like that in the ocean. As with those terrestrial mammals that lack dens and cannot carry their babies—for instance, horses—cetaceans give birth to precocious offspring, able to swim immediately after birth and so to follow their mothers through their fluid world. For a cetacean, the one dependable source of safety in the ocean is other cetaceans. So the cetaceans became very social, with those using the most open and, perhaps, most dangerous waters showing the largest group sizes. In the words of Richard Connor, a biologist who has studied dolphins for decades: "Perhaps no other group of mammals has evolved in an environment so devoid of refuges from predators. Many cetaceans, especially smaller open-ocean species, have nothing to hide behind but each other."[32] Although a few other land-evolved animals have made a partial transition back to the ocean—seabirds, sea turtles, and sea snakes—none has done so as completely as the marine mammals, particularly the cetaceans. And none have been anywhere near as successful, transforming ocean ecosystems with their presence.[33]

What Mammals Bring to the Ocean

Marine mammals bring to oceanic ecosystems a range of unusual, and sometimes unique, attributes—attributes that make sense on land but are hard to evolve in the ocean, presumably because they are handicaps when half-formed or not combined with other characteristics. Breathing air may appear a disadvantage to an oceanic animal—all those round trips to the dangerous surface—but there are substantial compensations. Whether coming from water or air, oxygen must enter the body through some hole or membrane. As the amount of air needed is proportional to the metabolic energy used by the animal, and metabolic rate goes up faster than surface area, it means that the size of the blowhole or gills is relatively larger in bigger animals. While blue whale blowholes are huge—a person could slip down one—they are still a rather small fraction of the surface area of the whale. In contrast the gills of a large fish, like a whale shark, take up a large part of its body. Thus air breathers are less constrained as they become larger. Whales can be very large. The efficiency of air breathing makes it easier for the mammal, as compared with the squid or fish, to maintain a fast metabolism and a high body temperature. This is particularly the case in the oxygen-poor mid-depths of the ocean where they have a huge advantage over gill-using prey, competitors, and potential predators. By bringing oxygen with them like a scuba diver, they effectively create an exclusive oxygen-rich niche for themselves in the depths. The other species forced to eke out their oxygen from the water around them don't stand a chance. It's as if the marine mammals arrived in their midst from the planet Krypton, like Superman.

Terrestrial vertebrates evolved an integrated set of structures that allow them to breathe air: noses, nasal passages, and lungs. They use adaptations of these structures to make their sounds, from human speech to birdsong to the echolocation of bats. The marine mammals brought these air passages into the ocean, where, it turned out, they were even more effective at producing sounds. This is because the very different densities of air and water mean that vibrations of the membrane between them are transmitted particularly well through the water. So the marine mammals are good at making sounds in an environment where sound travels fast and far.

Manatees are vocal, as are some seals, but the cetaceans have gone furthest.[34] The echolocation clicks of porpoises, the songs of humpback whales, the screams of orcas, the thrumming of blue whales, and the clicks of sperms are the loudest animal sounds in their frequency range and are a nearly omnipresent part of the ecology of the oceans. In some parts of the world, when we lower a hydrophone (i.e., an underwater microphone) from our boat we invariably hear the sounds of cetaceans. The whales and dolphins, and to a lesser extent the pinnipeds, use their hearing to map their environment and their social world.[35] As we have noted, the toothed whales, including the dolphins and porpoises, also produce sounds that echo off potential prey, social partners, predators, and oceanographic structures. These biological sonars give the animals a detailed picture of their surroundings and so a much more proactive role in the ocean than most other oceanic animals who can only sense, often poorly, their immediate surroundings.

This sound production is possible because of the air passages that the air-breathing marine mammals brought into the ocean. These air passages are used in a variety of ways, some of which we don't fully understand, to make sound.[36] The sound-producing organs of their terrestrial ancestors, their larynxes, evolved to make louder and more complex sounds. For instance, dolphins have two nasal passages and two sets of sound-producing organs and can simultaneously produce two different sounds. The importance of sound underwater gave natural selection considerable power to shape the vocal apparatus of the whale. Other structures appeared, such as the click-producing *museau du singe* (literally, "monkey's muzzle") of the dolphins, culminating in one of the strangest, and certainly the largest, sound-producing devices in the animal kingdom, the sperm whale's nose. The nose, formally called the spermaceti organ, takes up a quarter to a third of the animal's body. It contains fine grade oil, is surrounded by a huge muscle, and is bookmarked by air sacs, connected by looping nasal passages. Clicks are made by the *museau du singe*, which is right at the prow of the sperm whale, in contrast to the top-of-head position in the other toothed whales and dolphins. These clicks are formed and focused by the oil and air sacs of the spermaceti organ.[37] The immediate result is the natural world's most powerful sonar system, a system that makes the sperm whale a supremely adept predator.[38] The spermaceti organ is central to the nature of the sperm whale and to all its other attributes, including, as we will discuss, its culture.

The terrestrial heritage of the marine mammals is evident in their air breathing and its consequences for metabolic rates, size, and sound production. There is one other characteristic of the marine mammals and especially the cetaceans that is remarkable among marine creatures, but is less obviously tied to air breathing: their brains.[39] If we look across species, then larger-brained animals are generally more cognitively able.[40] This makes sense; we know from designing computers that more processor usually means more processing power. Furthermore, across the animal kingdom, larger brains are intrinsically more complex, with proportionally more wiring.[41] However, big brains tend to be in big species. There have been two reactions to this. The first and predominant perspective is that bigger animals just *need* big brains to run their big bodies, so when we are comparing brains across species we should do so as a ratio of brain weight to body weight, or an "encephalization quotient," which represents how much bigger or smaller an animal's brain is compared with other animals of the same body size.[42] Humans have the highest brain-to-body ratio—and encephalization quotients—of any species, thereby making us the brainiest creatures on Earth, which seems to be right. After all, no other species have apparently spent any time at all thinking about this issue. However, there are actually few good data, or much theory, as to why *relative* brain size is the best indicator of cognitive ability, other than a general feeling that large animals need large brains. Instead, there is increasing evidence from structural analyses of brains, as well as from attempts to test species with different-sized brains on comparable tasks, that absolute size may be a better general measure of cognitive ability.[43]

As we said, larger processors generally give more functionality. However, we also know from computers that getting the same amount of computing power into a smaller volume is difficult, and, if we are at the retail end of clever devices, more expensive. So, we can think of the absolute size of a brain as indicating, in a very general way, its cognitive power, while its relative size is a measure of how hard evolution had to struggle to get it that big. Hence, from this perspective, it is easier for larger animals to be smarter, and they generally are.

The human brain is big in absolute terms. But it is much more remarkable in relative terms; we devote a lot of our energy to maintaining it, and

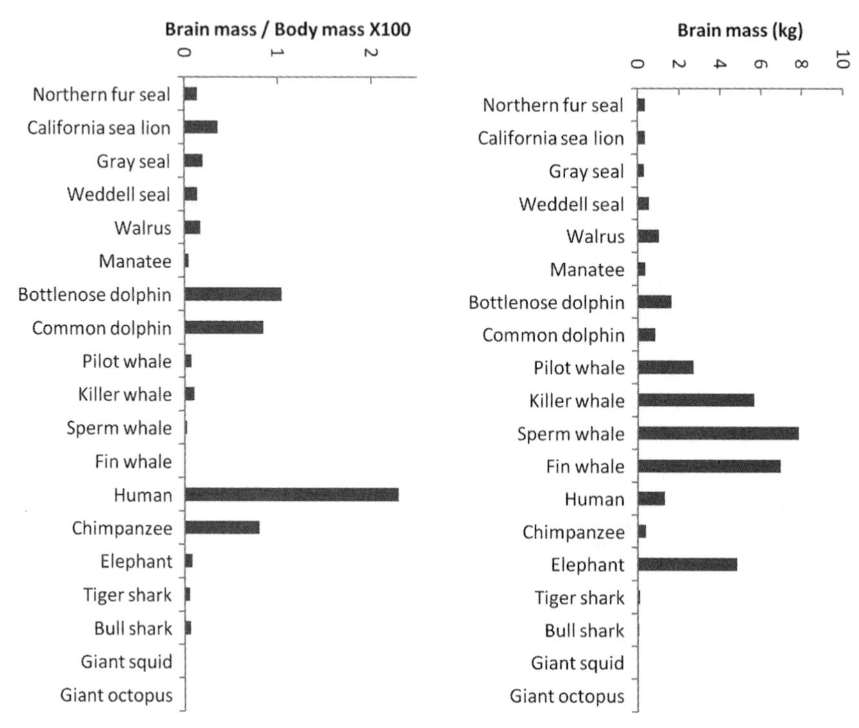

3.3. Relative and absolute brain sizes of some large marine and terrestrial animals (references: Douglas-Hamilton et al. [2001]; Marino [1998]; Northcutt [1977]; Würsig [2008]; and, for giant octopus and giant squid, from estimates made by Matt Wold [personal communication, April 2, 2014] using a variety of published sources).

human females don't always have an easy time getting that large-brained baby through their pelvises. It seems that large brains were tremendously important to humans during our evolution, and it was worth paying considerable costs to possess them. In contrast a large whale can have an even bigger brain with potentially even more cognitive power, at relatively little cost. The whale brain, although very large in absolute size, takes up a small fraction of the animal's energy needs, and the whale has no real pelvis through which to push a big-brained baby.

So how do brains measure up between species in both absolute and relative terms? In figure 3.3 we show brain weights and body weights for a number of larger species among mammals and fish. In relative brain size, humans are top, by quite a long way; as a proportion of body size, human brains are over twice that of their nearest competitor, the bottlenose dolphin, with the largest whales and elephants having tiny brains relative to their huge bodies. The seals and sea lions have brains of somewhat similar relative size to comparably large terrestrial mammals but that are quite a lot smaller than dolphins. Fish, as well as squid and octopus, have smaller brains, although both sharks and squid are bigger brained than most fish of their size.[44] In absolute terms the picture changes, with the largest whales, and then the elephants, topping the list. Sperm whales have the biggest brains on the planet, six times larger than those of humans.

As we look at the sizes of the brains of these large animals in figure 3.3, there are several strong patterns. The first is that mammals have the largest brains, both absolutely and relatively.[45] But is this because a larger brain is needed to control either the systems that keep the blood at a constant temperature or its collateral attributes, such as satisfying a much bigger appetite? Or is it because a stable body temperature helps in the development of a large and complex brain? Well, both are probably true, and more. It is complicated; mammalian adaptations lead to other mammalian adaptations and there are feedbacks. Figure 3.4 shows a simplified version of the linkages among mammalian adaptations, as depicted by John Allman.[46] Central is homeostasis, the maintenance of body temperature, but it leads, directly and indirectly, to brain development. So mammals and birds, the two groups of warm-blooded animals, have the largest and most complex brains on Earth, both in relative and absolute terms. As we will see in chapter 10, culture adds additional links and feedbacks to this network, likely further increasing brain size and complexity. This may explain a second pattern: the relatively larger brains of the cetaceans compared with pinnipeds and other marine mammals. A third remarkable feature of the data in figure 3.3 is the size of the human brain. For the size of our bodies, we have enormous brains. Brains seem to have been very important in human evolution over the last two million years. We pay large costs, particularly in energy costs and birthing difficulties, for our cognitive apparatus. Is this another cultural consequence?[47]

Moving to reproduction, the cetaceans are also most unusual marine animals. A mother cetacean gives birth to one offspring every one to five years, whereas some of her fish competitors annually spawn eggs by the millions. Cetacean young, therefore, are precious. However, unlike the fish eggs, which are nearly all doomed, that single whale or dolphin baby has a high probability of survival because it is fed, mainly through lactation, and protected by its mother and, often, by others in its social group. The young cetacean becomes a part of its community's social network, sometimes a central part of it, and learns through exposure how it works.[48] Philosopher Thomas White suggests that "dolphins may need this network of relation-

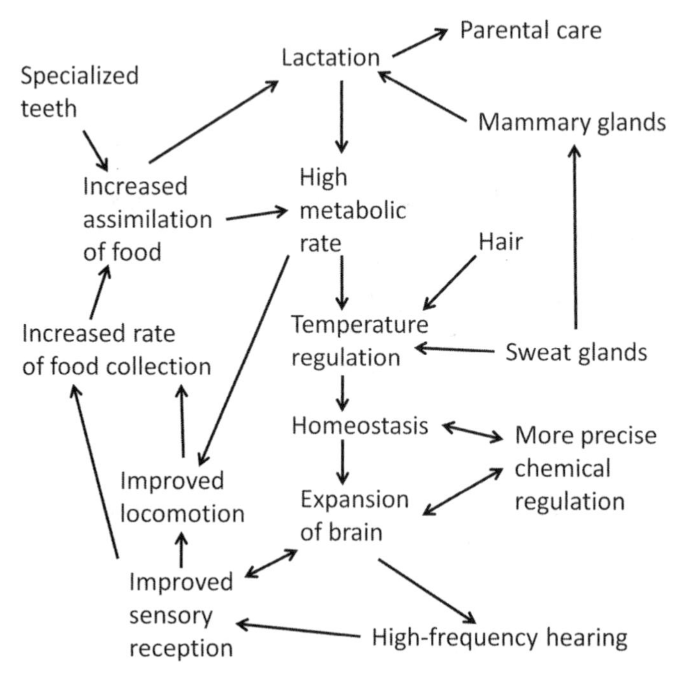

3.4. The roles of homeostasis (temperature stabilization) in mammals (based on Allman 2000, 105).

ships far more than humans do."[49] Such social networks are the ideal social substrate for culture. During this long and intensive period of maternal care, the young cetacean is exposed to information about where members of their community go, what they eat, how they hunt, and how they manage their social relationships. They receive, we would argue, their culture.

So, while the mammals carried 350 million years of terrestrial evolutionary baggage when they entered the ocean, they also brought some useful attributes, which the animals that had spent the intervening millennia in the ocean had not evolved. They breathed air and so could become large and move fast in oxygen-scarce waters. The air passages were well suited to produce loud and complex sounds, and they came into the ocean with an excellent sense of hearing so they could well exploit the significance of sound in the ocean. They arrived in the ocean with relatively large and complex brains. In the case of the cetaceans, these brains became much larger, both in absolute size and relative to body mass. In gestation and lactation, the females were well equipped to provide excellent care for a very few offspring. This mother-calf bond, in the case of the cetaceans, turned out to be a good foundation for the sometimes complex societies they would need in order to prosper.

Taking Culture to the Ocean

As we discuss in chapter 9, the ocean is habitat where culture is potentially extremely useful. The resources of the ocean vary hugely over scales of space and time so large that it is hard for one animal to make sense of. The accumulated knowledge of others is potentially a wonderful resource. But to make good use of this knowledge pool, an animal needs several attributes: a social structure with important bonds, a longish life, and an effective decision-making system. While long-lived animals are not uncommon in the ocean—the orange roughy, or deep sea perch, can live over a hundred years, as may some other fish and sea turtles—the other prerequisites for an important culture are much rarer. Rather fortuitously, the physiology and lifestyles of the mammals provide good cultural substrate, and the marine mammals brought these attributes with them into the ocean.

Culture needs a community of social relationships over which the knowledge flows. At least some of the cetaceans have complex social structures, built around the long-term care of calves. This good parenting, or more accurately, mothering, is itself a consequence of the basal mammalian attributes of gestation and lactation, enhanced by the "one large offspring at a time" reproductive strategy of the marine mammals.[50] This constraint against litters containing many small altricial babies—think a dozen puppies or piglets—is imposed by a heat-sapping ocean without refuges. Few creatures in the deep ocean outside the cetaceans possess this social substrate for culture. Away from land, there is little parental care outside the cetaceans. Pinnipeds and birds that feed in deep waters do provide parental care but raise their young on land or ice and usually soon leave them to deal with ocean life on their own.[51]

Cetaceans also have large and complex brains, another mammalian heritage, which may be useful in making sense of lots of socially learned information. As we discuss in chapter 10, these brains may even have evolved, in part, in order to use all this information efficiently.

The cetaceans vary enormously, in habitat, in size, in how they make their living, in their social systems, and in their brains. Within this diversity, our knowledge is not evenly distributed. It is overwhelmingly concentrated on the four best-known cetaceans, the humpback, killer, and sperm whales and the bottlenose dolphin. These four are iconic representatives of the cetaceans in most peoples' minds; say "whales and dolphins" to someone and the picture that pops into his or her mind is almost certainly one of these four. They also represent well the diversity of cetaceans: they have very different body sizes, environments, social lives, and cultures. In the next three chapters we will consider first the baleen whales, with most information on the humpback whale; then the dolphins and porpoises, where our information is greatly biased toward the bottlenose dolphin; and finally the larger toothed whales, especially the killer and sperm whales. For each group we will describe their ways of life, their social structures, and the evidence for culture.

4 | SONG OF THE WHALE

The Lives of Baleen Whales

The baleen whales are extremes among animals. Some, like the fin whale, strike most of us as extremely beautiful. In contrast, the physical form of the gray whale is an acquired taste. Their most arresting feature though is their size. The blue whale, at twenty-six meters long and 105 metric tons is the largest animal that has ever lived. Adults of all the baleen whale species, except the six-meter pygmy rights, are as big as the largest land animal, the African elephant, and most are much larger.

Their baleen, a unique adaptation for filtering food from the water, allows extraordinary feats of feeding. While whale sharks use gill rakers, and crabeater seals have specially adapted teeth to sieve prey, neither system has the flexibility or utility of baleen. These whales use their baleen filters in several ways. Blue, fin, minke, and humpback whales "lunge feed." Approaching a school of prey, they accelerate, and then, just as they reach the school, they open their mouths by dropping their lower jaws. The pressure of the water on the lower jaw balloons out their pleated throats, breaking the momentum of the whale, and engulfing a huge volume of water and prey. The whale, now nearly stopped, closes its mouth around this potential bonanza. However, the mouth is not quite closed, so that water is squeezed out through the baleen plates, while the prey is trapped inside the mouth. One gulp of a fin whale may contain sixty to eighty metric tons of water and food, more than the volume of its whole body.[1] The food may be almost any schooling marine creature: including krill, herring, mackerel, capelin, or squid. Lunge feeding is an exceptionally efficient way of feeding on schooling prey, which individually may be anywhere from half a centimeter to half a meter long. Lunge feeding by humpback whales, and almost certainly the other lunging baleens, comes in many variants. It can be done fast or slow, at the surface or at depth, singly or communally, and with or without bubbles, tailslaps, or sounds.[2]

The other major feeding method of the baleen whales is "skim feeding," the preferred method of right and bowhead whales. The animals swim steadily along with their mouths partially open. Water enters at the front and leaves through the baleen at the sides, with food being caught on the mesh formed by the baleen. Skim feeders have finer, longer baleen than the lunge feeders. The baleen plates of bowhead whales can be over four meters in length, and the pronounced arch of the mouth needed to accommodate this large filter gave the bowhead its name. These whales generally catch smaller prey. For instance copepods, just a few millimeters long, are a favored food of right whales.[3] In another innovative use of baleen, gray whales swim along the bottom on their right sides, sucking sediment through their short coarse baleen to filter out amphipods and other benthic creatures. These are all exceptionally efficient ways of making a living in the ocean, especially in productive areas where the prey schools are common and dense. As bulk feeders, the baleen whales need dense aggregations of prey. But when they get them, life is good. As big animals, they can store the energy from this bounty in their blubber. This permits another remarkable characteristic of the baleen whales: their ocean-spanning migrations.

For most baleen whales, life is highly seasonal. Generally, the summer is for feeding, and the winter for breeding. Summer is spent in temperate or arctic waters, and winter in, or near, the tropics. Why? Good question. Spending the summer in temperate or polar waters is easy to explain—that's where the food is, the waters there are just more productive. But then why spend all that energy migrating to the tropics and, even worse, fasting half the year, instead of sticking where the food is, sucking up all the calories, and pumping out more calves? The default answer here is that warmer/calmer waters are better for calf survival, but this is a hard idea to test directly (a study where you tried to keep half the population from migrating? Better you than us . . .), and theoretical analyses are somewhat equivocal. Also, plenty of small toothed whales seem to do just fine in high latitudes all year round. Another related suggestion is that they migrate to avoid killer whale predation on vulnerable calves—killer whales are much more common in the high latitudes—creating what can be called a mobility refuge.[4] There are exceptions. The tropical Bryde's whales don't seem to migrate much, and the arctic bowhead whales, although they do migrate seasonally, never go near the tropics.

Even within species, the general seasonal migrations do not always hold. Humpback whales are the archetypical migrating whale. They retain the distance record for a migrating mammal, with some humpbacks feeding during the southern summer around the Antarctic Peninsula but spending their winters off Costa Rica, eighty-three hundred kilometers to the north and the other side of the equator.[5] However, there are humpbacks off the Arabian Peninsula that do not make any substantial seasonal migrations, as their habitat naturally changes with the monsoons. The Arabian Sea becomes productive in the summer months with southwest monsoon winds inducing upwelling of nutrients, leading to plankton blooms and dense patches of food for the humpbacks to feed on. The same waters are relatively barren in the winter when the winds blow from the north, and the humpbacks use this time for breeding.[6] Other Northern Hemisphere humpbacks have roughly the same schedule, feeding from about May to October and breeding in the winter months, but they make substantial migrations between feeding and breeding areas.[7] It is not just humpbacks. Gray whales in the North Pacific move between the Bering Sea and lagoons of the west coast of Baja California, Mexico. Right whales in the northwest Atlantic feed in the Gulf of Maine during the summer and give birth off Florida in the winter. Blue whales move seasonally between the waters off Alaska and Central America, tracking peaks in prey abundance.[8]

While summer is feeding time for all baleen whales, there are two rather different ways of spending the winter. Humpback, gray, and right whales generally congregate during winter in well-defined locations, some of which have become famous as whale-watching destinations: San Ignacio Lagoon in Mexico for grays, Maui for humpbacks, and Hermanus, South Africa, for southern right whales. The waters are shallow and clear, indicating little phytoplankton and ocean productivity, and there are *lots* of whales. Because of the density of whales and scarcity of other life, the whales have little to feed on, and it seems that most animals fast through the winter months, even though they are active. Females give birth while males compete for mates physically and acoustically. During these months, they live off the blubber reserves built up during the previous summer. These are fat whales, especially at the start of their winter fasts.

In contrast, the rorquals—blue, fin, sei, Bryde's, and minke whales—are slim, wonderfully sleek, streamlined creatures. Apart from the tropical Bryde's whale, most rorquals, like the fat whales, migrate seasonally to warmer waters. However, unlike the fat whales, the rorquals do not seem to gather in winter in traditional breeding areas (with the important exception of a recently discovered blue whale nursing ground off southern Chile).[9] Their winter migrations seem less determined. Although we know rather little about where, or how, the rorquals spend their winters, the information that has been gathered suggests broad wintering grounds within which they wander.[10]

These two types of wintering behavior are what biologists call "alternative strategies." Each has benefits and costs. The fat whale strategy is to congregate in particular places. These are usually relatively calm, warm, and quite well protected, good places to raise a young calf. There are lots of whales around, so they are also good places to find a mate. But, on the downside, they are not good places to find food—thus the need for fat bodies and thick blubber to survive the winter. The alternative rorqual strategy is to disperse in winter, to roam widely, presumably giving the advantage of occasional meals. A thin, streamlined body makes this searching more efficient, and with this winter sustenance they do not need the energy stored in thick layers of blubber. But they are usually in rougher waters, so we might suppose that their calves have a harder time. However, the most obvious disadvantage of the rorqual "dispersing" strategy is that suitable mates are not necessarily at hand. How do these animals scattered over large areas of ocean find mates? That is an important part of the story to come.

The lives of the baleen whales, especially the fat whales, are framed by the seasons and their migrations between the summer feeding grounds and the winter breeding areas. It is during the winter that the calves are born and suckling starts. The fetal baleen whale grows fast in the womb, 2.7 centimeters per day for an unborn blue.[11] After birth, the calves suckle on the rich milk of their mothers, 20–50 percent fat, the fast growth continues, so that the calf has added about 60 percent of its mother's length by its first birthday.[12] During the last weeks of gestation and the first few months after birth all that energy is coming from a fasting mother. In springtime the calf follows its mother on her migration to the cold-water feeding grounds, where, over the summer, the calf starts to feed itself. During the female's subsequent trip to the warm-water wintering grounds, she may become pregnant, the beginning of an eleven-month gestation. This leads to a two-year interval between births, the most common pattern for the baleen whales. However, there are exceptions. Some females, perhaps especially female minke whales, become pregnant soon after giving birth and so produce a calf annually, while others, such as right whales, may have calves every three or four years. The young baleen whale becomes sexually mature at about eight to eleven years of age and goes on to live into her eighties or nineties, if a blue or fin whale, or forties, if a much smaller minke whale, assuming she evades the attentions of the Japanese or Norwegian whalers.[13]

The life histories of baleen whales are rather unusual. Ecologists sometimes arrange species on a rough continuum. At one end are species that live short lives, breeding early and often—think rabbits. Their lives have been shaped by natural selection to get lots of offspring into the next generation as quickly as possible.[14] At the other end of the spectrum are organisms that live slowly, give and receive prolonged care from one or both parents, mature late, reproduce slowly, and live long—think elephants.[15] The baleen whales live longer than any other mammals, but, compared with the larger toothed whales like sperm and killer whales—as well as large land animals, like elephants, that give birth every five years or so—most baleen whales reproduce relatively rapidly, though still nowhere near at rabbit speed. All this is possible because they get so much energy out of the ocean. They can then nourish fast-growing babies, as well as their own huge bodies. This size gives them safety. While killer whales are capable of killing a blue whale, this is rare.[16] Much smaller minke whales are rather more vulnerable, but still have a pretty good chance—if they don't encounter human whalers—of making it through from one year to the next .[17]

The life histories of species and individuals within species vary somewhat from the general pattern we have outlined. But in one case, the deviation is major. The bowhead is a huge and fat baleen whale of arctic waters. These whales can reach extraordinary ages for a mammal. The way we know this is incredible: artifacts that are of human origin are occasionally found embedded in the carcasses of bowheads killed in a regulated hunt by Alaskan Eskimos. The artifacts are in the whales because of previous unsuccessful attempts to catch them. The dates of these earlier encounters are seldom known precisely but in 2007 an explosive harpoon tip of a type known to have been manufactured between 1879 and 1885 (and likely used by whalers out of New Bedford before 1890) was recovered. This gives a minimum age of 117, but the animal was likely older as the New Bedford whalers would not have wasted expensive harpoons on very young animals.[18] The bowhead lives its life extremely slowly, becoming sexually mature at about age twenty-five, with females giving birth every seven years or so; if it avoids the whalers, a bowhead has a good chance of living well past a century.[19] It is definitively the end of the slow life-history spectrum.

Baleen Whale Communities

What about baleen whale societies? Behavioral ecologists like us assume that social life has evolved generally to maximize the reproductive success of individuals—basically, the number of grandchildren.[20] Evolution works on a stage built from the biology of the animal and its environment. So we think of the social life of baleen whales in the context of their extraordinary feeding abilities, their general safety from predators, their seasonal migrations, and the nature of their breeding grounds.

Let's start with feeding abilities. How can sociality help an animal get its jaws around a large, dense school of small, slow-moving prey? At the most basic level, social organization can minimize disruption when several animals are feeding on the same school. As each animal lunges, the school scatters, making for a less dense and less profitable mouthful for subsequent lungers. Lunging together therefore reduces the negative consequences of feeding on the same school together, and we find that several species of baleen whale form groups of about two to six animals that lunge simultaneously through prey schools.[21] For skim feeders, like bowhead whales, feeding in a coordinated echelon formation may channel food into each others' mouths.[22] Minimizing prey school disruption using synchronized lunges or echelon feeding requires coordination among neighbors, but the coordination is relatively simple, as the behavior of the prey is not particularly challenging. Long-term relationships between individual baleen whales are rare, at least as far as we can tell.[23] However, there are indications of some social structure on the best-studied feeding grounds. While we think of baleen whale prey as small schooling animals, and small animals cannot swim fast because of the viscosity of the ocean, sometimes baleen whales go after prey that can move pretty quickly. For humpback whales, the waters of southeast Alaska are a major feeding ground. There are a number of prey species there, but a particularly important food is the Pacific herring. These fish are no easy catch. Adults are about thirty-three centimeters long, can swim fast horizontally, and because they have good control of their gas bladder, can also move quickly up and down the water column. They have good vision and hearing. Pacific herring are nutritious and do form schools, but, unlike much other baleen whale prey, they can make strenuous and effective efforts to escape the humpbacks' lunges. In response, the humpbacks have developed a range of techniques to corral the herring. They blow encircling nets of bubbles, they use sounds, and they make use of their long flippers. And they work cooperatively, in groups of up to twenty animals. Much of this was uncovered by Fred Sharpe: "I remember my first bubble net. It was 1987 and my first time researching whales in Alaska. Sitting quietly on the waters of Chatham Strait, a circle of bubble [*sic*] began to form at the surface. Over the hydrophone came a wild cacophony of trumpet blasts. As the whales burst through the surface, I leapt up and cheered. That was the moment I became captivated with the Alaskan humpbacks."[24]

He has been studying them ever since. Chatham Strait is a great place for Pacific herring. The whales there form tight, and enduring, associations and work together to corral the herring using bubbles, sounds, and flipper waves.[25] There is some evidence that Chatham Strait individuals specialize in the different elements of cooperative feeding, with particular individuals being bubble blowers or trumpeters.[26] Fred Sharpe notes: "Small numbers of roving whales also enter Chatham Strait, some of whom will temporarily join the core community. However, these visitors do not seem to deter the community from its highly coordinated mass attacks on herring shoals. Rather, the presence of these recruits appears to provide more herders for the effective containment of herring schools."[27] Roughly fifty kilometers away in Frederick Strait, a larger assemblage of humpbacks feeds off a more diverse range of prey, including a lot of relative easy-to-catch krill. These humpbacks have the transient, impermanent social relationships that are more typical of baleen whales worldwide.[28] Interestingly, and important for our discussion of baleen whale culture, there is no evidence that baleen whales continue to feed in the same group as their mothers do in the years following weaning, even among the tight, enduring groups of cooperative feeders in Chatham Strait.[29]

Of course we may easily be missing other types of social structure among feeding baleen whales. The great majority of what is known concerns humpbacks, and their social life has only been studied on a few of their many feeding grounds. Our hunch is that the bowhead whale, that extraordinary creature of many surprises, is a good candidate for undiscovered foraging-based social structure.[30] Also, we do not know whether baleen whales help each other to locate prey. Roger Payne, among others, has speculated that they use their sounds to alert each other to feeding opportunities.[31] Although he suggested this thirty years ago, it remains speculation. However, from accompanying their mothers during their first summer, the young learn about foraging locations and, perhaps, also about foraging methods.

When behavioral ecologists investigate a species' sociality, usually the first factor they consider is predation: how working with others, or just being around others, can reduce the risk of being killed and eaten by a predator.[32] The second factor is feeding. As, we will discuss in the next two chapters, scientists believe that predation is at the heart of dolphin sociality and is important even for sperm whales. With the baleen whales, we have taken the factors favoring group formation in the reverse of this usual order, looking at feeding first. This is because they are so large and, thus, safe from most predators. But they are not completely safe, particularly when faced with killer whales.[33] When Hal followed humpbacks in his sailboat off the northeast coast of Newfoundland in the 1970s and 1980s, he noticed that, when the whales were feeding, their groups were noticeably variable in size. Sometimes the humpbacks were alone, at other times in groups of ten or more. Using sonar traces, he found that these group sizes were quite closely related to the size of the fish school on which they were preying, with solitary whales preying on schools a meter or two across, while the largest groups fed on schools spanning one hundred meters or more.[34] But when the humpbacks took a break from the feeding, they preferred to form pairs. This pairing may have some defensive function, as when attacked by killer whales, humpbacks, like some gray and right whales, can form tight clusters.[35] In contrast, minke whales typically flee by themselves when attacked.[36] The baleen whales may also use social groups as defensive measures on the warmer-water breeding grounds, although there is less evidence for this. The females and their newborn calves stay close to one another but usually do not interact much with others.

Mating is, by its very nature, a social activity. However, for some baleen whales the mating process goes well beyond consensual sex between one female and one male. Most obvious are the "surface active groups" of right whales, containing one female and a number of males trying, very obviously, to mate with her.[37] There may be anywhere from two to over forty males in a surface active group. The males compete by trying to push each other out of the way and getting themselves into a location where they can mate with her. They use the horny callosity growths on their heads as weapons in these shoving matches.[38] A female may incite this competition, using particular sounds to attract the males, and by lying on her back for much of the time, thereby frustrating the males who cannot get near her genital areas.[39] These surface active groups last an average of one hour but can continue for much longer. Breeding humpback whales also form surface active groups with several males surrounding a female, although there are some differences from the right whale pattern.[40] While right whales involved in these surface active groups move little, it is different with humpbacks. We have followed a surface active group of breeding humpbacks charging across forty kilometers of Silver Bank off the Dominican Republic, with our small boat straining to keep up. In surface active groups of right whales, waving penises are prominent, and mating itself is seen quite often, while the matings of humpbacks are much more discrete, being rarely witnessed.

The other baleen whales don't form surface active groups when breeding in any obvious way, but this does not mean an absence of sociality. As we will discuss in the next two sections, sounds seem to be an important part of the mating process for baleen whales, and the whales can hear each other over large ranges. The animals may well be interacting in significant ways but be too far apart for humans to consider them part of the same group—or even to see them together. This mismatch between what is likely meaningful for the whales and what we humans can see is a problem for our understanding of all the cetaceans but perhaps especially for the baleen whales, whose sounds are so loud and carry over such long ranges.

Our current image of the lives of these huge animals suggests three principal classes of baleen whale social relationship. These social ties may form conduits for cultural information. The relationship classes are: the mother-calf bond; the feeding aggregation; and breeding males.

Mothers and their calves are very tight for the first six to twelve months after birth, with the calf rarely straying far from its mother.[41] During those months calves learn about migratory routes—we will discuss the evidence for this later in the chapter—and perhaps also the structure of feeding grounds and what is good to eat.

While feeding, a baleen whale lives in a community of animals using the same ground, and sometimes also in a tighter nested community of those using the same feeding techniques. The feeding whales form groups, usually not very stable, but when cooperation becomes very profitable, as with the herring-feeding humpbacks of Chatham Strait, important bonds can develop and endure. There may be roles for particular individuals. These aggregations look to us like good substrates for the flow of information about prey distribution and abundance, as well as foraging techniques.

Breeding males interact socially, both in surface active groups, where they are often in physical contact, and over much greater ranges—perhaps up to tens or hundreds of kilometers—using their sounds. We usually imagine these social interactions between breeding males as competitive and sometimes as aggressive, but, as we will see later in the following section of this chapter, there may also be cooperation. The breeding males learn their songs from each other—and perhaps much more.

So what does it mean to speak about a "community," our context for culture, when it comes to baleen whales? For some of the less social species, this is tricky. We have very little evidence for prolonged social bonds, social delineations, or other social structures that we might use to define communities. It also gets tricky when one considers the geographic scales over which these animals can hear each other and so potentially interact. Some fascinating insights from military bottom-mounted hydrophone arrays, designed to listen for Soviet submarines, indicate the potentially huge scales such acoustic communities may encompass. Acousticians given access to the arrays were able to track a single blue whale on a forty-three-day, 2,500-kilometer circuit in the mid-Atlantic that spanned the latitudes of Miami almost to Washington, DC, by triangulating on signals picked up over hundreds of kilometers.[42] Does this mean there is nothing we could meaningfully call a community in baleen whale society? Not really. We are about to discuss humpback whale song, where the cultural community is an entire breeding population, in acoustic contact throughout their migration. In other cases, it seems that baleen whale communities are more circumscribed, incorporating only those animals in a particular area that might, through proximity, be able to sense or communicate with each other. There is also good evidence for cultural transmission within such subsets.

The Song of the Humpback Whale

As we described in chapter 1, the song of the humpback whale is a prominent part of the acoustic environment of the ocean (see plate 3). But the humpbacks' songs have now reached well beyond the ocean—they have left the solar system on board the *Voyager* spacecraft.[43] They have become part of human culture and, in one respect, have had a major impact on human behavior. This all began in the late 1960s when recordings of humpback sounds made from U.S. Navy hydrophones were heard by the American scientist Roger Payne. He then made his own recordings. It was a profound experience: "The first time I ever recorded the songs of the humpback whales at night was off Bermuda. It was also the first time I had ever heard the abyss. Normally you don't hear the size of the ocean when you are listening, but I heard it that night . . . That's what whales do; they give the ocean its voice, and the voice they give is ethereal and unearthly."[44]

Then, Roger Payne and his colleague, Scott McVay, tried to make sense of these recordings. Their story of discovery is told in *Thousand Mile Song* written by David Rothenberg.[45] Payne and McVay's paper in the influential scientific journal *Science* includes the statement: "Humpback whales emit a series of surprisingly beautiful sounds."[46] Rothenberg, a philosopher and musician, notes—rather sadly, it seems—that no scientists had used the word "beauty" in a technical paper about whale song before. Or since. We, as whale scientists, know why. But Payne and McVay had the guts to include the *b* word. Indeed, they did much more than note the song's beauty—they used it.

Roger Payne put out the long-playing record *Songs of the Humpback Whale* in 1970. It swept up the charts. According to the jazz musician Paul Winter, "*Songs of the Humpback Whale* is a timeless classic of the earth's music. It deserves a place in our cultural pantheon, alongside the music of Bach, Stravinsky, and Ellington."[47] The songs were incorporated into human music: into jazz by Paul Winter and into classical music by Alan Hovhaness, among many others. But when the pure-voiced folk singer Judy Collins sang the old whaling song *Farewell to Tarwathie* accompanied only by humpback whales, the public was entranced. As people heard these extraordinary, and beautiful, sounds of the deep, their image of whales changed, from being sources of margarine in the 1960s to icons of the burgeoning environmental movement in the 1970s.[48] This shift in attitude had a large role in the 1982 International Whaling Commission's declaration of a moratorium on commercial whaling and its near-elimination in 1986. During the 1960s, population biologists had warned that whaling was unsustainable.[49] But, as we know from so many fisheries, dire warnings from population biologists are almost always trumped by commercial interests, and the destruction continues.[50] However by the mid-1970s the whales had become different, they had become singing beings. People listened to whale songs, and politicians listened to the whale-song listening people and, then, finally, acted on the warnings of the population biologist Cassandras. The whales were "saved" from the immediate threat of large-scale commercial whaling.[51]

In retrospect, the scientific discovery of the humpback whale songs, and Roger Payne's promotion of their musicality, came at just the right time. The whale populations were mostly in desperate shape, or heading that way fast, but not yet gone. It was also fortuitous that although humpback whale song has been heard and recorded in many parts of the world over the past fifty years, the Bermuda songs of the 1960s that Roger Payne recorded and publicized are considered by human musicians to be among the most beautiful ever heard from humpbacks.

So what is this song?[52] First it is very loud indeed. The hydrophone that we tow behind our sailing boat can pick up humpback whale song at ranges of about fifteen kilometers, and the whales can almost certainly hear each other much farther apart, especially in deep water where sound travels better.[53] The song is also long. It cycles with a period sometimes up to thirty minutes, but whales can sing continuously for many hours.[54] It is made up of sounds varying in frequency from thirty to four thousand Hertz, so from three octaves below middle C to four octaves above it, roughly the range of hearing of a middle-aged human. The notes, or units, of the song can last anywhere between 0.15 and eight seconds and take various forms, including the purest whistles, low moans, coarse grunts, and prolonged rattles.[55] These notes are arranged in a very definite way, and this structure is what makes a "song" of the song of the humpback whale.[56]

A song cycle contains about eight "themes," and these have a distinctive, and invariant, order: theme I is followed by theme II which is followed by theme III, and so on. When the cycle is complete, theme I follows theme VIII, usually without a pause. Then we hear theme II, and so on. Because it is a continuous cycle, designating one of the themes as "theme I" is largely arbitrary, although the whales usually choose the end of a particular theme in order to come to the surface to breathe. Each theme consists of a number of nearly identical "phrases." The most flexible part of the song is in the number of phrases in each theme. There can be two or more than twenty, and a single whale may change the number of phrases in a theme between consecutive cycles of a song: for example, two "theme III" phrases in one cycle, and ten "theme III" phrases the next, but "theme III" is almost always followed by "theme IV."

To illustrate all this, we'll take Payne and McVay's "Song type A" from Bermuda in April 1964, which has six themes, illustrated in figure 4.1 and verbally described by David Rothenberg:

> *Theme I*: "Phrases of three units, beginning with a sound like a motor rumbling, followed by clear, sustained moans, first warbling, then dipping, then steady, each time a bit higher . . . that builds toward a conclusion with two high, sustained notes."
>
> *Theme II*: "A long series of [quite variable] piercing *buweeps*."
>
> *Themes III and IV*: "Great sweeps morphing into low, slowly rising pure tones. In between the sweeps are astonishingly high cries. Theme IV seems to end with increasingly strong, deep bellowing units."
>
> *Themes V and VI*: "Evenly spaced, rigid in time, ending with two low grunts. The main difference between them is that in theme VI, the low tones are steady and hold to a single low pitch, while in theme V, they wooze and waver."[57]

Later research found even more structure in the song. For instance parts of the song *rhyme* in the sense of repeating patterns "in highly rhythmical contexts," ending phrases with similar notes, as with human rhymes.[58] Humans use rhymes as mnemonic devices, and maybe the whales do too, as rhyming sections are most likely to be present in the most complex songs. The songs contain recurring patterns and structures, "rules" one could say, that persist even though the actual content of the song has moved on.[59]

Payne and McVay, after analyzing many songs from Bermuda recorded over a decade or so, concluded that "there seem to be several song types around which whales construct their songs."[60] Variation among songs became a major focus of the research in Roger Payne's laboratory in Lincoln, Massachusetts, during the 1970s and 1980s.[61] Katherine Payne, who was married to Roger during that period, took the lead role in this painstaking research.[62] She and her colleagues found that in any area at any time, the humpbacks virtually all sing the same song but that the song evolves with time:

> Basic units change in frequency, contour, duration, and the ways they are organized to make phrases. Phrases change in the numbers and types of

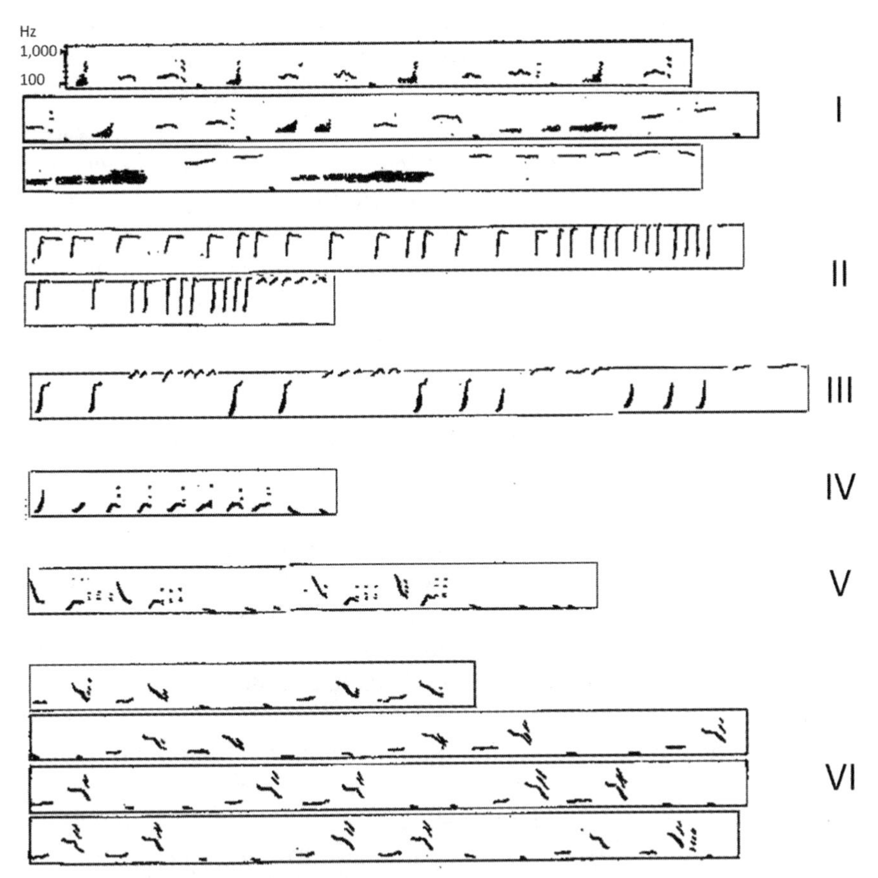

4.1. The "score" of the song of humpback Whale 1 from April 28, 1964, off Bermuda. This tracing of a spectrogram plots frequency (*vertical axis*) against time (*horizontal axis*) and shows the six themes of the song. These are sung in order, and theme I follows theme VI. This cycle of the song lasted about ten minutes (adapted from Payne and McVay 1971, fig. 2).

units they contain and in their rhythmic patterning. Themes gradually occupy a larger or smaller percentage of the song on average. . . . After some five or ten years, every theme is either much changed as a result of many little changes, or it has become obsolete and dropped out of the song, or both. At the same time, new phrase types have been introduced, imitated, and developed into new themes.[63]

Katherine Payne and her colleagues also found that songs in different ocean basins follow the same general rules. But they have different content and distinct trajectories of change.[64]

Humpback song has clearly had its effects on human culture—influencing both our music and whaling practices—but what of our interest in whale culture? These discoveries are particularly important for us because there is only one way large numbers of animals can sing the same song that evolves over periods of time that are much less than an individual's lifetime: culture. Genes cannot do this, nor can individual learning. The icing on the cultural cake comes, however, from the discovery of another kind of song change, in which the song being sung at any location can change dramatically, into an entirely new form, with new units, new phrases, and new themes, within less than a year. A revolution, rather than an evolution.

The first description of such a "cultural revolution" in humpback song was by Michael Noad and his colleagues who were studying humpback whale song off Peregian Beach on the east coast of Australia.[65] Here, the coastline forms a kind of chicane for migrating humpbacks heading between Antarctica and the Coral Sea, so they are predictably present in large numbers during defined periods of time as they pass through. This means they can be intensively and exhaustively studied during northward and southward movements, which Noad started doing during the 1990s. In 1996, having become familiar with the song of the east Australian humpback, he heard one whale singing a new and totally different song, with new phrases and themes. In 1997 this novel song took over; by the end of the season virtually all the humpbacks passing Peregian Beach were singing it. Noad was puzzled—this challenged the pattern of gradual evolution described in other humpback breeding locations around the world. It was not until his colleagues played him a song they had recorded off the *west* coast of Australia that he realized what had happened. This "new" song was actually the same as that sung by the population of humpbacks that migrate up the west coast of Australia into the Indian Ocean. The song had somehow jumped across the continent. How? A reasonable scenario is that a few Indian Ocean humpbacks, on their way north from the Antarctic feeding grounds, had gone the wrong side of Australia and into the Pacific. They kept singing their Indian Ocean song, and for some reason this proved irresistible to the east coasters. This was the first scientifically described nonhuman cultural revolution. It was likened to the Beatles crossing the Atlantic in February 1964 and changing the course of North American music, as American bands imitated the style of the wildly popular Liverpudlians.[66]

When you track song over time within ocean basins, fascinating patterns emerge. Ellen Garland, Michael Noad, and their colleagues have painstakingly traced the evolution of humpback songs across the South Pacific. They used recordings made between 1998 and 2008 in locations from the east coast of Australia to French Polynesia, about six thousand kilometers to the east.[67] This expanse of ocean includes several humpback breeding grounds, whose members migrate seasonally to the Antarctic to feed in the southern summers. But as humpbacks are generally faithful to their mother's breeding ground, individuals rarely make east-west movements between these South Pacific grounds. Garland and colleagues traced the movement of song types *eastward* across this huge ocean. So, for instance, the "blue" song—they identify the different song types with colors—was heard off eastern Australia and New Caledonia in 2002, American Samoa, and the Cook Islands in 2003 and off French Polynesia in 2004. By then it had been superseded by the "dark red" song on the east coast of Australia (see plate 5). The authors liken the pattern to "cultural ripples" propagating across the ocean. It remains a puzzle why the South Pacific songs should move from east to west, rather than the reverse, or both ways. But think of the spread of human cultures across Eurasia. Eurasia is about as wide as the South Pacific. During most of human history, ideas generally spread from China westward, but sometimes, as in the nineteenth and twentieth centuries, the predominant flow was from Europe eastward. The reasons behind these largely one-way fluxes of human culture are complex, and so may be the forces driving the eastward progression of humpback song culture in the South Pacific. These are extraordinary findings. The rate and level of cultural change in the humpback song is, in the words of Garland and colleagues, "unparalleled in any other nonhuman animals and thus involves cultural change at a vast scale."[68]

The gradual progression of humpback songs eastward across the South Pacific contrasts with the picture in the North Pacific, where songs being sung near Hawaii and off Socorro Island, Mexico, changed in synchrony through the winter breeding season, even though they are forty-eight hundred kilometers apart, farther than the span between adjacent grounds in the South Pacific.[69] This is quite a feat, suggesting either that there is movement between the breeding grounds or that humpback singers may alter their own songs in response to songs they have heard from other, very distant, humpbacks. Perhaps the different shapes of the oceans in the Northern and Southern Hemispheres lead to different geographical patterns of song evolution. The North Atlantic and North Pacific are narrow at their polar extremities, so the humpbacks that generally forage at high latitudes in summer are thrust together. Then, as they sing on their southward migration in early winter, they keep in touch, and the songs of the different breeding grounds are synched. In contrast, Southern Hemisphere humpbacks have the entire Antarctic in which to feed, so are not as clustered when feeding or on migration. Thus it may be that Southern Hemisphere breeding grounds end up being less connected and their songs more independent.[70]

Once Payne and McVay had described the humpback song in 1971, scientists were fascinated. The quite remarkable discoveries about the evolution of the song increased this fascination. They asked: why, who, and when? Why do humpbacks sing such extraordinary songs? Why do the songs evolve? Who sings the songs? When do they sing?

The "who" and "when" questions were answered clearly in the 1980s. Only males sing, and they sing on the winter breeding grounds and during their migrations.[71] Tellingly, there is a hiatus in the evolution of the song during the summer feeding months. The whales start singing again in the late fall with the same song that they were using the previous spring.[72] Using the obvious parallels to birdsong, these results strongly suggested that the songs are involved in mating, either attracting females or repelling other males. It only remained to work out which. This has not turned out to be as straightforward as was hoped.

Scientists have watched and listened to singers as they interact with other humpbacks, and they have played humpback songs back to the whales using underwater speakers—which would usually induce nearby whales to move away—but still it is not clear what the song is used for.[73] Females don't obviously approach singing males, so it does not seem to be a mate attractant.[74] Jim Darling has studied humpback whales off Hawaii for decades. In a 2006 paper, he and his colleagues summarize 167 interactions between singers and other whales.[75] The singers often meet with each other, and nonsinging males, in an apparently cooperative manner, and then go off to encounter females together. Darling and colleagues then interpret the song as "providing a means of reciprocity for mutual assistance in mating." This, for behavioral ecologists, is pretty radical—breeding males may collaborate but are not expected to use elaborate, lengthy, and presumably expensive displays to organize this.

In 2008, two years after the publication of Darling's hypothesis, a group of Australian scientists interpreted their observations of singing humpbacks off Queensland in a more conventional framework. Joshua Smith and colleagues describe males joining female-calf pairs and continuing their songs if they had been singing or starting singing if they had not been singing. They conclude with the suggestion that singing with a female helps a male's prospects of mating with her, increasing her receptivity.[76] Smith and his colleagues think that males may join singers because, given that males sing when with females, as observed so often by Darling, other males assume that the song indicates the presence of a female. Sometimes the male's assumption is wrong, because males frequently sing when alone, but the costs of the mistake are small. Why do the males sing when alone, then, if the song is about stimulating females to mate? Smith and colleagues point to a little bit of evidence that females will sometimes join singers. But perhaps song could also generally make the population of females in the neighborhood more receptive, which would be to the male's general advantage.

There are other theories for the function of the humpback song.[77] However, the most recently published hypotheses by Darling and colleagues that the songs organize male collaboration, and by Smith and colleagues that the song stimulates female receptivity, seem most plausible to us. Both theories have difficulties. Why do males need a display as elaborate as the humpback song to organize a rather cryptic form of collaboration, when males of other species who form clear alliances, like the bottlenose dolphins that we consider in the next chapter, get by with short simple signals? And, in the case of Smith's hypothesis, why do the males often sing alone? Both hypotheses may be wrong. They may also both be right, as nature loves efficiency. Time and again studies of animal communication have identified signals that perform multiple functions, providing information along several dimensions simultaneously, information that is used in different ways by different receivers. Consider the information different listeners could extract from you simply saying "Hello." This signal can contain information on your identity, your location, your disposition to whoever you are addressing, and your current mood. Listeners could use your utterance for a variety of purposes depending on their interests, and male and female listeners would often have quite different ones.

Both hypotheses suggest functions for three of the most compelling features of humpback songs, their complexity, evolution, and beauty. According to E. O. Wilson, perhaps the world's most prominent zoologist, the song of the humpback whale may be "the most elaborate single display known in any animal species."[78] It changes almost continually. And it is, to many human ears, beautiful. Darling and colleagues suggest that because of the songs' dynamic complexity they provide "a real time measure of association between individuals," like an ever-changing "secret code" among the collaborating males.[79] It is not as clear, though, why this bonding signal should be beautiful, if indeed it is perceived in such a way by male humpbacks. If the song is intended to improve the sexual receptivity of a female, then the male's goal should be to produce a song that is beautiful—that is, stimulating—to her ears.[80] One of the origins proposed for human music is precisely this, courtship, and many birdsongs have a similar role.[81] We then only need to suppose some correlation between human and humpback concepts of beauty to explain the success of the *Songs of the Humpback Whale* in the human charts.[82] In this scenario, complexity and change may be driven by female choice.[83] Female humpback whales like a complex song, one that everyone sings, but one that has some novelty. The males comply. Humans have similar preferences: we like the typical, but with a little novelty thrown in.[84]

As behavioral ecologists we then ask the question: Why? Why do female humpbacks—or male humpbacks, if Darling's hypothesis is correct—like the complex, ever-changing, yet stereotypical song? The characteristics of the dynamics of humpback whale song, evolution at a rather steady rate, with occasional revolutions, match those of human art, music, and literature. In his 1990 book, *The Clockwork Muse*, Colin Martindale shows that trends over time in human art and music fit with laws derived from what we know of human psychology and the principles of cultural evolution.[85] Similarly, we expect that the dynamics of humpback song will be the product of both their neurobiology and their cultural pressures. But we expect both of these to be shaped by evolution. One possibility is that the song characteristics are set up as a kind of test. Singing a well-known song with a personal twist signals two things simultaneously—that you "belong" (because you know the song), but that there's something a bit special about you (because you have the smarts to introduce your own inflections). A male who gets the challenge of singing the complex ever-changing song "right" is a "good" male, and worth mating with, for a female, or collaborating with, for another male.

This is all speculation, as are our ideas about the overall functions of the songs. However, what we do know about the humpback song is that it is an important part of the acoustic ecology of the ocean; that it is loud, long, complex, beautiful; and that it is culture.

The Songs of Other Baleen Whales

The beautiful, complex, and ever-changing song of the humpback whale has made the leap into human culture. But other baleen whales sing, at least by the criteria used to identify birdsong.[86] Bowheads, blues, minkes, and fin whales sing. Sei and Bryde's whales seem to sing, although the evidence is sparse. We know little of pygmy right whales: they may or may not sing. Only in the cases of right and gray whales do we know enough to be fairly sure that they do not sing.

We will start with the bowhead, whose song seems to be most similar to that of the humpback. Bowhead song also covers a wide range of frequencies, is loud, consists of units within phrases that are sequentially repeated, is sung mainly in the winter, evolves over months and years, and differs among the waters off Alaska, north of the Pacific, and off Greenland in the northern North Atlantic.[87] The bowhead song is simpler than that of the humpback. Cycles last a minute or so rather than up to thirty minutes. Typically they have fewer phrases and so are more akin to folk songs than classical sonatas. However, unlike the humpbacks, some of the units in a bowhead song can consist of two quite different sounds made simultaneously.[88] Also unlike the humpbacks, the populations of bowheads wintering off Greenland and in the Chukchi Sea can sing two or three quite different songs at the same time. We don't yet know whether this means that *individual* bowheads sing two or more songs in the same season. We do know that the songs change completely between years. As with the humpbacks, this makes it certain that cultural transmission is playing a role. Speculations about the function of bowhead song follow similar lines to those debated for humpbacks and are even more uncertain.[89] The bowhead song has been dwarfed by that of the humpback in popular and scientific interest, the latter likely because the bowheads remain in, or near, Arctic waters in the winter, and the wintertime Arctic is not an easy place to do research. However, bowhead song is fascinating in its own right. The underlying evolutionary and cultural processes between the songs of the two species may not be that different, but we do not know.

The dwarf minke whale, the smallest of the rorquals, may have the quietest song, but it is still quite something. Described from studies off the Great Barrier Reef and dubbed the *Star Wars* vocalization because it "sounds almost synthetic, metallic, or mechanical" and uncannily like a laser blast from the science fiction movie, it is "complex and finely structured."[90] It is made up of three units, A, B and C, in each of which the minke makes two distinct sounds at the same time. Unit A, which is sung three times in a row, is a complex pulse with a high-frequency down-sweep on top of a low-frequency rumble. Units B and C are longer, complex "pulsed sounds." These *Star Wars* sounds are repeated in a rigid, patterned sequence. In the North Pacific, another species of minke whale makes "boing" sounds, quite similar to the *Star Wars* vocalization.[91] There are two versions of the boing, one heard in the waters around Hawaii ("the central boing"), the other off California (the "eastern boing").[92] Although we do not know their function, or which sex makes them, both the boings and *Star Wars* sounds fit definitions of "song" and are made primarily in the winter months when the animals are breeding.[93] However, unlike the humpback song, the minke songs do not seem to change much with time. This is important because it means that the minke songs do not have the "synchronous change in complex characteristics" property that makes us certain that humpback and bowhead song are culture.

The fin whale makes extremely loud, extremely low sounds, the "20-Hz pulses."[94] These are perhaps the simplest of the baleen whale songs.[95] The pulses are pure tones that sweep down from about twenty-three to sixteen Hertz over about one second. They are repeated at regular intervals of about seven to twenty-six seconds and go on for hours, interrupted briefly as the whale surfaces to breathe.[96] Sometimes the pulses are made in "doublets," with alternating short—say, eight seconds—and long—say, eleven seconds—intervals between pulses.[97] Roger Payne and Douglas Webb speculated in 1971 that fin whales oriented to each other at ranges of hundreds, or maybe thousands, of kilometers using their calls.[98] More recent studies confirm the potential range of hearing, but point to the fin whale song as a mating signal. All singing fin whales whose sex has been checked were males, and fin whales usually sing in the winter breeding season.[99] The song differs consistently between ocean basins in the average interval between pulses, how much the pulses sweep down, the patterning of the doublet call, and whether the two pulses in the doublet differ consistently in intensity—the first is usually a little quieter.[100] In some regions, such as the western Mediterranean, it seems that there are distinct populations, identifiable by their song, in different areas.[101]

Blue whales have a song that is proportionate to their size, similar to the fins' but, on average, maybe a little lower, a little louder (plate 6). Blue whale song can be heard by hydrophones at ranges of hundreds, and sometimes thousands, of kilometers.[102] The blue whale song, like the minke and fin whale songs, contains only one theme, which is repeated. Each phrase of the theme is made up of one to five different units in a prescribed sequence. Some of the units are pulsed or throbbing, others are pure tones. All are low pitch, too low for us to hear properly unaided. A whale can sing for days, taking only short breaks to breathe.[103] Currently, blue whale scientists recognize eleven song types worldwide from different ocean basins, suggesting distinct populations, although in some cases populations have overlapping ranges with adjacent song types.[104] All whales within each of these populations sing virtually the same song, but there are major and obvious differences between the different types and a little individual variation.[105] The structures of these songs are remarkably stable. So the songs recorded in 1964 and 1997 off New Zealand sound very similar.[106] Unless you have perfect pitch . . .

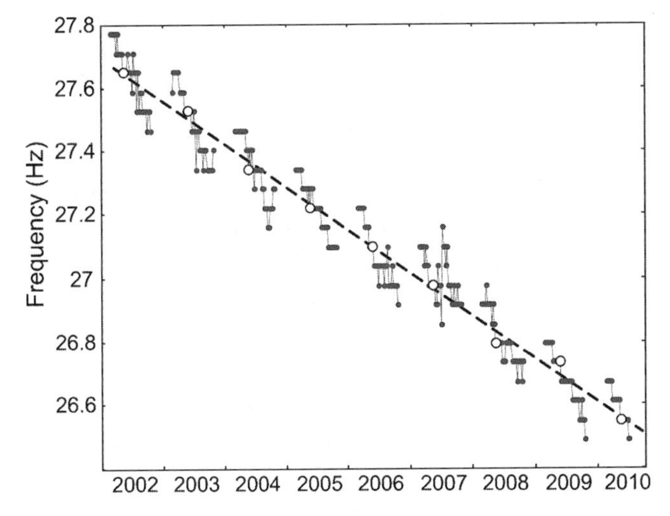

4.2. Songs of blue whales off Australia decreasing in frequency year by year and through each winter season (adapted from Gavrilov, McCauley, and Gedamke 2012, fig. 5a).

In the late 1990s, Mark McDonald and his colleagues were developing an automated detector of blue whale songs. They wrote computer routines to pick out the songs of the blues from among all the other sounds propagating through the oceans, whether from other whale species, fish, waves, earthquakes, or humans. They noticed that each year they had to tune the detector a little lower. They looked into what was behind this annoying trait, and then, in 2009 published their extraordinary discovery: the blue whale songs everywhere in the world's oceans have gradually, but steadily, gotten lower in pitch over the last thirty years.[107] The Pacific songs got deeper, the Atlantic songs got deeper, and so did the Indian Ocean songs. The rates vary a little, decreasing at a rate of from 1.8 percent per year in the North Pacific and 0.8 percent per year in the South Pacific, but they imply a decline of about an octave over a span of fifty to a hundred years. At any time, at any location, the singing blues are remarkably faithful to a particular frequency. But the following year they have all gone down a little. Alexander Gavrilov and his colleagues examined these trends in more detail for the Antarctic blue whales singing off the southwest tip of Australia.[108] While these blues decreased their singing frequency every year at about the same rate that McDonald had found in his global study—0.135 Hz per year—the song frequency declined at a much faster rate during the winter singing season: 0.4 Hz per year (figure 4.2). At the start of each season the frequency rebounded to about the average rate of the previous year before starting its within-season decline, giving an overall saw-toothed pattern of frequency with time. This high conformity and change in unison over weeks and years among whales that can hear one another, at least within oceans, means that social learning has to have a role, as with the humpback songs.

McDonald and his colleagues considered a range of potential drivers for their remarkable discovery. We will summarize the explanations that they consider most plausible.[109] For behavioral ecologists, the most obvious potential explanation is sexual selection. If the song is used by males as a mating call—there is some evidence for this—and the females "prefer" lower sounds, then there will be pressure for decreasing song frequencies through the process of cultural evolution that Robert Boyd and Peter Richerson call guided variation.[110] The rate of change of blue whale song frequencies is much too fast for standard genetic sexual selection, in which the more successful males have more offspring who tend to sing like their fathers, so in this scenario it has to be a cultural process. However, selection needs variation—we can envisage singing blue whales noticing that other singers who had lower songs seemed to be more successful and adjusting their own frequencies to match. But the conformism by which every singing blue whale at any time uses very nearly the same frequency leaves exceedingly little variation and, thus, little basis for a sexual selection hypothesis.

Some animals, including birds and other whales, change the pitch of their vocalizations as noise increases, usually away from the frequencies that dominate in the noise, seemingly to improve the audibility of their acoustic signals. The oceans are getting noisier in the frequency ranges used by blue whale songs, primarily due to increased shipping, so this is a potential explanation for the change in frequency, but because shipping produces very low frequencies the prediction is that blue whales should increase the pitch of their songs to improve audibility. Another source of "noise" for blue whales is other whales, particularly fin whales whose frequencies overlap those of blue whale song and who are also very loud. Fin numbers have probably been increasing since the end of whaling thirty years ago, so perhaps the decline in blue whale song frequencies reflects an increasing communal desire to distinguish themselves from fins or, more simply, to avoid the interference of the calls of the much more numerous fin whale. McDonald and colleagues think not, noting that "fin whale calls and songs, however, change seasonally and/or geographically, and blue whale songs occur above, below and within the fin whale frequency bands."[111]

The most intriguing of McDonald and colleagues' potential explanations for the decline in pitch of blue whale song is that it is a consequence of increasing numbers of blue whales themselves. Blue whale populations were devastated by whaling around the world, and there is evidence that in some areas, such as the eastern North Pacific, they have begun to recover. Higher-pitched blue whale songs are louder than lower-pitched ones, presumably because higher frequencies are easier to make and, thus, have longer range. If we further assume that there is a preference among listening blues for lower sounds, as in the sexual selection hypothesis, then at low population numbers a singer uses a higher frequency to reach as many potential mates as possible, but as numbers build up, sexiness becomes more important than range, so frequencies decline. This is an intriguing hypothesis, but it depends on a number of untested assumptions (about, e.g., the song function, and the preferences of listeners) and is confounded by factors such as increasing noise in the ocean.

Let us pose another hypothesis. Cultures can change arbitrarily in a particular direction without much variation or an external cause. Between 1910 and 1970 the hemlines of women's skirts rose, or occasionally fell, arbitrarily over decades, with very little variation among the skirts around at any time.[112] There was a cultural drive in one direction for one attribute, hem length, in skirts of all kinds of styles. Maybe blue whale song culture, which is clearly conformist in form and in the pitch used in any year, is in the thrall of such a cultural drive, which has spread all around the world courtesy of the long-range propagation of blue whale sounds. The hypothesis is incomplete in that we don't have a convincing explanation of what happens when the frequency bottoms out—songs can only go so low, just as hemlines can only go so high. Perhaps then, like hemlines going back down again when the skirt has almost disappeared, the cultural drive can reverse. Although McDonald and his colleagues do not discuss cultural drive as a potential explanation of the declining frequencies of blue whale song in their landmark paper, they had thought of it. Mark McDonald explains: "One of the reasons we did not mention a hypothesis similar to what you suggest is that it implies all the world's blue whales are in acoustic communication. I don't have a problem with that implication, but the reviewers no doubt would. Either they are all communicating and have adopted the same direction of frequency shift or the seven song types studied just happened to all go the same direction."[113] The scientists reviewing McDonald's paper were fine with a discussion of a frankly tenuous hypothesis that ocean acidification could affect the frequencies of blue whale song, but would not, he felt, be open to an explanation that would be near the top of the list were this the behavior of humans, rather than blue whales: cultural drive propagating around the world.

We have summarized some of what is known of the songs of the baleen whales in table 4.1. Our level of knowledge varies hugely between species.

Table 4.1. Characteristics of the Songs of the Baleen Whales

Species	Vocalization	Source Level (dB re 1μPa@1m)	Duration of Cycle (min)	Frequencies (Hz)	Structure	Variation with Place	Variation with Time	Sources
Humpback	Song	171–89	8–16	30–4,000	~8 themes; >1–20 phases; >1–20 units	Across ocean basins	Over months, years	Au, James, and Andrews 2001; Payne 2000
Bowhead	Song	158–89	~0.5–1	20–5,000	Several songs; >2–20 units	Across Arctic	Over years, seasonally	Cummings and Holliday 1987; Delarue, Laurinolli, and Martin 2009; Stafford et al. 2008; Tervo et al. 2011; Würsig and Clark 1993
Dwarf minke	*Star Wars*	150–65	~ 0.5	50–9,400	1 theme; >3 units	Unknown	None detected	Gedamke, Costa, and Dunstan 2001
North Pacific minke	Boing	150	~0.5	1,000–5,000	1 theme: brief pulse, long call	Across ocean basins	None detected	Rankin and Barlow 2005
Fin whale	20-Hz pulses	186	~0.25	18–42	1–2 units; "doublet call"	Between ocean basins	None detected	Thompson, Findley, and Vidal 1992; Watkins et al. 1987
Blue whale	Song	188	~2	16–100	1 theme; >1–5 units per phrase	Between ocean basins	Change in frequency, not character	McDonald, Mesnick, and Hildebrand 2006

There seem to be some universals. The songs are built on phrases that themselves consist of units sung in a prescribed order. The phrases can be sung many, many times, and the songs can go on for hours or days. The song structure varies between ocean basins and, sometimes, across ocean basins. In humpbacks and bowheads there are several types of phrase, grouped into themes that are sung in a given order. In these species the song structure evolves with time. In the rorqual species—blue, fin, minke—the structure, which is simpler, is stable over decades. Apart from a few aberrant individuals, all songs made by any species in a particular place at a particular time are nearly the same or, in the case of bowheads, follow two or three types. All sexed singers have been males, and song is usually not heard in the summer when the whales are feeding.

Together, these facts strongly suggest that baleen whale song is involved in mating. For the rorquals, the thin whales that do not aggregate in winter, the obvious interpretation is that the songs mainly function to attract females for mating, that they are courtship songs. As McDonald and colleagues put it, "Such sounds are ideal for communication between individuals of a widely dispersed and nomadic species."[114] But we do not know. We do not even know the function of the so-much-better studied songs of the humpback whale, or those of the bowhead, another species that seems to aggregate in winter. But the hypothesis of Joshua Smith and colleagues that male humpbacks sing to stimulate female receptivity seems to fit in this framework.[115] In the dispersed species, attracting the female is the major task for a male; in the species that gather on breeding grounds, finding a female is not so hard; turning her on may be the challenge. Female right and gray whales, the nonsinging species that congregate in winter, have other ways of choosing mates.[116]

Humpback, bowhead, and blue whales in the same place at the same time all sing the same song or, in the case of bowheads, one of two or three stereotypical songs, but these songs change, without losing conformity, over timescales much shorter than population turnover could account for. To produce such effects, social learning must have a role, so the songs are culturally transmitted.[117] In the other species where there is no record of temporal change in the stereotypical song, we cannot be as certain. The geographical variation in song type suggests culture, but perhaps different genes in different places lead to different songs. However, if song were subject to major genetic influence, one might expect hybrid songs to pop up occasionally like we hear in gibbon duets—say, between the eastern and central boings of the North Pacific minke whale—but they do not.

Nonvocal Cultures of the Baleen Whales

Baleen whale behavior is not limited to lolling around singing. Most of them migrate. Born on the low latitude breeding grounds, young whales follow their mothers away from the equator to the summer feeding areas. Humpbacks, and probably many of the other species as well, such as right whales, learn this route, or at least its endpoints, from their mothers' behavior and repeat it pretty much every spring and autumn for the rest of their lives.[118] At a population level, this results in matrilineal feeding grounds, which can be detected in the genetic profile of the species.[119] Suppose that you, a humpback whale, were conceived on Silver Bank in the West Indies by a humpback mother who feeds each summer off Newfoundland and a father who spends his summers off Iceland. A year later you are born, also on Silver Bank. After a few months you follow your mother to Newfoundland and learn to feed there with her: the good places, the different techniques for lunging on krill, capelin, or squid. Throughout your life you migrate back and forth between Newfoundland and Silver Bank, never visiting your father's summer home.

Thus, for a baleen whale, both the feeding ground and breeding ground are learned, usually from the mother's behavior. So these are vertical cultures, propagating predominantly from parent to offspring, but maybe not entirely. During the winter months, the waters around the central Hawaiian Islands throng with humpback whales, many thousands of them. Sit on any beach or headland in Maui between January and March, and you will likely see them. But precontact Hawaiians, although highly attuned to the marine world, made no reference to the species. It seems that the humpback whales only began to use Hawaiian waters in any significant way about two hundred years ago.[120] The numbers have increased particularly dramatically over the past forty years, likely faster than the reproductive capability of the population. Louis Herman, who documented this increase, thinks that animals not born off Hawaii were "enlisted" to use Hawaiian waters by the sounds and behavior of others—horizontal social learning.[121]

Although humpback whales almost always use the same breeding and feeding grounds as their mother, unlike the matrilineal whales that we will meet in chapter 6, they do not stay grouped with mother on small scales within a given ground. However, in the Gulf of Maine, another feeding ground of the whales that breed on Silver Bank, animals from the same matriline associate more frequently than we would expect by chance.[122] This seems to be because individuals learn feeding styles and prey preference from their mothers and so share them with their maternal kin, who are likely to be their associates.

Particularly dramatic evidence for this appeared after a deadly incident that began in late 1987 and continued on into early 1988. At least fourteen humpback whales died in Cape Cod Bay, a part of the Gulf of Maine, after eating mackerel that were contaminated with a natural, red-tide-like poison. Geneticists analyzed mitochondrial DNA (mtDNA) from ten of these fatalities. This mtDNA controls the basic chemical power plants of cells, and the mitochondrial genes are passed down through the mother, via her egg cell, and are not recombined with the father's genes when babies are made, so mtDNA traces one's maternal lineage. The scientists were startled to find that eight of the ten had one of two particular mtDNA profiles, called haplotypes, that were rare in the wider Cape Cod population.[123] The odds of this genetic pattern showing up by chance were thus vanishingly low without some other factor pulling these whales together. It turns out that nine of them had been sighted feeding together during the preceding two months, so they were not strangers.[124] Feed together with your maternal kin, face the consequences together. This particular feeding habit, which proved fatal, was likely passed down from mother to daughter over several generations, and thus came to be shared and characteristic of these matrilines.

If we have interpreted these discoveries correctly, knowledge about feeding grounds and some aspects of foraging are culturally transmitted vertically—from mother to calf. In the case of the selection of feeding grounds, this is probably through local enhancement. But the other elements of foraging that the calf learns from mother could be through almost any of the social learning mechanisms. It will be hard for us to distinguish which, but this does not speak against a cultural hypothesis. In the development of foraging techniques by human hunter-gatherers, it has also been difficult to pin down social learning mechanisms, and few would argue these are not culturally inherited.[125]

Humpbacks forage in a range of ways, from a straightforward lunge of one animal through a krill swarm, through to the complex cooperative herding of herring schools off southeast Alaska. Humpbacks are the most watched baleen whale, and a lot of this watching has occurred on the summer feeding grounds, particularly the waters off Cape Cod and southeast Alaska, on opposite sides of the North American continent. In both areas the humpbacks blow bubbles when feeding. While bubbles are used by quite a few foraging marine animals, the humpbacks are the bubble artisans.[126]

Probably the first scientific record of the use of bubbles by humpbacks was in the writings of the whaler M. A. Ingebrigtsen, who watched them feeding near Bear Island in the Norwegian arctic in 1905. A humpback went "a short distance below the surface of the water, swimming in a ring while at the same time it blew off. The air rose to the surface like a thick wall of air bubbles and these formed the 'net.' The 'krill' saw this wall of air bubbles, were frightened into the centre." The whale then "came up in the centre to fill its open mouth with 'krill' and water, after which it lay on its side, closed its mouth, and the catch was completed."[127] Off Cape Cod, the humpbacks have two general types of bubble feeding: bubble columns and bubble clouds.[128] Each bubble column is made up of lots of bubbles forming a rough cylinder about one to two meters in diameter and is blown from a depth of perhaps three to five meters. There are usually four to fifteen columns in a series, which can be arranged in a line, as a semicircle, a *J*, a complete circle about the length of a humpback in diameter, or a spiral comprising up to two complete turns. Having blown the pattern, the humpback then lunges up against the line of bubbles or through the center of the circle or spiral. Bubble clouds seem to be made of single large exhalations, perhaps a bit deeper in the water column. The bubble cloud expands as it rises toward the surface. The humpback lunges through the cloud, which seems to be blown directly beneath a prey school, usually made up of small fish like herring or sand lance, the humpbacks' primary food in the Cape Cod area.[129] It seems that the bubbles disorient the prey, inhibiting their escape responses, thereby giving the lunging whale a better meal than it would have had without the bubbles.[130]

Things are more complex for the bubble-feeding humpbacks off southeast Alaska. Bubbles are used on both herring and krill schools and may be combined with slams of flukes, flipper movements, and trumpet-like calls.[131] Up to twenty whales can feed together on herring schools (plate 4). During a single feeding attempt, they can use several of the bubble structures employed by the Atlantic humpbacks, making a curtain, a cylinder, or bubble clouds. Individuals may specialize on different parts of the complex social foraging—"bubblers," "trumpeters," and "herders"—although we don't know how consistent these roles are.[132]

All this bubbling and other behavior may be socially learned and therefore culture. The associations among maternal kin on the feeding grounds support this hypothesis. But we don't know for sure, and it's possible these behaviors are not culture. Perhaps each animal is discovering bubble feeding on its own or—though this seems improbable to us—bubble feeding is entirely coded in the humpback genome. There is one variant of bubble feeding where the case is stronger for its being culturally based, and this is because, as with the songs of humpback, bowhead, and blue whales, there has been a change in the population's behavior over timescales of less than a generation. The behavior is called lobtail feeding, which is a variant on the bubble-cloud feeding that we described a little earlier. The whale lifts its tail into the air and then slams it down onto the surface: a lobtail. It lobtails up to three times, then dives, emits a bubble cloud directly underneath the disturbance caused by the lobtails, and finally lunges up through it, often with a mouthful of fish. The lobtails may help concentrate or confuse the prey, increasing the profitability of the subsequent lunge.

In the waters off Cape Cod, lobtail feeding was first seen in 1980, by just one of the 150 feeding whales observed that year.[133] In 1981, of the fifty-one known whales seen feeding at the surface that year, two were lobtail feeding. The behavior spread through the 1980s. By 1989, forty-two of the eighty-three whales seen surface feeding that year were lobtail feeding. Lobtail feeding had become common through the population, with similar proportions of males and females, but it was far from universal. A substantial proportion of the population was never seen to lobtail feed. Lobtail feeders would feed in the same area, sometimes on the same fish aggregation, as whales using other methods, primarily standard bubble-cloud feeding, without the lobtails.

Mason Weinrich, a scientist who has dedicated decades to studying this population, tried to work out how the animals were learning this new feeding technique.[134] Lobtail feeding was rare in the older animals. The whales seemed to pick up the technique from age two, although two-year-olds sometimes appeared to be "playing" or "practicing" at lobtail feeding, with smaller, less-dense bubble clouds and no evidence of actual food. Many of the lobtail feeders had mothers who were known not to lobtail feed. Thus, the evidence accumulated by Weinrich suggested primarily oblique or horizontal transmission of lobtail feeding among immature whales, or from elders to the young.[135] Recently, Luke has been working with Weinrich, a student named Jenny Allen, and other colleagues in an analysis of the data records from the beginning of lobtail feeding in 1981 right through to 2008. This analysis highlighted several important points.[136] The first was that the use of this foraging tactic was strongly related to the abundance of sand lance in the habitat. The incidence of lobtail feeding in the population waxed and waned roughly in line with how much sand lance was around, so lobtail feeding appears to be a specialization that takes advantage of a particular reaction shown by this species of fish. The second was that, as in Weinrich's original study, having a mother that did this made very little difference to whether her calf would go on to develop the behavior. Third, Luke and his colleagues found that the data on the spread of the behavior (who took up lobtail feeding and when they started) massively supported a role for social learning: those humpbacks with many associates who were lobtail feeding themselves were much more likely to take up the habit than those with few lobtail-feeding associates.[137]

This analysis demonstrates how ecology and culture can interact with each other—ecologically, the availability of a particular prey item, the sand lance, was varying over time. At some point, one bright, or lucky, humpback figured out that hitting the water with his or her tail did something to the sand lance (perhaps causing them to bunch together more, making the shoal easier to enclose with a bubble net), and since then this trick has been spread and maintained in the population by cultural transmission. The lack of relationship with maternal inheritance is interesting. It is a strong contrast to the pattern of mother-offspring transmission of foraging techniques among dolphins that we shall explore in the next chapter and, also, to the inheritance of migration routes in humpbacks themselves. Instead, the behavior is mostly learned after weaning, and can therefore spread rapidly within generations. This is fortunate—or maybe the result of smart use of social learning by the humpbacks—since the abundance of sand lance can vary manyfold within intervals of just two to three years.

As with songs, scientists studying baleen whale foraging methods have concentrated on the humpback whale. Humpbacks are generally both more visible and more available than other baleen whales—the whaler Ingebrigtsen believed them "far more intelligent than other species of whale"—and their attractiveness to whale watchers gives scientists opportunities that would be hard to come by otherwise.[138] Whether singing or foraging, humpbacks seem to have it in their nature to behave in elaborate, demonstrative ways. However, other baleen whales have varied foraging methods. For instance the minke whales in Puget Sound have at least two individually distinctive foraging styles, lunge feeding and "bird-association" feeding.[139] Some consistently use one, some the other. The authors of the study, Rus Hoelzel and his colleagues, believe that these foraging methods are learned, but are they socially learned, and therefore culture?

We have concentrated on singing, migrating, and foraging as candidates for baleen whale culture because these are the parts of their behavior we know most about. But culture could well be involved in almost anything the animals do. For instance, southern right whales sail.[140] They tip themselves up, head down, with their tails sticking out of the water. They orientate themselves so that their tails are perpendicular to the wind. After sailing for a while, they may right themselves, swim back upwind, tip up, and do it all again. They only do this when the wind is between fifteen and thirty-three kilometers miles per hour.[141] Roger Payne notes that while sailing is an excellent way of getting around the ocean—we use it extensively ourselves—right whales seem to only use it as play, rather like the Incas inventing the ever-so-useful wheel and then only using it in toys.[142] We do not know whether sailing in southern right whales is culture, but it is telling that the very similar northern right whales do not sail. Then there is the "tail-up" behavior, a frequent habit of the humpback whales on Abrolhos Bank off Brazil, and occasionally elsewhere.[143] The whales, of both sexes and all ages except calves, hold their tails in the air, but, unlike the right whales who usually position their tails perpendicular to the wind as a sail, the Abrolhos humpbacks often hold their tails parallel to the water, frequently rotate about their axes, and sometimes move upwind.[144] Why? The tail-up of the Abrolhos humpbacks is another of the many cetacean behaviors whose function we really do not understand.

The baleen whales are extraordinary animals, with a wide range of intriguing behavior, both vocal and physical. Some of this behavior is culture, some may be. We have outlined the evidence here. We will evaluate it in chapter 8. Now it's time to meet some dolphins.

5 | WHAT THE DOLPHINS DO

We are moving on in this and the next chapter to the toothed whales, the members of the second suborder of cetaceans, the Odontoceti. We split this group in two because in terms of society and culture there is an important division that does not follow the suborder taxonomy of the cetaceans. In this chapter we will talk about dolphins in the common usage of the term and as epitomized by the bottlenose dolphin. In chapter 6 we will consider several species that, despite belonging to the same Delphinidae family, carry the word "whale" in their common name as a reflection of their larger size, for example, killer whales. As we shall see, despite their taxonomically close kinship, evolution has taken the societies and cultures of the "dolphin" Delphinidae and the "whale" Delphinidae species in quite different directions. We include the small, but nondelphinid, porpoises in this chapter—although we have little to say about them—and the large, also nondelphinid, sperm whale in the next. Here we will introduce the societies of dolphins and describe variation in their behavior around the globe, much of which we argue is a result of cultural transmission. Our discussion will dwell on two extraordinary pieces of behavior—sponge carrying by dolphins in Australia, and dolphin-human fishing cooperatives in Brazil and Burma. Culture in dolphin societies often seems to revolve around the bond between mother and calf and the life skills and knowledge passed on during their time together. In some cases it looks like the possession of these skills and knowledge can come to define social divisions within dolphin communities.

Dolphin Societies

Dolphins live in about as wide a variety of aquatic habitats as we can imagine. Ganges River dolphins reach the foothills of the Himalayas, and hourglass dolphins circle the Antarctic continent, while rough-toothed dolphins live mostly far offshore in the tropical ocean. They eat all kinds of fish, squid, and sometimes other things, and the diversity of their foraging reflects the diversity of their prey. Larger species have generally few predators when adult, except, of course, killer whales—the ultimate predators—but for smaller species, and for calves, sharks can be deadly enemies. We humans also cause them problems, by catching them, sometimes directly for food and often indirectly as bycatch in fishing nets or insidiously by altering or invading their habitat. Increasingly, with dark irony, we are also at risk of loving them to death with uncontrolled dolphin watching. While human and dolphin societies interact in multiple, complex ways, these interactions are not always negative. Cultural transmission in both parties seems to play a large part in these interactions.

Despite the fascination many people have with dolphins, the prominent role they play as symbols of nature within our own societies, and the values this leads us to project onto them, we still know virtually nothing about the societies of the vast majority of dolphin species. Of the thirty-four species of large-brained mammal that most people would commonly call dolphins, our knowledge is immensely skewed toward just one: the bottlenose dolphin, "Flipper." This makes us the poorer in the struggle to understand our own society. Since dolphin and human societies are products of evolution, their diversity offers us a tremendous natural experiment to help understand those evolutionary processes. Despite this, the little we do know shows that with all the variety in habitat comes a range of social systems. We best know the species and populations that conveniently live their lives close to shore, especially in areas amenable to our presence like sheltered seawater lagoons, bays, and estuaries.

The bottlenose dolphin is an iconic . . . species? Actually there are two recognized species covered by this common name, both belonging to the genus *Tursiops*: regular bottlenose (*T. truncatus*) and Indian Ocean bottlenose (*T. aduncus*), and there may be more.[1] We also recognize "ecotypes"—for example, "onshore" and "offshore"—that are probably subtly adapted to the pressures of the different habitats.[2] For convenience, from here on, unless we specify otherwise, "dolphin" will refer to the bottlenose genus.

These are the dolphins we know most about, for three main reasons. First, they are highly successful, using almost the whole range of subpolar coastal habitat, which means we see them a lot. Second, they are highly adaptable, which means they can adjust relatively easily, at some level, to captivity. Finally, they seem possessed of a natural and bold curiosity, which leads them to a rich diversity of interactions with humans. We think that this combination is not coincidental and that their curiosity and adaptability has enabled them to acquire rapidly the skills and knowledge necessary to thrive in a diversity of habitats. With the ability to learn new skills comes the selection pressure to be able to transfer that knowledge efficiently to those that matter to you—especially your children. As we discuss in chapter 9, the sea is a dynamic place, so a general learning ability that incorporates social information from others as well as the products of your own learning can match the timescales of change far better than an accumulation of genetic mutations. A lot of dolphin behavior makes much more sense as a product of the former than of the latter. We include in our evidence the extraordinary rapidity with which dolphins have learned to take advantage of the opportunities afforded by our own activities in their habitats.

So, even within one genus of closely related species—bottlenose dolphins—there is much variation in habitats, from deep fjords, to shallow, sandy estuaries, to the deep ocean. This is reflected in their societies, even though we only really know a lot about, again, coastal populations. A general, but not universal, picture is that coastal bottlenose dolphins live in quite well delineated communities, generally defined in geographical terms and separated by distance and/or physical barriers. For example, there is a community that inhabits the shallow waters of the Little Bahama Bank off the Bahamas, members of which can roam its approximately two hundred kilometer extent but do not cross the deep waters to nearby banks.[3] This is likely because the deep ocean is dangerous for small groups of dolphins, which also probably explains why fully oceanic dolphins often form groups numbering the thousands. In some areas with long contiguous coastlines, such as Florida or Australia, the barriers between communities can be porous and the picture is more of a stepping-stone series of communities between which individuals can interact and sometimes mate.[4] There can also be social structuring within communities, where certain dolphins tend to stay in particular subareas of the community range and tend to associate with other individuals that share their habitat preference, and this can be reflected in the genetic structure of communities.[5] Even populations that at first glance look homogenous turn out on deeper examination to have subtle but significant structuring.[6]

Within these communities, individual female dolphins form flexible networks of associates, any of whom they can be sighted with. They may have preferred associations with certain others, and even preferred associates for particularly activities such as socializing or foraging, just as we might have cycling friends and drinking pals.[7] These relationships are important. For a female, success at raising her calves to independence, a gold standard of evolutionary achievement, depends heavily on how successful her closest associates are at doing the same thing. For example, biologist Celine Frère and her colleagues studied the calf-rearing records of fifty-two females from one of the best-known dolphin populations in the world, in Shark Bay, Australia (to where we shall be returning).[8] They used animal pedigree models to link success at rearing calves to both genetic variation, representing the genetically inherited component of successful mothering behavior, and the success of each female's closest social associates. The result was provocative: the success of a female's social associates was over twice as important as her genetic inheritance in explaining why some females were better than others at raising calves. The dolphins probably inherit both their genes and some part of their social networks from their mothers. This is consistent with females learning mothering skills *socially* from other females, which is fascinating from our cultural viewpoint, but other explanations, such as good habitat selection by groups of associates, are also feasible. This work highlights both the importance of long-term studies of these long-lived animals and the vital significance of social networks for female dolphins.

Young dolphins have prolonged associations with their mothers. They generally stay close to their mothers until weaning at three to four years old. A calf spends much of her time in what is called the "infant position"—swimming alongside her mother at touching distance behind her pectoral fin, tucked in close to the mammary slits from which she takes her nourishment. From this position, the calf is exposed to everything her mother does during their time together, and it seems that this is a crucial period in which the young dolphin learns about its neighborhood, its social community, and the skills it will need to make a living later on.

For males life is different. They play no part, as far as we are aware, in rearing calves. Their goal is paternity. With females only giving birth every three to six years, mating opportunities are scarce and much male social activity appears dedicated to the maneuverings required to secure and maintain access to females against competition from other males, whether the females wish it or not. What we have learned about male sociality in recent years has laid the mythical image of the peaceful dolphin to rest. Male competition is fierce. There is documented evidence that fights can lead to unconsciousness, so it is entirely possible they could result in fatality.[9] There is also evidence that male dolphins sometimes kill calves, for reasons that are poorly understood.[10] Males sometimes live fairly solitary lives, but it seems that in many situations this is not the primary route to successful paternity. To fight off other males, and successfully pursue his intentions with sometimes reluctant females, a male dolphin needs help. He needs allies.[11] Richard Connor, a biologist who has spent twenty-five years studying the alliances of male bottlenose dolphins, puts it this way: "If you're going to run into your enemies, you better be with your friends, or have some that are close by, willing to be recruited."[12] Alliances between males are known from well-studied populations of both *Tursiops* species in Florida, the Bahamas, and Australia.[13] The sizes of the alliances vary between populations, and in some populations alliances appear not to occur at all. We don't understand the drivers of this variation, but whenever they occur they function the same way. Males use alliances to outcompete other males and to engage in a behavior called herding, where alliance members will aggressively guard a single female in a highly coordinated way. This serves to prevent other males getting to her and also allows the alliance to "guide" her into, for example, shallower waters where it might be easier for them to force their attentions on her. In Shark Bay, Australia, Richard Connor has described how these alliances reach a kind of Machiavellian peak, as dyads and triads of highly associated males form what are called second-order alliances (alliances of alliances) with one another in a rapidly changing picture richly evocative of human politics: "It's those guys you rarely see. What have they been up to since the last time you met them? Are they still on your side?"[14]

In 1982 the primatologist Frans de Waal published a book about the social life of a chimpanzee community housed in seminatural conditions at Arnhem Zoo in the Netherlands. He provocatively titled the book *Chimpanzee Politics* because he saw direct parallels between the power struggles and social maneuvering among the chimpanzees in the zoo and the behavior displayed by politicians all around the world.[15] Given sufficient observation, someone—probably Richard Connor!—could doubtless write a similar account of the Shark Bay dolphins that would, like de Waal's book, cross the boundary between ethology and history.

In recent years, life in Shark Bay has been dominated by what Connor calls a superalliance, a loose group of fourteen males that back each other up at various times, according to political dynamics we don't yet understand.[16] Alliances are fascinating in their own right for understanding social evolution, but what particularly attracts our interest with respect to culture is that they are often characterized by extreme synchrony in the movements of their members—dolphins in alliances produce identical, or even mirrored, movements with pinpoint timing.[17] The synchrony is so extreme that some suggest it is a signal in itself. But the fact that dolphins can do it speaks to their ability to track the actions of others and, it seems, reproduce them. Even though not in alliances, captive dolphins can be trained to behave with this level of synchrony to the delight of audiences around the world.[18] This mimicking ability is an important part of the case for cultural transmission in dolphins.

This picture of dolphin societies means that we can think of bottlenose dolphin communities at the level of the geographical habitat, such as an estuary or lagoon, or at the much smaller scale of the matriline or male alliance. The bottlenose dolphin social system, with open networks of females, and sometimes alliances of males, may be representative of those of other dolphin and porpoise species, especially those living inshore. But it may not. River dolphins, inshore porpoises, and other species living in convoluted habitats seem more solitary and may have simpler social systems. Offshore, oceanic species like common, striped, or right-whale dolphins form groups sometimes numbering in the thousands. We sometimes assume that within these huge groups social relationships are fluid, and mating promiscuous, but we do not really know—there may be permanent social units embedded within some of the vast aggregations. Some sections of the population seem to become attached to certain places, such as oceanic islands like the Azores, producing distinctions between resident and nonresident dolphins.[19] Intermediate in a different way are the spinner dolphins that spend the nights resting in structured groups over the shallow waters off the Hawaiian archipelago and then feed offshore at night, perhaps using very different social structures.[20] Our current picture of dolphin societies is currently a thin torch beam of light in a dark hall, a beam aimed at Flipper. We have much to learn.

The Diversity of Foraging Strategies of Bottlenose Dolphins

Bottlenose dolphins have a wide range of ways of making a living. In the coastal communities where we know them best, the diversity of strategies is astounding (see table 5.1). They drive fish onto beaches and then leap onto the beaches to eat them. They use sponges as nose guards when foraging in rough bottoms. They can "whack" fish with their tails, up to nine meters into the air, and "kerplunk," slamming their flukes on the water surface and so startling fish out of sea grass beds.[21] They use a specific sequence of actions to catch, kill, and prepare cuttlefish for consumption. In some instances they appear to show division of labor in cooperative groups. They engage in "carousel feeding," a combination of horizontal and vertical circling that works to bunch prey against the surface, through which they swim to create confusion and break up the shoal.[22] Individual dolphins specialize, so that in many study areas scientists have found subgroups of dolphins that forage in distinctively different ways, such as the "kerplunkers" and "spongers" of Shark Bay, Australia. This is just the start: we are a long way from mapping the true diversity of ways in which dolphins make their living.

"Beach hunting" involves individual dolphins chasing single fish into

Table 5.1. Some of the Foraging Methods of Bottlenose Dolphins

Type	Description	Where	Communal Foraging?	References
Deep water	Feeding on deep-water (>500 m) fish/squid (it is unknown how exactly)	Various	?	Klatsky, Wells, and Sweeney 2007
Whacking	Hitting fish with their tails, up to 9 m in air to stun them	Various	No	Wells, Scott, and Irvine 1987
Hydroplaning	Moving very fast in very shallow water to catch fish that have sought shelter near beaches	Shark Bay, Australia	No	Sargeant et al. 2005
Strand feeding	Driving fish schools onto beaches and then stranding, temporarily, to catch them	Various	Yes	Silber and Fertl 1995
Sponging	Placing a sponge on the beak to forage in crevices on the bottom	Shark Bay, Australia	No	Smolker et al. 1997
Conch carrying	Dislodging prey from conch shells	Shark Bay, Australia	No	Allen, Bejder, and Krützen 2011
Kerplunking	Startling fish from refuges with tail slap	Shark Bay, Australia	No	Connor et al. 2000
Bottom grubbing	Flushing out prey from sea grass with snout	Shark Bay, Australia	No	Mann and Sargeant 2003
Crater feeding	Diving into sand bottom to catch buried fish leaving crater	Bahamas	No	Rossbach and Herzing 1997
Mud-plume feeding	A U-shaped plume of mud is created by fluke strokes, which the dolphins lunge through to catch fish	Florida Keys, USA	No	Lewis and Schroeder 2003
Mud-ring feeding	Swimming in a circle around fish school striking the muddy bottom with tail, stirring up a ring of fine sand and silt around school	Florida Bay, USA	Yes	Torres and Read 2009
Rooster-tail foraging	Fast lunges at the surface followed by dives	Shark Bay, Australia	No	Mann and Sargeant 2003
Carousel feeding	Horizontal and vertical circling to concentrate fish school against the surface, where it is broken up	Various	Yes	Bel'kovich et al. 1991
Barrier feeding	Driving fish against barrier of other dolphins	Cedar Keys, USA	Yes	Gazda et al. 2005

Killing and preparing cuttlefish	Removing less favorable parts of prey using complex methods	South Australia	No	Finn, Tregenza, and Norman 2009
Trawler feeding	Feeding on discards behind prawn trawlers	Moreton Bay, Australia	No	Chilvers and Corkeron 2001
Cooperative fishing	Working with human cast-net fishers using signals	Laguna and Imbe-Tramandai, Brazil	No	Simões-Lopes, Fabián, and Menegheti 1998
Provisioning	Being given food by humans	Shark Bay, Australia	No	Mann and Sargeant 2003
Begging	Begging for food from boats	Various	No	Donaldson et al. 2012
Depredation	Stealing fish from fishing lines and fish farms	Various	No	Read 2005; Lopez 2012

Note. These foraging methods exclude those only known to be used by only one or two individuals. Also, bottlenose dolphins everywhere swim along near the surface and grab prey!

very shallow waters on a specific beach in Shark Bay, chasing them for hundreds of meters parallel to the beach and then using a specialized "hydroplaning" technique to catch the fish in water that only just covers their pectoral fins. Research by Janet Mann and Brooke Sargeant highlights social learning as the most plausible explanation for the limited numbers of dolphins using this technique, as the only calves ever observed hydroplaning in this way were born to mothers that also did it, but the technique is not reserved to specific genetic lineages.[23] This technique is similar in concept, but different in execution, to the strand feeding techniques used by dolphins in the estuarine mudflats in the southeastern United States. Here the dolphins often act collaboratively to create bow waves that wash fish up onto a mud shore, from where they are picked off by the dolphins that hurl themselves up the shore after them.[24]

Another striking cooperative feeding tactic is the "mud-ring feeding" of dolphins in Florida Bay in the United States.[25] When a group finds a shoal of fish, one dolphin swims in a circle around it, striking the muddy bottom with its tail and thus stirring up a ring-shaped cloud of fine sand and silt. The other dolphins gather around the outside of the ring, knowing that when the circle is almost closed, the fish inside it will panic and start jumping out of the water to try and escape—right into the waiting gapes of the dolphins (plate 7). It seems that individual dolphins become specialized into specific roles within this cooperative framework.

Dolphins in Cedar Keys, Florida, use another cooperative hunting technique where one dolphin, the "driver," herds fish against a wall of companions. The fish again attempt to escape this trap by leaping, but the dolphins are impressive fielders, so it does not go well for the fish. In one study, researchers led by Stephanie Gazda observed two different dolphin groups perform 145 separate bouts of this behavior and in all of them just a single individual from each group did the driving.[26] This degree of specialization is not expected unless it takes some practice to drive fish well. It is not clear why certain dolphins become specialized like this—they don't obviously seem to get more fish and surely use more energy herding than do their comrades, waiting for their food to arrive.

In South Australia, dolphins go through an extraordinary sequence of behaviors after catching a cuttlefish, all designed to increase the suitability of the prey for eating. Julian Finn and his colleagues documented this processing from underwater video. They observed that: "Cuttlefish were herded to a sand substrate, pinned to the seafloor, killed by downward thrust, raised mid-water and beaten by the dolphin with its snout until the ink was released and drained. The deceased cuttlefish was then returned to the seafloor, inverted and forced along the sand substrate in order to strip the thin dorsal layer of skin off the mantle, thus releasing the buoyant calcareous cuttlebone."[27]

Removing the ink, containing the pigment melanin that inhibits secretions in the digestive system, as well as other chemicals that apparently impair taste and smell, would, the scientists state, "improve palatability and internal digestive processes."[28] Removing the calcareous cuttlebone before eating the cuttlefish has obvious advantages. This technique allows the dolphins to take advantage of a short-lived glut of cuttlefish, which occurs every year from May to August when the cuttlefish participate in a synchronized spawning period. The prey is easy to catch because the cuttlefish are suffering their version of postcoital stupor—the dolphins apparently time their arrival to exploit this ruthlessly. Finn noted that social learning is highly likely to be involved in the development of this complex behavioral sequence, but because we currently do not know the details of how they develop it, or which individual dolphins do and do not perform this processing technique, it is currently impossible to definitively pin on the "social learning" label. Still, a highly involved and intricate sequence of techniques like this, apparently unique to a subpopulation of coastal dolphins, puts a cultural hypothesis front and center.

Even in the same general habitat, it seems dolphins will find ways to specialize into distinctive foraging niches if the opportunity presents. These specializations can be at the individual level, shared by just a few animals, or practiced widely, so the putative foraging cultures of bottlenose dolphins form a kind of continuum with, and may be based on, individual innovations.[29] One of the best-studied bottlenose dolphin populations in the world—the *aduncus* version—lives in Shark Bay in Western Australia. One reason why we and other scientists find the Shark Bay population of bottlenose dolphins so fascinating is the diversity of foraging strategies that have been described in this one location. Janet Mann and her students have spent many years documenting them and have described twelve, some of which are listed in table 5.1.[30] We have already mentioned the hydroplaning technique with which dolphins swim up beaches into water just inches deep to chase fish against the shoreline. Hunting for golden trevally, a species of marine fish, is apparently practiced by only a single dolphin in the population and involves catching these large (up to fifteen kg) fish that then take over an hour to break up and consume. A few dolphins are "provisioned" with fish provided by people when the dolphins swim into knee-high water around the Monkey Mia resort in Shark Bay. "Rooster-tail" foraging is so called because it involves fast lunges at the surface that make a spray of water fly off the dolphin's dorsal fin, followed by an immediate deeper dive. In "bottom grubbing," the dolphins flush out prey from sea grass with their snout. Mann and Sargeant have shown that in the last two examples there is an important relationship between a mother's use of these techniques during the first year of life and the probability that the calf will subsequently develop the tactic in later years: "No calves develop these foraging techniques unless their mothers engaged in them."[31] This puts cultural transmission between mothers and calves among the leading hypotheses for explaining the development and persistence of these types of behavior. But of all the Shark Bay foraging specializations, one has attracted more study than all the others combined. The argument over whether it is a cultural behavior or not strikes to the core of this book. We turn to the famous Shark Bay spongers.

The Sponger's Tale

What is a sponger? This is the name given to dolphins that show one of the most intensely studied single pieces of cetacean behavior: bottlenose dolphins in Shark Bay are quite commonly seen carrying conical sponges, apparently gathered from the sea bed, on their noses (plate 8). The level of scientific interest is sparked by two things—our tool user's natural interest in tool use in other species and the suggestion that this behavior might be culturally transmitted from mother to calf.

Shark Bay is large (ten thousand square kilometers), relatively shallow (about ten meters deep), and is formed by the Edel Land peninsula and Dirk Hartog Island and bisected by the Peron Peninsula. It is a place of natural splendor, a world heritage site with comparatively little human impact, and well populated with dolphins, as well as dugongs, turtles, sea snakes, whales, and, of course, sharks. The dolphins have quite small home ranges, about six kilometers across.[32] East of the Peron Peninsula, sponging is seen in relatively deep channels (about eight to twelve meters) roughly five kilometers from shore, but only by a specific subset of dolphins: fourteen females and one male. The carrying of sponges is only visible at the surface, but it is likely that waving sponges in front of scientists' cameras is not the primary purpose of the behavior. In a frustrating irony, despite the fact that we know the dolphins dive with the sponges on their noses, what they actually do when down there has never been observed because the water gets murky toward the bottom of the channels where sponging mostly occurs. The leading hypothesis about why they do it is that the sponges are used to protect the dolphin's nose as it roots around in the sandy bottom looking for fish—some of which are endowed with nasty defensive spikes. Scientists have tested how plausible it is by rooting around in the sand with sponges themselves—it seems to work.[33]

Sponging is worth thinking about in some detail because it illustrates many of the challenges and sources of controversy in studying social learning and culture in wild cetaceans. If the bottom-grubbing hypothesis is correct, then this technique is a rare example of a nonhuman using tools to access an ecological resource that is too risky otherwise. A venomous fish spine in the nose is no trivial thing. So the sponger's tale is bound up with ecology—deepwater channels with sandy bottoms in which various species of fish take refuge from the many predators of Shark Bay, offering a new opportunity to anyone smart enough to get at them safely. It is a specialized behavior and, as a consequence, only occurs in areas where the ecology is right—where prey and sponges are available.[34] Is this then just a local response to a local ecology, with dolphins growing up in the channels learning all by themselves how to forage there? Not really. Dolphin home ranges in Shark Bay, although small, overlap considerably, and there are also dolphins that forage in the same channels without carrying sponges.[35]

We'll never know who the first sponger was, or how they came on the idea, but at some time in the past, there must have been one. How did one dolphin become many? Did that dolphin have some kind of genetic mutation that meant only its descendants would be able to carry sponges? Reaching the bottom of channels, although they are deep relative to the surrounding habitat, is well within the diving performance envelope of these animals, so it seems unlikely. However we can look at the present-day genetics of spongers for clues. Geneticist Michael Krützen and his colleagues have done just that. In the eastern gulf of Shark Bay fourteen of the fifteen spongers all belong to the same genetic female line, as determined by studying the mitochondrial DNA.[36] That single dolphin not belonging to the genetic line allows us to exclude the idea of a single "gene for sponging," but it is still technically feasible that sponging results from more complex genetic interactions. Unfortunately for anyone trying to figure this out, the pattern is consistent with two distinct possibilities: that sponging is related to some genetic complex transmitted down the female line or that it is passed on by learning from mother to calf, in which case you would get the same link with maternal lineage genes, as they also descend by the same route—mother to calf. We can add to this, though. There is another group of fourteen spongers in the western gulf of Shark Bay, completely separate from those in the east.[37] Like them, the western spongers all share a common maternal genetic ancestry, but, importantly, it is a different one from those in the east. Furthermore, there are also members of both of these maternal lines that don't do any sponging at all. The more we look, the harder it is to base the sponger's tale purely on genetic causation. We will return to this question in chapter 8, where we weigh up the evidence for culture in cetaceans.

The transmission of sponging seems to be from mother to daughter—and occasionally to a son. But, if all else is equal between spongers and non-spongers, this does not work in the long term. On average each mother in one generation generates another mother in the next, but if, as with sponging, some offspring don't pick up the trick, in each succeeding generation there will, on average, be fewer sponging mothers.[38] Eventually there will be no sponging mothers at all, and no sponging. To keep sponging going over the long term, something else must be going on. Perhaps the sponging mothers have relatively more offspring than other dolphins, or sometimes sponging is invented independently by a dolphin whose mother was not a sponger, or there is a little horizontal transmission, with occasional learning of sponging outside the mother-calf bond.[39] All of these processes are feasible and will undoubtedly be targets of the Shark Bay researchers as they delve deeper into the lives of the dolphins.

Like the provisioned dolphins a few kilometers away, the spongers mainly associate with each other, forming a kind of subcommunity.[40] The spongers have other characteristics, for instance, they are more solitary than the average eastern Shark Bay dolphin, usually foraging alone or with their calves. While sponging was first seen in the well-studied eastern gulf, it is actually much more common in the western gulf, which has more deep channels, as well as more sponges. Are these independent innovations, or did some wide roaming female (by Shark Bay standards) bring sponging from west to east, or vice versa? Genetic studies now underway by Michael Krützen and his colleagues should give us a good idea. The sponger's tale looks set to fascinate for a while yet.

Human-Dolphin Fishing Cooperatives

Perhaps the most extraordinary of dolphin foraging specializations is to fish cooperatively with humans. These cooperatives have been described from Australia, India, Mauritania, Burma, and the Mediterranean, but best known are those in Brazil.[41] Off Laguna, in southern Brazil, a group of human fishers and a group of bottlenose dolphins have been cooperatively fishing for many generations of each species (plate 10). The origin of the cooperation is lost to living memory—it has been going on for longer than any of the humans or dolphins now taking part have been alive. Treading very carefully so as not to disturb the distinctly real living people make from the activity, Brazilian scientists Paulo Simões-Lopes and Fábio Daura-Jorge have attempted to build up a picture of this remarkable behavior.[42]

The fishing cooperatives are based along the mud shores of a lagoon system, where fishers can carry their nets into knee- or thigh-deep water ready to cast them out toward the deeper channels. When they arrive at one of seven specific spots along the shore of the lagoon, the fishers make a commotion by slapping their hands and nets onto the surface water. This, they say, lets the dolphins know they are there. Following this, if dolphins and appropriate prey are around (the main one being mullet), the dolphins will herd fish, much as they do in other populations, except here the barrier they herd against is the mud bank with the fishers on it. When this happens, the fishers watch the dolphins closely. They say they are watching for a particular signal, a distinctive type of dive, in which the dolphin arches its back high out of the water. This is the cue for the fishers to throw their nets in front of them, catching the herded shoal. The success of the enterprise demands careful attention to the dolphins' behavior. These fishers are adept at recognizing the dolphins individually, so they know that only some dolphins do it. The ones that do cooperate get bestowed with names of Brazilian ex-presidents, soccer players, and Hollywood stars—the fishers are particularly fond of Scooby and Caroba (the name of a medicinal plant), who have been involved in the cooperative for over fifteen years. They also know which ones don't, and won't bother casting when these "bad" dolphins are around. This kind of cooperative fishing also occurs at another Brazilian site, Imbe-Tramandai, 220 kilometers south of Laguna, and there are two known cases of dolphins moving between the two sites.[43]

For the humans, the payoffs are clear: fishers catch more when dolphins are involved.[44] It is less clear if or how the dolphins benefit. The current working hypothesis of the Brazilian scientists is that the net cast results in isolated or injured fish, which the dolphins can easily pick off, but it is only a hypothesis at this stage. All this raises some intriguing questions. How did this cooperation arise originally? Probably we will never know. How do dolphins come to be cooperators? Do the fishers train the dolphins, perhaps inadvertently? Is the distinctive dive an intentional signal on the part of the dolphins, or just a cue that the fishers have picked up on (they themselves have no doubt whatsoever that it is an intentional signal)? There is so much to be learned in this Brazilian estuary. Why, for example, does less than half of the local dolphin population take part in the cooperative?

Another fishing cooperative between cetaceans and humans developed many generations ago on the Ayeyarwady River, Burma, involving Irrawaddy River dolphins.[45] The interaction and the dolphins have been given special protection by the Burmese government and are beginning to become a tourist attraction.[46] Unlike the Brazilian fishers who wade into the water, those on the Ayeyarwady River throw their cast nets from small canoes. They usually do this near the river bank that the dolphins use as a barrier into which to drive the fish. The Burmese scientist Tint Tun studied the cooperative fishing in three villages. Here is his description:

> During the cooperative fishing, fishermen and dolphins communicated [with] each other by means of audio and visual signals. Fishermen sent audio signals to dolphins by making sounds with . . . a conical wooden pin, lead-weights, paddle, cast-net, and making guttural sound by mouth. Dolphin[s] communicated [with] the fishermen with body and fluke signals. Dolphins herded fish against sand banks and drove them towards the fishing canoe's direction. Then, fishermen threw their cast-nets timely and orderly onto the fish. Dolphins preyed upon fish, which stunned or darted away from the sinking cast-net or some fish protruded from mesh aperture of the net. Fishermen reported that dolphins sometimes spy-hopped to check surrounding.[47]

Tint Tun found that when dolphins were assisting, the fishers caught an average of eighty grams of fish per cast as opposed to twenty-five grams per cast when fishing without dolphins.[48] So the dolphins clearly help the fishers, but, as in Brazil, we cannot be sure that the fishers help the dolphins.

There is also a possible human-dolphin fishing cooperative in the West African country of Mauritania, but we know less about it. On one side are members of the small, impoverished Imragen tribe and, on the other, bottlenose dolphins—or possibly humpback dolphins. In the middle are schools of yellow mullet that the dolphins drive toward the nets of the Imragen fishers. However, in this case it is not clear whether the dolphins are responding to signals from the fishers or vice versa—or neither.[49]

Prior to European contact, aboriginals seem to have formed fishing cooperatives with dolphins in several locations along the east coast of Australia.[50] Like the Brazilian cooperatives, they appear to have involved bottlenose dolphins, often with mullet as the prey, and both humans and dolphins seem to have benefited. However, in addition to nets the aboriginals fished with spears—sometimes allowing dolphins to pick impaled fish from the spears—and the cooperatives held a spiritual, as well as a sustenance, significance for the aboriginals. The cooperatives seem to have ended with the advent of Europeans along the coast, Europeans who on occasion killed the cooperative dolphins. There was also a cooperative between killer whales and whalers in Twofold Bay, Australia, that we will describe in chapter 6. This may have grown out of the aboriginal experience of cooperating with cetaceans and persisted for many years in the European-led small-scale whaling industry.[51]

These cooperatives are unusual, perhaps unique, in that both cetaceans and humans change their behavior to cooperate, both appear to benefit from the cooperation, neither trains the other, and the cooperative transmits intergenerationally in both species. The transmission is probably mainly matrilineal, from mother to offspring, in the dolphins. Again, a cultural hypothesis is a strong contender for explaining this transmission in the dolphins. We know the transmission is patrilineal in the humans—because they said so! We don't even have to ask if the transmission is cultural in our species because nobody would seriously argue otherwise.

These fishing cooperatives are extraordinary interactions between wild dolphins and humans. Perhaps the only other example of such cooperation that does not involve capture or training is the way that greater honeyguides—small birds that live in sub-Saharan Africa—lead humans to bee colonies using special double-noted calls and posture signals that indicate where the hive is. Once the humans have extracted the honey that they want, the birds come in and feed on the remains. There is little doubt that the signaling on the part of the bird is intended to have this effect, but it is also true that the birds are not very selective as to whom they signal.[52] For example, they have been seen signaling to baboons and mongooses, neither of which paid them the slightest attention.[53] Every bird in the species does it—the behavior is not confined to a subgroup. These facts suggest that there is a largely genetic basis to the honeyguiding. By contrast, only a subgroup of the Laguna dolphins work with fishers (around half of a population numbering at least sixty), and there is no suggestion of any kind of mating barrier between those that do and those that don't.[54] Likewise, there is no record of any attempt to herd fish toward other mammals on the shore. It is much harder to explain this pattern as having a largely genetic basis.

Foraging Strategies as Lifestyles

In human societies, the way we make our living affects more than just the amount of money in our pockets—it guides our social interactions. We develop friends in the workplace, or struggle to reconcile night shifts with relationships. The way an individual dolphin makes a living, the foraging tactics it employs, also has implications. One can talk of sponging, for example, not just as a foraging tactic but as a lifestyle with social and other consequences.[55] Spongers are more solitary, for example. They also spend relatively more of their time foraging than do other dolphins, they dive for longer periods, and they spend more time in the deep channels apparently apt for sponging. As a consequence, they associate more with other spongers than any of the other dolphins in the Shark Bay population. Different approaches to socializing, longer hours, different neighborhoods. Their lifestyles are different. They form a sponging community.[56]

A particularly insightful long-term study of dolphins in Moreton Bay, Australia, illustrates how these lifestyles come about. Moreton Bay is a large shallow bay off the city of Brisbane in Queensland. In the 1990s, biologists Louise Chilvers and Peter Corkeron started to study the dolphins there, identifying the individuals using photographs and watching their behavior.[57] A majority of the dolphins, 154 individuals, spent a fair amount of their time following prawn trawlers working the bay and foraging off the detritus that this fishing method would bring to the surface, as well as discards tipped off the sterns of the boats.[58] However, eighty-eight other dolphins used the same areas but did not associate with trawlers. The dolphins that did and those that didn't follow trawlers avoided each other, segregating into two communities. Despite sharing the same physical habitat, the communities had almost nothing to do with one another, except possibly mating. Writing in 2001, Chilvers and Corkeron concluded that "the feeding opportunities for trawler dolphins that are created by trawling have become part of their habitat requirements, although trawler dolphins are capable of foraging on other food sources (e.g. at weekends when trawling is banned). Recent fisheries management planning is calling for increased temporal and spatial closures of the trawl fishery in Moreton Bay. . . . The behavioural responses of the trawler dolphins to these closures should be investigated."[59]

The fisheries managers took pretty radical action, and by 2005 most of the study area in the southern part of Moreton Bay was closed to the prawn trawlers. What happened to the dolphins? Many of us wondered. The answer came from Ina Ansmann who repeated the surveys in 2008–10, after trawling had ended.[60] Happily, she has shown that the trawler-feeding dolphins didn't all starve to death. In fact, they went back to doing what the others had been doing all along—foraging for themselves in the shallow waters, ranging more widely over their habitat searching for food, which was now much more patchy. Tellingly, the social division has almost completely disappeared as well—the population is now well mixed. The trawler and nontrawler communities had merged, with members of the previously distinct groups now associating frequently.

This story, played out over decades, raises absorbing questions. It seems that some part of the population cottoned onto the easy meal, but others, almost ostentatiously, refused it. Were these two preexisting communities within the Moreton Bay population, with different cultures, one exploitative and one conservative? Or, as seems indicated by the posttrawling merger, was the community division driven by individual personalities, with some dolphins accepting the trawler bonanza, others rejecting it, and each foraging style coming to define a distinctive community? While the studies indicated possible mating between the two Moreton Bay communities, if mating between the communities slowed down, then it is not hard to see how this could lead to the development of "ecotypes," subgroups of a species defined primarily by what or how they eat but sufficiently separated over time by their lifestyles that evolution can potentially lead to genetic differences between them.[61] In chapter 6, we explain why we think this has happened with killer whales, who appear to be culturally conservative creatures. But bottlenose dolphins are almost the epitome of adaptability. They quickly worked out how to exploit the trawling, which began in the 1950s but only became common in the 1970s, and shaped their social system around it but then adapted both individually and socially to the closure of the fishery.

Similar processes seem to have been at work among the bottlenose dolphins involved in the Brazilian fishing cooperatives, as well as the spongers in Shark Bay. Recent studies have shown how social divisions—between cooperative dolphins, who take part in the joint fishing effort, and noncooperative ones, who never do, or between spongers and nonspongers—have apparently built up in these populations, too.[62] Because culture is learned socially, we generally think of social structure as a driver of culture, but these examples suggest the reverse, that cultural behavior can shape society.[63] This kind of flexibility, adapting quickly both to new ecological opportunity and to the loss of a niche, must surely be crucial to the success of the species and helps explain the diversity of dolphin lifestyles.

This flexibility can be double-edged, though. It often creates problems when dolphins, like those in Moreton Bay, learn about the opportunities provided by our methods of getting food from the sea. While in Moreton Bay the interactions were largely benign, dolphins and porpoises around the world have learned to take fish from a variety of fishing gear. This can be fatal if they get caught in the nets or lines. Even if they don't, it can lead to conflict because the animals get so good at taking fish from gear that it has an impact on the livelihoods of the fishers whose nets and lines, to their chagrin, are providing à la carte menus of fish that need no chasing.[64] In a typically human paradox, sometimes we willingly give the fish away, such as at the Monkey Mia resort in Shark Bay, where the feeding of dolphins is a big tourist draw. Despite the apparent generosity, this can also create problems only understood when careful long-term study reveals the lifestyle consequences.

It used to be a bad roll of the dice in Monkey Mia to be born to a mother who takes part in the provisioning, as provisioned dolphins were less successful in bringing up offspring than the dolphins who did not come into the beach. Calves typically don't follow their mothers into the very shallow waters where provisioning takes place and are left vulnerable, particularly to shark attack, as a result.[65] The feeding of dolphins got started in Monkey Mia, forty to fifty years ago when it was a small fishing camp. Shark Bay dolphins sometimes come into the beach to herd and catch fish, as part of their natural diversity of feeding tactics. When they approached the fishing camp, some fed from the fishers' discards. The fishers would occasionally toss fish to the dolphins, getting similar pleasure to the tourists who now visit Monkey Mia specifically to take part in this ritual. The dolphins grew to appreciate this. Then a boater, Alice Watts, found a dolphin who would take fish from the hand. The news spread, among humans, as well as, it seems, among the dolphins. People came to Monkey Mia to feed the dolphins, dolphins came to Monkey Mia to be fed. At the start, the feeding was erratic and indiscriminate, unmanaged. It quickly got out of control.

The speed with which knowledge about these fish giveaways can apparently spread through dolphin populations can rapidly turn a low-level interaction into a bigger problem. In Cockburn Bay, Australia, around 550 kilometers south of Monkey Mia, sport fishers have also begun amusing themselves by feeding bait and discards to local dolphins, who will now approach boats of all kinds apparently to "beg" for provisions. This is a risky thing for dolphins to do if the boaters are not expecting it—propellers can be dangerous. How does this knowledge spread? Bec Donaldson and her colleagues have shown that in Cockburn Bay, the propensity for begging appears to follow the social network in that local population, which is just what we would expect if the dolphins were learning the behavior from each other.[66]

In Monkey Mia, humans, in the collective form of the state government of Western Australia, have acted. Based on the advice of the scientists, they limited which dolphins got fed, when they got fed, and how they got fed. Now the feeding is professionally controlled by rangers of the Shark Bay

Marine Park, a salutary story of how our interactions can be managed positively, and the protocol seems to work quite well for both parties. Individual dolphins are only fed three small fish three times a day—volunteers helping the rangers weigh the fish going into each bucket—which makes up just a small proportion of their daily food requirements, so they cannot become dependent. The tourists have close encounters with the dolphins, which they enjoy. The dolphins have close, and controlled, encounters with the tourists, which they at least put up with for a decent snack. There has also been learning between the species, the dolphins picking up on the cues of the humans, such as the volunteers rinsing their empty buckets in the sea—when this happens, the dolphins know the feeding session is over—as well as quickly adapting to the changes in feeding protocol. But humans also learn from dolphins. Rangers and managers pick up on the cues of the dolphins—when they are ready to be fed, when they have had enough of humans—and so adjust their procedures as well as the information that they give the tourists. Feeding happens no more than three times a day, only in the mornings. Only five dolphins are ever fed, all females. The provisioning lifestyle, while still different from others in Shark Bay, is now perhaps less fraught. It has also spanned dolphin generations, just like sponging. Since around 1995, the rangers have begun selecting which dolphins to provision, but prior to that the lifestyle was maintained some other way. How do young dolphins learn about provisioning? By themselves, independently? Most provisioned dolphins had mothers who were themselves provisioned, and most calves receive their first fish while accompanying their mothers into the provisioning area, so this is implausible. A gene for provisioning? There are three different matrilines in the provisioned dolphins, all containing numbers of nonprovisioned dolphins, so this doesn't work either.[67] By imitating their mothers? Or simply by being with her and thus experiencing the interactions with humans and learning the rewards themselves? These latter two seem the most plausible explanations, and they fit with the detailed observations of the scientists who have watched the provisioning interactions develop.[68] Both point to the importance of social information flowing from mother to calf: culture.

In another part of Australia, provisioning of bottlenose dolphins appears to have led to a kind of culture that seems remarkable to us but probably not so remarkable to the dolphins themselves. Beginning in 1992, at the Tangalooma Resort, Moreton Bay, wild dolphins have been offered fish in the evenings. But since 1999, the dolphins have, occasionally, given fish or cephalopods that they themselves caught to the Tangalooma staff, "likely a manifestation of the particular relationship between the provisioned dolphins and the human participants in the provisioning."[69] One of the interesting aspects of this behavior is that it is occurring in an area where there used to be a cooperative fishery between the dolphins and aboriginals, as well as the trawler dolphins that we described earlier.[70]

So from foraging tactics flows a lifestyle, a set of social relationships, and a society. What are the longer-term consequences of this kind of specialization, where individual foraging tactics are honed to target specific ecological resources? We think we can see them in the mosaic of patterns revealed by genotyping dolphins over large areas, entire oceans in fact. Geneticist Rus Hoelzel has done this. He describes the pattern as "consistent with local differentiation based on habitat or resource specialization."[71] So dolphins can specialize into localized ecological niches, and these specializations, with their attendant lifestyles, can lead to degrees of population separation that we can now detect through measuring the dolphins' bodies and probing their DNA but that do not entirely cut off gene flow.[72] If the foraging tactics on which such resource specializations are based spread through social learning, then social learning is, down the line, affecting the genetic structure of the species.

Communication and Play among Bottlenose Dolphins

Like other cetaceans, dolphins communicate mostly using sounds. They make many whistles, which seem to be an important way of conveying information. Over decades, a persuasive body of evidence has been built up (largely by Vincent Janik, Stephanie King, Laela Sayigh, and Peter Tyack, following pioneering suggestions by David and Melba Caldwell) showing that each dolphin has a distinctive "signature" whistle, which they use rather like we use names.[73] When isolated they make these whistles, presumably to get in touch, and may use each others' signature whistles to initiate communication. They do this mainly among close associates, so mothers with calves, and male alliance members with one another.[74] Copying other dolphins' signature whistles seems to be mainly about maintaining social bonds. Some scientists have doubted the signature whistle story, so it is still controversial in some quarters, but evidence has been steadily accumulating in its favor.[75] The signature whistles of bottlenose dolphins have been rather well studied, and they are special because we know that they develop through learning—a young dolphin is not born with its signature whistle.[76] They have also been reported, but not studied in anything like the same depth, in four other dolphin species (spotted, white-sided, common, and humpback dolphins).[77] But the whistles are not shared—each individual has his or her own—so, at first glance, they are not culture by our definition.

Dolphins do not, however, only produce their own signature whistles—they can adeptly mimic the signatures of other dolphins and produce a variety of whistles that appear to have nothing to do with signatures.[78] It remains to be seen just how much of an individual's repertoire is given over to its signature during its daily life, but when an individual dolphin gets isolated, it makes a lot of its own signature whistle.[79] They can remember and recognize individual whistles, and we've begun to understand how they use these whistles when meeting each other at sea, sending them back and forth to each other as they approach one another.[80] Painstaking detective work by Stephanie King and Vincent Janik has shown that dolphins remember and produce copies of the signature whistles of at least those other individuals in their community with whom they have strong social bonds.[81] Furthermore, captive dolphins can remember, and strongly react to, the signature whistles of dolphins they lived with twenty or more years earlier but had not seen since.[82] It follows from this that each dolphin develops an awareness of the signature whistles of multiple other dolphins in its community, and this knowledge must be similar across many dolphins within a given community. It has to be learned socially—you cannot learn another's whistle without hearing it being made by them—and to the extent that dolphins in one community carry knowledge of the unique collection of signature whistles belonging to that community and, more important, which individuals the whistles "belong" to, then this is shared information. Perhaps, then, we should describe this aggregated and shared knowledge that dolphins build up about their community as culture.

Bottlenose dolphins also make other types of whistles, as well as other sounds, such as "brays" or "pops," which are usually also presumed to be communication.[83] These sounds vary geographically—brays have only been reported from dolphins feeding on salmon in Scotland, and pops only from the dolphins of Shark Bay. Similarly, when all the whistles made by a community, whether signatures or not, are measured and compared with those from other communities, many dolphin populations are shown to have regional dialects.[84] If learned, and they almost certainly are (we will present the evidence for vocal learning in chap. 7), these sounds are also part of the vocal cultures of dolphin communities.

For a dolphin, sound is the primary mode of communication, and communicative sounds usually emanate from the vocal system. But they do communicate in other ways, although because most of their behavior is underwater, it is hard for us to pick up on these nonvocal signals. Bottlenose dolphins living in Doubtful Sound, a New Zealand fjord, occasionally make side flops—leaps from the water, landing on the side—that appear to be used as signals to initiate travel. They also perform upside-down lobtails—slamming the top of the flukes onto the water—that appear to signal the end of a period of travel.[85] Such signals are not known from other bottlenose dolphin populations. However, the Doubtful Sound bottlenose dolphins have a particularly cohesive group structure, with lots of animals traveling together, so using demonstrative signals to regulate group behavior makes sense. As they have not been reported from other populations, we strongly suspect that these signals are socially learned, and so culture.

Play is a particularly hard type of behavior to identify rigorously, although, like pornography, most of us know it when we see it.[86] Dolphins can appear to be highly playful—most obviously in the characteristic way they approach vessels to "surf" the bow waves generated by their passage through the water, as well as the waves arriving at beaches. Because play does not usually directly relate to ecology, the way that play behaviors develop and spread within populations can be highly informative if we want to know more about how dolphins might learn from each other.

Take the case of Billie the tailwalker.[87] Billie was a female bottlenose dolphin from South Australia who, in 1988 when approximately five years old, was taken into a commercial aquarium for medical treatment for about three weeks after being found trapped in a canal lock. While there, she was housed with other dolphins who were permanent oceanarium residents and had been trained to perform a variety of behaviors on demand during shows put on for visitors to the facility. One of these behaviors was tailwalking, where a dolphin beats its tail to emerge vertically from the water and then maintains that position while moving backward, appearing to walk over the water. Once recovered, Billie was released back into the wild, and her progress after release was monitored by a group of volunteers led by Mike Bossley of the Whale and Dolphin Conservation Society in Australia.[88] Mike was amazed to see that Billie started to perform tailwalking after her release. Billie, despite never receiving any training herself while recuperating, must have learned this spontaneously from the trained dolphins. The behavior has no known utility in the wild and, as you can imagine, uses up a fair amount of energy, but Billie kept it up. Not only that, her enthusiasm was apparently infectious, as Mike observed: "Another female dolphin called Wave began performing the same behavior (plate 9), but does so with much greater regularity than Billie. Four adult female dolphins have also been seen tailwalking."[89] Mike also reports that several calves have also been seen producing recognizable approximations of tailwalking. So, a behavior learned in captivity, with no obvious function beside play, has apparently gone on to become something of a hit in the wild and persists to this day, twenty-five years following Billie's release and after Billie's own death in 2009.

How did the tailwalking spread? We can perhaps get some idea by examining the synchronous behavior we touched on earlier in this chapter, when we mentioned that males in alliances sometimes show extreme synchrony. We don't think this is play, in fact it may be quite serious business for male dolphins hoping to contribute genes to future generations, but we think there is a connection between Billie learning to tailwalk and the role that synchrony plays in dolphin society. Richard Connor, the world expert on dolphin alliances, has described this in some detail for a simple form of synchrony—surfacing to breathe at the same time. The degree of synchrony is extreme. In one study, the average time difference between two dolphins surfacing in a "synch surfacing" was 140 milliseconds—less than one-fifth of a second.[90] He has also described how alliance members take part in a variety of synchronized behavior when they are "consorting" with a female (a neutral term used when males guard potentially fertile females in a way that would be illegal between humans in most countries) including synchronized leaps, surfacing in the same and opposite directions, and turning into, or away from each other.[91] These are combined into complex sequences of synchronous display. Connor describes the following sequence as a "simple display": "Two males surface synchronously behind a female, swim forward on either side of the female, turn outward synchronously and swim parallel in the opposite direction to the female before turning in synchronously behind her and surfacing side by side synchronously again."[92]

Alliances also engage in synch surfacing when there are no females around, although they appear to hold back on the more elaborate display elements. Connor suggests this kind of synchronous behavior is an adaptive signal, indicating to all parties present the current health of the alliance relationship. If he is right, then synchrony plays a functional role in male alliances and thus, since alliances seem vital in getting access to females, to success at the mating game. It may act, for example, to reduce tensions between alliance members, which explains why the displays are more intense when there are females about, because both males are hoping to get lucky, and this creates some tension between the friends that needs diffusing. *Plus ça change.*

Whether the tension-reducing hypothesis is true, there is little doubt that this kind of synchrony is really important to male dolphins. In order to match each other through such a complex display, dolphins must have an incredibly acute awareness of the relationship between the actions of another—in this case, the alliance partner—and their own.[93] If synchrony improves reproductive success, then dolphins may well have evolved a psychology that both gives them the ability to copy others and also "rewards" them for achieving it. We will go into this more in chapter 7 when we think about how whales and dolphins learn from each other.

Returning to the topic of play, it is often thought to be a way to rehearse and develop skills that will be important in other times, so one might expect dolphins practicing synchrony through play to end up producing the same behavior as each other.[94] Couple this with psychological rewards, and you could get a very powerful social learning mechanism that leads to seemingly functionless behaviors spreading through populations simply because dolphins like being synchronous. Thus Billie, placed suddenly in an unfamiliar social setting, would have a powerful motivation to synchronize behavior with her new tank mates, and it could spread through the wild population after her release through synchrony-play.

What makes this even more interesting is that we humans are often partial to a bit of synchrony. It also appears to play some pretty crucial roles in human coalitions. Anthropologists Edward Hagen and Gregory Bryant have suggested that this kind of coalition signaling might have been a factor in the evolution of music and dance.[95] Fans of rugby union will know of the haka displays performed by the New Zealand All-Blacks, where the team performs a synchronized Maori war dance before a game, reproducing in often grim detail the intended fate of their opponents.[96] Modern armies spend inordinate amounts of time and effort drilling cadets to march and maneuver in lockstep, and dictatorial regimes especially have an apparent need to advertise the strength of their military coalitions with displays of synchronous goose-stepping. As Richard Connor suggests, these similarities in the use of synchrony in coalitions may go more than skin deep and, in fact, constitute a fascinating evolutionary convergence.

The Oceanic Dolphins

Apart from a brief excursion up the Ayeyarwady River to meet the fishing cooperatives between the Irrawaddy dolphins and Burmese fishers, this chapter has so far largely stuck with the bottlenose dolphins of bays, estuaries, and other coastal waters. We know much less about the foraging and communication of the other dolphin species and porpoises. It seems likely that the social and foraging behavior of inshore species like the humpback dolphin is not too dissimilar to that of that of the bottlenose dolphin.[97] But are there similar specializations within and between populations?

One exciting study of wild dolphins is that of the spotted dolphins on the shallow banks off the Bahamas. This study has been going on since 1985 and is led by Denise Herzing.[98] Herzing and her colleagues can swim with the dolphins and observe them closely in ultraclear waters, giving unparalleled insight into dolphin life. We will refer to some of their observations when considering social learning in chapter 7. In their publications, Herzing and her colleagues hint at variation in behavior that might be cultural, but so far they have not published a systematic study of behavioral variation in the spotted dolphins of the Bahamas.[99]

As we move offshore, to deeper waters, even less is known. The dolphins of these habitats, whether "offshore" bottlenose or members of the twenty or so deepwater dolphin species, live a very different life from their coastal or freshwater cousins—a more dangerous one. They typically form large schools, sometimes thousands strong, most probably primarily for protection against a suite of open-ocean predators.[100] In their featureless deep ocean world, other members of their school are the only constant physical presences these dolphins experience, and losing one's school may be a death sentence. Additionally, their food tends to come in huge unpredictable shoals, rather than the "bite here, bite there" of the coastal dolphins.

Compared to the coastal dolphins, we know very little about the societies of these oceanic animals and how behavior may or may not spread within them. However, a few in-depth studies of semicoastal dolphin species give us something of a feel for how lifestyles change in deeper waters. Bernd Würsig, Melany Würsig, and their colleagues have studied dusky dolphins for decades, first off Argentina and more recently in the waters off New Zealand.[101] Dusky dolphins, like some other semicoastal dolphin species, usually spend the day huddled quietly in shallow waters near the coast, where they are pretty safe from predators, and then go out into the deeper waters at night to feed individually on creatures that migrate up from the depths in the dark. This is what they do off the peninsula of Kaikoura on the South Island of New Zealand.[102] However, each winter a subset of the Kaikoura duskies, mostly males, travel to the waters of Admiralty Bay about 275 kilometers away where they completely change their lifestyles, feeding cooperatively in daytime on small schooling fish that become seasonally available.[103] How do young males learn about this seasonal opportunity? The most plausible driver is that they learn from older unrelated males since their mothers mostly don't do it and dusky fathers have very little to do with their progeny. Toward the end of a recent book on dusky dolphins edited by the Würsigs, authors Heidi Pearson and Deborah Shelton conclude, in a summary chapter, that "the seasonal movement between Admiralty Bay and Kaikoura is likely an example of inter-generational or oblique culture."[104]

It was while thinking about life in huge oceanic dolphin schools that Ken Norris, who perhaps more than anyone launched modern cetacean science, first proposed that dolphin societies may possess something we could identify as culture. He offered a typically pertinent description of what he meant by the term: "By *culture* I mean the collective concourse of ideas or concepts within a species; their origin, passage, and storage. Because ideas are constructed in the mind of the conceiver, they cannot be directly linked to the genome. Instead, they pass from individual to individual by instruction and observation, and they may change with each transmittal. They can be stored as memory, as traditions, or in iconography. I perceive the elements of culture in many higher mammals and some birds."[105]

Norris spent many years studying oceanic spinner dolphins around Hawaii, who, like the dusky dolphins, tend to spend the daylight in large inshore social groups, going offshore at night to feed. Spinners, as their name suggests, also parallel the duskies in their elaborate leaps. Norris documented how their behavior was organized into bouts, how the schools maintained cohesion by synchronizing their switching between behavior bouts, and what it must be like to live within such a large three-dimensional society. Along with his colleague Carl Schilt he came up with the radical notion of a "sensory integration system" by thinking about what information would be available to an individual dolphin within such a large school (see plate 11).[106] It would have its own information, from vision and echolocation, but there would also be a wealth of information available from the movements and vocalizations of other members of the school. This information is not private, it is available to all, and it is potentially hugely valuable. For example, sudden rapid movements from schoolmates could warn you about the appearance of a predator, while a cacophony of echolocation from the other side of the school could alert you of a nice bait ball from which to feed. Norris suggested that the school would function to integrate these sources of information—hence, sensory integration system—because dolphins within the school would repeatedly both receive and transmit such information, perhaps inadvertently simply by acting on it or perhaps through mimicry or matching. If you are a dolphin in such a school you should pay attention to what others around you are doing. Paying attention to others facilitates the flow of public and social information. Norris suggested that this sensory integration would allow dolphin schools to respond collectively and effectively to both threats and opportunities no matter which part of the school first encountered them, using mental maps of the dynamic layout of the school and the dolphins within it. In this system, with constant attention paid to public information and the benefits of spreading information fast through a school, Norris saw the foundations of culture: "I contend that spinner dolphin society involves both socioecology and culture, and that both have had major roles in shaping the general genic evolution of the group."[107] His vision was so far ahead of his time that twenty years later we still have not been able to make much progress in understanding the complexities of oceanic dolphin societies. Today the paucity of data on these species still prevents us from properly testing Norris's contention, although recent simulation studies have shown just how well such a process works in "virtual" animals.[108] Oceanic dolphin societies are, in general, still as inaccessible to us as the dark side of the moon, as they wander in the pelagic deserts that have consumed thousands of unfortunate human seafarers. They will keep their secrets quite a while longer.

We have seen how dolphins discover and pass on to their calves knowledge about diverse ecological resources, how those discoveries in turn shape the lifestyles and societies of the dolphins, and how this seems to be underpinned by a drive for synchronous behavior. To what extent do these features of the dolphins' lives relate to our notion of culture? Are dolphin communities, with their diversities of foraging strategies and unique collections of vocalizations, repositories of significant amounts of cultural information? With these questions we leave the dolphins and proceed in the next chapter to consider their larger toothed cousins, especially killer and sperm whales.

6 | MOTHER CULTURES OF THE LARGE TOOTHED WHALES

Matrilineal Social Systems

In contrast to the loose "fission-fusion" social lives of the dolphins, the societies of some of the larger toothed whale species are strongly structured. They are focused on mothers. Formally, we call them "matrilineal" and mean by this that most females spend their lives in the same social group as their mother while both are alive—daughters stay with their mothers. This is not the same as "matriarchal," which means female elders have power or influence. Elephant societies are both matrilineal and matriarchal.[1] The large toothed whales live in matrilineal societies, but we currently have no evidence that they are matriarchal, though we suspect that they are. This matrilineal structure takes a range of forms. The most rigid is the hierarchically arranged social system of the "resident," fish-eating killer whales off the northwest coast of North America.[2] Both males and females spend their lives in their natal social unit. The social units are thus completely matrilineal in structure. Using the same waters are the mammal-eating "transient" killer whales. Their social system is more flexible, with occasional transfers of males, and sometimes females, between units, but it is still largely matrilineal.

The other well-studied species of large toothed whale is the largest of all—the sperm whale. Here again the basis of society is the matrilineal unit.[3] Females within units suckle and babysit each others' young and defend themselves communally. However, unlike the resident killer whales, males permanently leave their natal unit, thereafter having more flexible social relationships with a network of other males while roaming between groups of females when they are in the same habitats.

The two species of pilot whale also live in societies characterized by matrilineal units, as may a few of the other less-studied large odontocete species, such as false killer whales and perhaps melon-headed whales.[4] Not all large toothed whales are matrilineal. For instance, the deep-diving northern bottlenose whales that we study in undersea canyons south of Nova Scotia are about as big as killer whales but have a social system more like that of the bottlenose dolphins: networks of females and stronger relationships among pairs of males.[5]

The matrilineal social systems of killer, sperm, and pilot whales provide a fine substrate for cultural traditions to develop. The young whale learns from its mother or other matrilineal relatives. In this way distinctive, and often communal, behavior patterns easily build up within the matrilineal social unit. Social units may split, or group with other social units, sharing and learning behavior. Thus shared cultures can develop at higher levels of social structure, such as pods or clans. The cultural contrasts between the social units, pods, and clans may be reinforced by conformity.

We scientists find it easier to identify culture in these matrilineal social environments than in the more fluid social worlds of the dolphins, porpoises, or baleen whales. Basically, we look for behavior that is distinctive to the social entities. Behavior characteristic to matrilines may not be culture. It could be the fruit of genetic combinations that only that matriline holds or result from independent effects of a common matrilineal environment on the behavior of individuals. However, a purely genetic explanation is exceedingly unlikely if there is mating between matrilines, or if females occasionally transfer between matrilines, while different responses to different environments cannot be the sole cause if matrilines with different behavior share the same environment. Thus much of our best information on whale culture comes from these matrilineal species and, especially, their sounds. The vocalizations of the matrilineal whales are in some respects our clearest route into the cultures of whales and dolphins.

These matrilineal whales include two of the most extraordinary species in the ocean, the killer whale and the sperm whale. We also know more about these animals than almost any other cetaceans—only the humpbacks and bottlenose dolphins are as well known. So we will structure this chapter around the killer and the sperm, introducing each of these species in separate sections, describing their communities and their vocal and nonvocal behavior that is, or may be, culture. At the conclusion of this chapter, we will consider what we know about pilot whales and the other matrilineal whales.

"Killer Whale: The Top, Top Predator"

Of all whale scientists, Robert Pitman has perhaps spent the most time at sea and seen the most cetacean species. He edited the 2011 special issue of the journal of the American Cetacean Society, the *Whalewatcher*, on killer whales, the title of which heads this section.[6] He writes, on behalf of the killer whale scientists who are authors of articles in the issue, "Killer whales are the most amazing animals that currently live on this planet."[7] We do not study killer whales. Our primary research species, the sperm whale, had its own special issue of the *Whalewatcher* and is a pretty strong challenger for "most amazing animal on the planet."[8] We will lay out arguments for this in the next section of this chapter. But, even though sperm whales are much larger, have bigger teeth, bigger brains, dive deeper, and are extreme in many ways, we have to admit that Pitman has a good point. We rarely see killer whales during our work with sperms, but each time we do is memorable. Even when they just cruise by, they give the impression of immense power and determination, and when they start attacking the sperm or humpback whales that we are studying—both larger than the five- to eight-meter-long killers—the power, planning, and cooperation is there to see.

Some, like Pitman, say "killer whale," others say "orca." Both are fine.[9] "Killer whale" emphasizes their predatory nature, and they sometimes kill large prey in dramatic and bloody fashion; "orca" has come to represent the social, cultural, and cognitively complex creature that has never, as far as we know, actually killed a human in the wild.[10] These two sides of the same animal are closely related. They are such devastating predators because they are social, clever, and cultural. And the evolution of their societies and intelligences has undoubtedly been driven, at least partially, by their predatory nature.

There is another conundrum with the name "killer whale." People see, describe, and photograph killer whales all around the world, from within the leads that fracture pack ice near both poles to warm waters on the equator. All these animals are robust, medium-sized whales, with the characteristically dramatic black and white markings and tall dorsal fins—huge dorsal fins in mature males. But there are differences among groups of 'killer whales'. Some groups of animals are a little smaller, or their fins are a little pointier, or the characteristic white eye patch is reduced. Furthermore, there may be two or more of these morphological types in the same area. But the most interesting aspect of these differences is that these are not just *morpho*types—different body shapes—they are *eco*types: animals with different ways of life. Killer whales often eat in a most spectacularly visible way—for instance, a minke whale being torn apart or a seal being snatched from a beach—and even when foraging more cryptically they may leave behind traces of their food. Compared with many cetaceans, we have a relatively good picture of killer whale diet. In a few situations, scientists following killer whales can be pretty certain they have observed every feeding event of each individual and so can make detailed quantitative studies of foraging behavior.[11] And so we know that the different visually distinguishable types of killer whales eat characteristically different prey.

Best studied, and most well known, are the killer whales of the northwest Pacific, where there are three of these ecotypes: the "residents," which eat salmon and whose males have slightly forward-tilted dorsal fins; the "transients," which are a little larger, have pointed dorsal fins, and eat marine mammals; and the "offshores," which eat deepwater fish, especially sharks.[12] Two or more ecotypes may use the same waters but not in the same way. In the northeast Atlantic there seem to be at least two ecotypes, one that specializes in hunting cetaceans, and the other, smaller "type 2" that has a more generalized diet.[13] In the Antarctic the orca-ecotype situation is becoming increasingly baroque.[14] There seem to be at least five ecotypes, with perhaps more to come; "type A" specializes on minke whales, the "pack-ice killer whale" on seals, and the "Gerlache killer whale" on penguins, while the "Ross Sea killer whale" and "sub-Antarctic killer whale" appear to eat fish.[15]

What is going on here with all these killer whale ecotypes? Are they species, or subspecies, or races, or . . . ? At first glance, the presence of a range of killer whale ecotypes does not seem that unusual. Lots of animal species, wolves and coyotes, for example, come in somewhat different ecotypes that eat somewhat different things.[16] But generally they develop these variations in different places, and so geographical barriers allow for independent ecological, morphological, and, fundamentally, evolutionary trajectories in the different groups of animals. Sometimes the barriers are breached, perhaps because of geological processes—the opening of a land bridge for instance—an unusual migration, or human intervention, and the types start being found together, retaining their differences because successful breeding only occurs within types. But it was the barriers that allowed the differences to develop.[17] However, killer whales don't face much in the way of barriers; individuals swim thousands of kilometers all over the oceans encountering killer whales of other ecotypes. Yet the differences between the ecotypes are profound. New genetic studies show that the maternal lineages of these ecotypes have been well separated for many, many generations, with very little, if any, interbreeding. Geneticist Phillip Morin and his colleagues believe that there are multiple species of killer whale out there.[18] So how did this happen?

Lance Barrett-Lennard, a researcher based at the Vancouver Aquarium, has spent a lot of time pondering killer whale evolution. He has also spent a lot of time at sea with the animals themselves. He thinks the keys to their evolutionary diversification lie in three powerful killer whale traits: they are picky eaters, they are xenophobic, and they are cultural.[19]

While the killer whales of the world eat a very wide variety of prey, from blue whales to stingrays to herring, each individual killer whale and each ecotype of killer whales has a much more restricted diet and sticks to it with remarkable persistence. This was graphically illustrated in 1970 when three mammal-eating transient whales were captured alive off British Columbia for the display industry.[20] They were kept in a netted pen and fed fish like the resident whales that were usually captured; at that time no one knew that there were different ecotypes of killer whale. For seventy-five days they were provided with fish but refused them. One died. Four days later the other two began to eat the fish, but they reverted to mammal food on being returned to the wild after a few months of captivity.

The pickiness goes beyond general types of food, such as fish or mammal. The resident killer whales of the eastern North Pacific focus on Chinook salmon and are sufficiently disdainful of other salmon species that during the 1990s when the Chinook salmon population declined so did those of the resident killer whales, even though other salmon species were reasonably plentiful.[21] The Antarctic pack-ice killer whales specialize in using waves that they make to dislodge seal prey from ice floes. This is the way they catch most of their food. However, they often spy-hop around the ice floe first to make sure the victim is their preferred species, the Weddell seal.[22] Crabeater seals, about half the size of the Weddell, but otherwise pretty similar, are consistently spared. Crabeaters form about 83 percent of the regional seal population, Weddells just 15 percent. The pickiness extends to parts of prey. Killer whales that eat baleen whales often just eat the tongue and lips, and the Gerlache killer whales, when eating relatively tiny penguins, often discard all of the carcass except the preferred breast muscles.[23] Killer whales brought into captivity retain this pickiness, unwilling to try unfamiliar food.[24] Selectivity about food is a general killer whale trait, and, Barrett-Lennard argues, is part of their overall cultural conservatism.

Barrett-Lennard's second characteristic killer whale trait, xenophobia, is harder to document. However, in the eastern North Pacific, the different ecotypes either studiously ignore or go to some lengths to avoid one another. This description from John Ford and Graeme Ellis's authoritative book on the mammal-eating transients illustrates the evidence:

> Seeing residents and transients in the same vicinity is not a common occurrence, but it has been witnessed often enough by ourselves and our colleagues so that a pattern is starting to emerge. Either the two forms pass as if neither notices the other or the transients actively avoid encountering the residents. In cases of avoidance, transients on a "colli-

sion" course with a pod of residents deviate from their path and skirt around the residents, or reverse their course in order to stay clear. We have noticed on occasion that transients begin to take evasive action once the underwater vocalizations of an approaching group of residents become audible, at a range of a few kilometers or so. As transients are most often silent when they travel and forage, they may typically go undetected by the larger, vociferous resident pods.[25]

Meetings between the ecotypes are therefore rare events, but they have been observed, and the details of one event bear repeating here because they speak so powerfully to the concept of xenophobia in killer whale societies, which in turn speaks to an aspect of culture that pervades human societies—the way individuals identify with their own in-group, which can be recognized by completely arbitrary symbolic markers, and prefer them to any individuals or groups identified as "other." Ford and Ellis describe an encounter that started with Ellis receiving a phone call about killer whales in Nanaimo Bay, on Vancouver Island, which was their cue to get aboard their boat and try to collect some data. They soon encountered the whales and recognized them as belonging to a well-known resident group, J pod:

Soon after encountering the group, the whales began "porpoising"—swimming at high speed—toward Descano Bay, some two miles distant at the northwest end of Gabriola Island. Graeme [Ellis] ran ahead to investigate where the J whales were headed at such speed, and as he approached the small bay, he sighted more whales, which were creating a great deal of splashing in the water.

Moments later, Graeme arrived in Descano Bay to find more of J pod in a tight group, and they appeared to be in a very excited state. Shortly thereafter these were joined by the previous J group, still travelling at high speed. At 11:00, as the whales began to charge together toward the head of the bay, a group of three transient whales, T20, T21 and T22, suddenly surfaced a few metres ahead of the residents. The transients were clearly fleeing from the larger resident group, and it appeared that the residents were attempting to drive the transients toward and possibly onto the beach. All the whales were extremely agitated, and intense vocalizations were clearly audible through the hull of the boat, even over the noise of the outboard motor. Graeme observed what appeared to be fresh teeth marks on T20's dorsal fin and T22's flank. All ages and both sexes of the resident whales seemed to be involved in the fray. Just as the whales drew near a ferry dock located along the shore, a ferry backed out of its slip and disrupted the interaction. The transient group immediately dove and surfaced near the far side of the bay, followed by the J pod whales about 200 metres behind. About 5 minutes later, the transients left the bay and swam steadily south, with the J's following slowly several hundred metres behind. At 11:35, the transients went through Dodd Narrows, a small tidal channel leading to the south, but J pod did not follow. Instead they milled about for 25 minutes, when they were finally joined by the female J17 and her newborn calf, J28, who had not been present during the earlier altercation with the transients.[26]

The degree of antagonism between the groups described here is extraordinary—after all, these animals eat different things, so they can't be competitors for the same resources. The most efficient thing to do, from a natural selection perspective, would be to simply ignore each other and minimize the risk of injury associated with this kind of confrontation. There must be something more going on, and Barrett-Lennard suggests it is xenophobia.

"Cultural" is the third killer whale trait, and the subject of this book. Their culture is notably conservative. Killer whales are, in Barrett-Lennard's words, "capable of learning practically anything by example, but not prone to experimenting or innovating."[27] So new ways of doing things don't occur often in killer whale society, but when they do, they can spread fast.

In consequence, Barrett-Lennard argues, a group of killer whales develops a specific way of capturing and using a particular prey, which they pass on culturally, use fairly exclusively, and stick to.[28] They have little to do with other groups that have other ways of making a living and soon cease breeding with them. This is the proposed origin of the diverse killer whale ecotypes and perhaps the beginnings of a rare form of culturally driven speciation that we will discuss further in chapter 10.[29]

These attributes have served killer whales well. While each individual killer whale and each killer whale ecotype may be xenophobic and picky, as a whole species—or species-complex, if you like—*Orcinus orca* has a profound impact on the environment. They forage so efficiently that killer whale predation is probably a major regulator of some prey populations.[30] Jim Estes and his colleagues have amassed evidence suggesting that off the Aleutian Islands the destructive effects of predation by just a few North Pacific transients has changed the state of a whole ecosystem.[31] This is how they think it happened. The inshore Aleutian waters were covered in, and structured by, huge kelp forests that provide habitat to all kinds of creatures. However, the kelp is threatened by hungry sea urchins. A healthy sea otter population has kept the sea urchins under control, thereby letting the kelp flourish. But then, in the 1990s a few hungry killer whales started eating otters. Because a sea otter is a pretty small bite for a killer whale, they ate a lot. They almost completely removed the otters from a large stretch of the Aleutian coast. Without otters, the urchins mowed down the kelp, and the forest changed to barrens, a completely different habitat and ecosystem.

It also seems very likely that killer whales have eradicated whole species, perhaps many species. The diversity of cetacean species has generally increased over time, but between the Tortonian geological stage, 11.6–7.3 million years ago, and the Messinian, 7.3–5.3 million years ago, the number of living cetacean species decreased from about eighty-five to thirty-eight.[32] The numbers of sirenian and pinniped species also halved at roughly the same time.[33] Did these extinctions have anything to do with the appearance of killer whales about ten million years ago?[34]

The threat that killer whales represent also affects, in some cases probably profoundly, animals that they do not regularly kill. Avoiding being a killer whale's lunch is an important goal for many marine mammals. To minimize the risk they may haul out on land or ice—although killer whales have worked out ways of capturing some hauled-out seals, as we will discuss—or change their daily schedules, or sleep well below the surface in the case of elephant seals, or migrate to fairly killer-whale-free tropical waters in the case of baleen whales.[35] A common strategy for combating predation is sociality—safety in numbers. Other cetaceans especially seem to use each other to counter the killer whale threat.[36] This could work in a number of ways. Heightened communal sensing (such as the sensory integration system discussed in chap. 5) can allow them to detect the approaching killers and flee; they can hope the killer whales take one of their schoolmates rather than themselves; they might confuse the killer whales with many bodies moving quickly; or, in one important case we will meet soon, they can mount a communal defense.

Why are killer whales such efficient predators, posing threats well beyond any other marine carnivore, other than the technologically enhanced human? Killer whales are big, powerful, and fast, and they have large jaws and teeth. But all this applies to some sharks, and other cetaceans are also well endowed in these ways. Perhaps the essential difference with killer whales is a highly sophisticated *communal* intelligence in how they make their living. They learn from each other and work with each other.

With the baleen whales and dolphins that we met in the two previous chapters, we had to think a bit to circumscribe their communities, and we are not sure we really described the communities as the whales and dolphins experience them. The sense of community among killer whales is much more obvious. They move, socialize, and often hunt together, with the same companions year after year, companions who are usually their relatives. The fundamental killer whale communities are matrilineal units, mothers, with their offspring and grandoffspring, up to four generations living and moving together.[37] And then for any resident, fish-eating killer whale matriline of the northeast Pacific there are higher levels of community: matrilines that spend more than half their time together and are related are part of the same pod.[38] Pods meet other pods, with whom they may share greeting ceremonies or parts of their dialect or from whom they may choose a mate.

Thus, an individual resident is born into a particular matriline, most of which contain two to ten whales, which is part of a pod (mostly one to three matrilines), which is part of a clan containing two to ten pods that share similar dialects, which is part of a *community* of one to three clans. This is a specific geographically based meaning of the word "community" for resident killer whales, which we will contrast with our general usage of the word by italicizing the resident killer whale *communities*. So there is the southern resident *community* based in Puget Sound and around southern Vancouver Island, the northern resident *community* based off northern Vancouver Island, the southeast Alaska *community*, and so on. A resident killer whale stays in the same pod, clan, and *community* for its whole life, whether male or female. This is an extraordinarily unusual social system—it is almost universal among other mammals, whether marine or terrestrial, for either males or females to disperse from their mother. The male residents have been dubbed "momma's boys."[39] But the females are also "momma's girls." The units, pods, and clans within a *community* have overlapping ranges, so the killer whales frequently meet and greet members of other social entities. The resident killer whales not only socialize with other matrilines, they mate with them. The matings that take place during these get-togethers are not random. The two parents of a resident killer whale are usually members of different clans but the same *community*, except when there is only one clan in the *community*, as among the southern residents who use the waters of Puget Sound and southern Vancouver Island.[40]

Killer whales of other ecotypes, such as the northwest Pacific transients, also live in matrilineal groups which have well-defined membership. But the mammal-eating groups are not as large or as stable as the resident pods, and scientists have not found clear signs of higher levels of social structure such as clans and *communities*.[41]

Overlaid on the units, pods, clans, *communities*, and ecotypes of the killer whale are characteristic behavior patterns: this unit/pod/clan/*com-*

munity does this, and that unit/pod/clan/*community* does that. We believe that most, perhaps all, of this variation in behavior is cultural, although the level of certainty varies. Remember, we are basically still working with the method of exclusion: it is culture if we can rule out genetic and environmental causation of the patterns in behavior. In the world of the killer whale, environment can often be dismissed when the different behavior patterns are performed in the same place; genetics is usually harder to rule out as the highly matrilineal structure of killer whale communities means that behavioral differences correlate with genetic differences. As with the baleen whales and dolphins, we are most certain that culture is driving things when all the animals in a community change their behavior over time.

In the 1970s, the innovative Canadian scientist Mike Bigg and his American colleague Ken Balcomb worked out how to identify killer whales individually using photographs. They and their collaborators went on to use photo identification to work out the multilevel social structure of the two *communities* of resident killer whales that used the waters around Vancouver Island and in Puget Sound.[42] As their picture of killer whale society came into focus in the late 1970s and early 1980s, the project broadened. John Ford, then a graduate student at the University of British Columbia, began to study their sounds. Like many toothed whales, killer whales make three broad types of sound. Clicks are the working sounds, the products of their sonars, which the whales use to sense their surroundings. At close ranges, the killer whales seem to communicate using pure-toned whistles.[43] It was the third sound type, the "pulsed call," that attracted Ford's attention. Pulsed calls are made up of many short pulses produced at high rates. They may sound like screams or squeals or squawks. The pulsed calls of the killer whales can be extremely complex—with different pulsed-call elements being made in intricate patterns or even at the same time. Ford found that each pod of residents had its own repertoire of pulsed calls.[44] These repertoires are sufficiently characteristic that he only needed to listen to unseen killer whales for a few minutes to identify the pod. Dialect mirrors social system at other levels, too. The calls of the matrilineal units within pods are also a little different, while each pod shares some of its repertoire with other pods, parts of their clan. But there are other pods that have completely distinct repertoires, members of other clans. So the different levels of social structure—matrilineal units, pods, clans, and *communities*—are mapped onto characteristic dialects. Potentially, one might think, this is all genetics. As the society of the resident killer whales is so clearly matrilineal, the number of genes shared by two whales will be greatest if they are members of the same matrilineal unit, high if they are members of the same pod, lower for members of the same clan, and so on—the same pattern as with dialect similarity. However, this reasoning does not work. If dialect was determined by genes of the cell nucleus, those that are inherited from both parents, then the killer whales' propensity to find mates with very different dialects—outside their own clan in the case of the northern residents—quickly destroys any firm correlation between kinship and dialect.[45] It is very unlikely, and probably impossible, that complex variation in pulsed calls could be directly determined by the matrilineally inherited mitochondrial genes, and in any case these mitochondrial genes seem to be identical for all whales in each clan.[46]

Another nail—we would contend the final nail—in the coffin of genetic determination of killer whale dialect was driven in by a remarkable study of changes in resident killer whale dialects by Volker Deecke, John Ford, and Paul Spong.[47] They examined the structures of two pulsed-call types made by two pods of northern residents over twelve to thirteen years. Members of the two pods made each of the call types in a predictably different way. The structure of one of the call types, "N9," was stable over the entire period of the study; the other, "N4," evolved, so that by the end of the study all the members of both pods were making the call differently from the way they had made it a decade earlier, but the same as other members of their pod at that time. As in the case of the songs of humpback, bowhead, and blue whales, this cannot happen if the sounds are only controlled by genes. We will also present direct evidence for vocal learning in the next chapter. So the killer whale calls are culture. As the pods often meet, and the calls travel well under water, the killer whales must be familiar with the other dialects of the whales within their *community*, but they stick to "theirs."

Deecke, Ford, and Spong's study also found that the killer whales actively maintain the distinctions between the dialects of the different social units.[48] The evolving "N4" call changed in the two pods, and changed in similar ways, but they seemed to maintain a certain preferred distinction between the two repertoires—not too similar, not too different. The whales apparently care about the differences between their dialects.

We know much, much more about the sounds and social structures of the resident killer whales of the eastern North Pacific than those anywhere else. However, it looks as though the social and vocal worlds of other fish-eating ecotypes in other parts of the world are similarly structured and complex.[49] In contrast, mammal-eating killer whale ecotypes are much less vocal. Why? Their prey have good ears and so could hear the killers approaching and take evasive action; consequently, these killer whales primarily vocalize when not foraging.[50]

The pulsed calls of the killers trace their social world. As we can see the whales much less well than we can hear them, we know rather little about nonvocal parts of the killer whales' social lives. An exception is the remarkable greeting ceremony of the southern residents. When pods meet, each forms a rank. When the ranks are about twenty meters apart, they halt facing one another. After a pause of a minute or less, the whales dive, "and a great deal of social excitement and vocal activity ensues as they swim and mill together in tight subgroups."[51] This is quite a contrast with the meeting between the residents and transients described earlier.

After vocalizations, the other area of killer whale behavior that we know a fair amount about is foraging. Even though each individual killer whale, each social unit, and each ecotype has a highly specialized diet, killer whales as a whole use an extraordinary array of foraging techniques. They catch and consume sperm whales, blue whales, sharks, and stingrays, as well as seals, dolphins, penguins, salmon, and herring. Some of their prey, such as the large baleen whales, face no other nonhuman predators. Some of their foraging methods are complex, others seem quite simple: grab it and eat it. Some require the elaborate coordination of a whole group, others are snapped up by a single killer. We will focus on a few techniques that span some of the range of behavior that these creatures use to catch their prey. Table 6.1 summarizes rather more of these techniques.

Look at a killer whale, or a picture of a killer whale. It seems odd that a top predator—*the* top predator—should be colored so vividly that it must bring attention to itself. Other species at the summit of the food webs, such as lions, sharks, and human hunters, adopt camouflage. That bold black and white, characteristically killer-whale pattern would seem to shout out "danger" to potential food, spurring evasive action. But that same pattern is ideal in assisting the killers to coordinate visually with each other during their attacks.[52] The highly synchronized movements that they manage appear to outweigh the warning that the contrasting colors give to prey. Here is Robert Pitman's description of how the Antarctic pack-ice killer whales get their dinner (see plates 12 and 13):

> When they get into an area of pack ice they fan out. Individuals, or cows with calves, begin spy-hopping—lifting their heads above water to have a look around as they swim by individual ice floes. They are looking for seals that often spend their days resting on the ice. When a whale finds a seal, it spy-hops several times around the floe, apparently to make sure it is the right species. They appear to prefer Weddell seals. If it is the right seal, the whale disappears for 20 to 30 seconds, and then begins spy-

Table 6.1. Some of the Foraging Methods of Killer Whales

Type	Where	Food	Cooperative Foraging?	Share Food?	References
Carousel feeding	Norway, Iceland	Herring	Yes	?	Similä and Ugarte 1993
Picking off salmon	Eastern North Pacific	Salmon, especially Chinook salmon	?	Yes	Ford and Ellis 2006
Chasing tuna	Strait of Gibraltar	Bluefin tuna	Yes	?	Guinet et al. 2007
Pinning rays	New Zealand	Rays	?	?	Visser 1999
Depredation from fishing gear	All oceans	Tuna, swordfish, turbot, sablefish, flounder	?	?	Dalla Rosa and Secchi 2007; Yano and Dahlheim 1995
Chasing penguins	Antarctica, sub-Antarctic Islands, Argentina	Penguins (gentoo, chinstrap, others?)	?	?	Pitman and Durban 2010
Cruising for seals	Eastern North Pacific and elsewhere	Harbor seals, and others	Yes	Yes	Baird and Dill 1995; Beck et al. 2011; Jefferson, Stacey, and Baird 1991
Knocking seals from ice floes	Antarctica	Weddell seals (mainly)	Yes	Yes	Pitman and Durban 2012
Beach stranding for pinnipeds	Peninsula Valdes, Argentina; Crozet Islands	Sea lions; elephant seals	Sometimes (teaching?)	Yes	Guinet 1991; Guinet, Barrett-Lennard, and Loyer 2000; Hoelzel 1991; Lopez and Lopez 1985
Chasing down minke whales	Eastern North Pacific; Antarctic	Minke whales	Yes	Yes	Ford et al. 2005; Pitman and Ensor 2003
Attacking large whales	Various oceans	Humpback, sperm, gray, blue, etc.	Yes	Yes	Jefferson, Stacey, and Baird 1991; Pitman et al. 2001; Whitehead and Glass 1985
Cooperative foraging with human whalers	Australia	Humpback whale	Yes	Yes	Dakin 1934; Wellings 1964
Scavenging carcasses of large whales killed by whalers	All oceans	Many species	?	?	Whitehead and Reeves 2005

hopping around the floe again. During the brief disappearance, the whale apparently goes down to call in the troops, because within a minute or two the rest of the group is there spy-hopping around the floe also.

After a minute or two of collective appraisal, the group decides either to move on or to move in and attack. If they decide to attack, the members begin to swim in formation, side-by-side; then they head away from the floe to a distance of usually 5–50 [meters] . . . As if on cue, they turn abruptly toward the floe, swimming rapidly with their tails pumping in unison—synchronized swimming. A deep trough forms above their tail stocks and a wave, approximately 1 [meter] . . . high, forms above the flukes. The whales charge the floe and dive under it at the last second. If the floe is small, the wave will break over it and usually washes the seal into the water. If the floe is large (ca. 10 [meters] . . . or more), the whales carry their wave with them under the floe to the opposite side. This often causes the floe to shatter into smaller pieces, after which one or two whales use their heads to push the floe with the seal on it out into open water where they can wave-wash it again.

When the seal goes into the water, the killer whales immediately close in and attempt to take it by its hind flippers to drag it underwater. Although they could at any time easily kill the seal with a single bite or a ramming charge to the mid-section, they choose to wear out the seal and drown it.[53]

Now move to the fjords of northern Norway. Here the killer whales have a different challenge, and once again they work cooperatively.[54] The problem is that the food, while nutritious and plentiful, are small. Small oceanic creatures cannot move very fast, but they can turn quickly. So the thirty-five-centimeter-long herring—the killer whales' prey in the Norwegian fjords—while no match for the killer whales' speed or endurance, are much more agile. Like small songbirds mobbing an eagle, herring would normally be pretty safe from the relatively huge jaws of a killer whale because of this nimbleness.[55] However, the Norwegian killer whales have worked out ways of overcoming the agility gap. Tiu Similä and Fernando Ugarte, the scientists who described it, call it "carousel" feeding.[56] There are two phases. In the herding phase all the whales swim "in the same direction in a highly coordinated fashion and gathering the fish into a 'ball' close to the surface . . . with the white part of their bodies towards the fish."[57] The diameters of the herring balls vary from about two and half meters to over seven meters. They slow down for the feeding phase but continue to swim tightly around the herring ball, making sounds and blowing bubble clouds. Individuals slap the herring ball with their flukes, stunning the herring, which they then eat. They can move their flukes much more quickly than their whole bodies, so tailslaps are a much better way to catch nimble herring than lunges.[58] These feeding bouts can last anywhere from ten minutes to more than three hours. Feeding on herring has been seen elsewhere in the North Atlantic but has been described most clearly in Norway. In recent years the herring have no longer been entering the Norwegian fjords so the feeding behavior is no longer observed there. In two other areas where these whales feed on herring, they seem to have innovated an acoustic element to the technique. Off both Iceland and the Shetland Islands north of the United Kingdom, researchers have noted a characteristic long, low frequency call produced by whales just before they slap at the herring schools with their tails, which they believe evokes a bunching response in the prey school.[59] The call has never been heard in Norwegian waters. Much of this is reminiscent of humpback whales feeding on herring in southeast Alaska in the northeast Pacific (described in chap. 4): cooperative herding, bubbles, tail slaps, and loud sounds. These two very different species of whales have worked out quite similar ways to catch herring, a most nutritious food, but not an easy capture.

Catching herring and tipping seals from ice floes are difficult maneuvers but not particularly dangerous. Let's go back to the Southern Hemisphere, where killer whales have developed a truly hazardous way of making a living. Pinnipeds make good meals for killer whales. Hauling out is generally a highly effective defensive strategy against an aquatic predator. We have seen how, in the case of the Weddell seals about to be knocked off ice floes, killer whales got round this strategy. But land seems safe. How can a killer whale catch a seal on land? By going on land itself. At Peninsula Valdes in Patagonia, Argentina, the killer whales lunge into the surf or onto beaches to catch elephant seals or sea lions. Juan Carlos Lopez and Diana Lopez describe what happens:

> Typically, the killer whale about to beach itself swam toward shore, actively directing itself toward the prey, and occasionally surfing on an advancing wave. Sometimes other whales milled to the sides of the beaching whale, possibly to keep the prey from escaping in their direction. The whale caught the prey in the tumultuous surf zone or within an advancing wave on the beach, but usually while the prey was still in water. Because the killer whale is much larger than the prey, the killer whale "beached" or grounded while water was still flowing around it, and while the prey was still capable of swimming. Once grounded, the whale arched its

body, with the head and tail lifted up, and rocked sideways. This motion usually oriented it parallel to the beach, and a subsequent wave helped to lift it off the bottom. It then swam back into deeper water, carrying the prey in its mouth if it had been successful at capture. None of the whales we observed became permanently stranded; it usually took only two to four large waves to help dislodge a grounded animal.[60]

Killer whales hunting pinnipeds also strand intentionally on the Crozet Islands in the southern Indian Ocean.[61] However, this behavior has never been seen in the Northern Hemisphere, where surf is usually much less heavy; the waves seem to help the whales attack quickly, as well as to dislodge themselves from the beach. As with other killer whale foraging behavior, the whale who catches the prey typically shares it with others.[62] At both Peninsula Valdes and the Crozet Islands the older animals seem to teach the younger ones this rewarding, but dangerous, way of making a living, a suggestion we return to in the next chapter.[63]

Killer whales do not always feed cooperatively, but their food is often challenging. For instance, off North Island, New Zealand, killer whales frequently forage on stingrays, an apparently profitable but potentially dangerous, food.[64] The killer whales try to "pin" the rays to the bottom and then grab them; the rays try to escape by swimming under wharves, into shallow water, or even onto beaches. During the chase, a miscalculation may be as fatal to a killer whale as it was for Steve Irwin, "the Crocodile Hunter" and Australian TV star killed by a stingray in 2006.[65] Ingrid Visser, who first described the ray hunting, believes that the whales overcome their natural aversion to the shallow waters where they catch the rays by emulating their elders.[66]

As you read the accounts of this extraordinary array of techniques that killer whales use to make a living, the awe of the observers keeps slipping out into their dry scientific prose. But one of the most remarkable of the killer's foraging methods was known by humans from the day that it started, sometime between 1828 and 1843, because it involved humans.[67] In 1828, shore-based whalers starting catching the baleen whales, mostly humpbacks, that migrate past the waters of Twofold Bay, off the town of Eden in southeast Australia. In 1843 Sir Oswald Brierly, a marine painter who had become manager of the whaling station, wrote in his diary:

> They [the killer whales] attack the [humpback] whales in packs and seem to enter keenly into the sport, plunging about the [whaling] boat and generally preventing the whale from escaping by confusing and meeting him at every turn. . . . The whalemen of Twofold Bay are very favourably

disposed towards the killers and regard it as a good sign when they see a whale "hove to" by these animals because they regard it as an easy prey when assisted by their allies the killers.[68]

Over the years the cooperation became stylized, somewhat in the manner of the dolphin-human fishing cooperatives that we described in the previous chapter. Here is how William Dakin and H. P. Wellings describe it.[69] The killer whales came to Twofold Bay each winter. They would wait outside the bay. When a humpback came along, some of the killers would swim over to the whaling station and start breaching and lobtailing, to alert the whalers. Others would herd the humpback into shallow waters, into the path of the whaling vessel, which could relatively easily harpoon the harassed humpback. As the humpback towed the whaleboat with the harpoon line, the killers would continue their work, leaping on the back and blowhole of the whale. This presumably made it easier for the whalers, who, in the traditional Yankee-whaler fashion, killed the tiring whale with a lance. The whalers then anchored the dead whale to the bottom, leaving it for a day or so to the killer whales, who would pry open the mouth of the dead humpback to get at their preferred delicacy, the tongue. They might also eat the lips. After twenty-four hours the meat began to turn putrid, and the killers lost interest. The human whalers would return, tow the carcass to the whaling station and start processing it. The whalers believed that they were very much more efficient in finding and capturing whales when the killers were part of the operation than when they had to do it all themselves.

The whalers grew to know, and name, many of the killers that worked with them: Hooky, Cooper, Typee, Jackson, and so on. These individuals were more than names to the whalers. They recognized characteristic behavior. The most famous killer, Old Tom, who worked at Twofold Bay for decades, had a habit of holding onto the harpoon line with his teeth as the humpback towed the whaleboat though the water, hitching a "Nantucket sleigh ride."

In the heyday of the Twofold Bay operation there had been twenty or more killers working with the whalers. However, by 1929, the final year of whaling, only one was left—Old Tom—perhaps because mechanized Norwegian whalers working farther north along the Australian coast had killed the rest. The body of Old Tom was found floating in the Bay in 1930, and his skeleton, complete with grooves in the teeth from gripping the harpoon lines, now resides in the Eden Killer Whale Museum, New South Wales.

While most of the foraging specializations of killer whales that we have described operate at the level of the ecotype, the Twofold Bay killers' co-

6.1. Killer whales and whalers cooperating in Twofold Bay, Australia, 1910. With the help of Old Tom (*foreground*) the whalers have harpooned a female humpback whale (*out of the picture*) who is dragging them on a "Nantucket sleigh ride." A humpback whale calf is between Old Tom and the whalers, following his or her mother. This is a still from an early documentary film, produced in 1910 and first shown to the public in Sydney in 1912.

operation with human whalers is an example of a foraging specialization at the level of the group. There are others. Robin Baird notes that groups of mammal-eating killer whales that work the coastal waters of southern Vancouver Island have two primary strategies.[70] Some groups visit the waters throughout the year and stay a kilometer or more from shore, catching seals and porpoises as they forage; other groups appear only in the seal breeding season and hang around the pupping colonies close inshore, snatching the seals as they transition from land to water. Another specialist group is the AT1 transients of southeast Alaska who have worked out ways to corner speedy Dall's porpoises.[71] The AT1s, which use the waters of Prince William Sound, lost nearly half their twenty-two members following the Exxon Valdez oil spill and have produced no new offspring since then. Their specialist way of life seems on the way out.

We have summarized just a very few of the many feeding methods used by killer whales. They also snap up birds, chase down tuna and minke whales, take fish from longlines and seines, and rip up sperm and humpback whales. They do this using size, strength, speed, stamina, sounds, bubbles, and cooperation. The methods often include several types of behavior, each of which is performed precisely and which are carefully ordered. Cooperation among killer whales is a key part of many of their feeding methods, and given their sociality it seems almost certain that the complex foraging strategies are learned partially or wholly from each other. After capture, they quite consistently share their catch with the other members of their group, whether it be salmon or seal, large or small.[72]

The killer whale that we have portrayed so far may seem a serious animal, focused on social relationships and food with a conservative, xenophobic attitude to life. But their behavior has sides that seem light, almost frivolous. Perhaps they are playing, although, as we noted in chapter 5, "play" is a difficult concept for ethologists. Members of the northern resident *community* have a habit of rubbing their bodies on underwater beaches made of small smooth pebbles, sometimes for an hour or more, sometimes several times in a day.[73] Certain beaches off the northern tip of Vancouver Island are highly favored. Maybe beach rubbing helps remove parasites, maybe they beach rub simply because it feels good or because it's the "done thing" in their *community*. Neither the transient whales that use the same waters nor the southern residents, a little to the south, beach rub. However, the southern residents have their own typical behavior. They are particularly aerial creatures, leaping from the water far more than the northern residents.[74] In 1987 the southern residents briefly became obsessed.[75] A female in K pod started pushing around a dead salmon with the top of its head. The fad spread to the other two pods in the southern resident *community* over a five- to six-week period, so that nearly everyone had their dead salmon toy. Then, just about as quickly as it arose, the fad stopped. Scientists saw dead-salmon carrying a couple of times the next summer but never again. The craze had died. The southern residents also "play" with harbor porpoises, chasing them, tossing them, and sometimes killing them.[76] But these whales are strict piscivores so the porpoises are not eaten. What is this about? Robin Baird noticed that the number of these encounters increased suddenly in 2005.[77] Another fad?

Beach rubbing, breaching, pushing dead salmon, and recreational hunting of harbor porpoises all seem nonfunctional—at least they are not directly about feeding or mating—but they can be fitted into a daily schedule framed by more serious matters. However, from time to time Southern Hemisphere pack-ice killer whales take real time off: they leave the Antarctic where they make their living tipping the seals from ice floes and head north. They swim steadily north at about seven kilometers per hour for about three weeks to reach the waters off Uruguay and southern Brazil—the closest warm waters—and then turn around and head straight back to the Antarctic. These movements are not seasonal—different groups do it at different times of year—and they do not seem to be related to feeding or breeding. The whales do not linger in the warm waters. So why do they do it? John Durban and Robert Pitman, who discovered these movements using satellite tags that they put on the whales, speculate that, as with some of the seasonal movements of modern humans, it is about temperature.[78] The very cold Antarctic waters are hard for the killer whales physiologically. In particular, to maintain their core temperature they have to restrict severely blood flow to the surface of their skin, so metabolism and self-repair is also restricted. Their skin becomes yellow as it gets covered with diatoms. They come back from Brazil clean and shiny, rather like the tanned humans returning to Canada or northern Europe after a few weeks in the tropics. Perhaps, as with humans, their holiday leaves them in better spirits as well as with more attractive skin!

We have outlined some of the things that we know killer whales do. There is much, much more that we don't know. The diversity, just in feeding behavior, seems extraordinary. But it is diversity across all killer whales. Each individual killer whale has a very precise repertoire, a way of foraging, interacting socially, playing. It will ignore, and may disdain, killer whales with other ways of life. These ways of life are, we think, cultural. Although only in a few cases can we be certain that the behavior is socially transmitted, such as the pulsed-call repertoires of the resident killer whales and the fad of pushing dead salmon about, the weight of evidence is strongly in favor of a cultural component driving nearly all the behavior that we have described. Try imagining how genes alone, or individual learning, could have led to the greeting ceremonies of the southern residents, or the ways in which Old Tom and his colleagues cooperated with the Twofold Bay whalers—hard to see. It is not unreasonable to ask whether killer whale call dialects could function as symbolic markers of culturally distinct groups, which would be remarkably close to the kinds of things that social scientists claim set human cultures apart. Like Lance Barrett-Lennard, we see culture as a defining characteristic of killer whales.[79] It gives them their nature, and it gives them their success.

Sperm Whale: Animal of Extremes

Roughly seven hundred years ago we humans took over as the most collectively massive mammal on Earth. Until then we were surpassed by fin whales, blue whales, and sperm whales. Each population weighed about twenty to forty million metric tons, but as the human population passed about 430 million we outweighed each of them.[80] That blue and fin whales should have been dominant in mass is perhaps not very surprising. They are huge animals that eat low on the food chain—krill and the like—where there is much energy available, and using their baleen to lunge- and skim-feed they can ingest enormous amounts of high-quality food very efficiently and quickly. Before we started killing them, there was a lot of blue whale and fin whale biomass around. Sperms are something else. They eat their food, mainly squid, one item at a time, and these items are themselves predators, so relatively scarce and not especially nutritious. Sperms catch the great majority of their food far beneath the surface, far from their source of oxygen, and these squid are not easy prey to catch. We humans have been so unsuccessful in our own attempts that much of what is known about many deepwater squid species comes from analyses of the contents of the stomachs of sperm whales that were killed by whalers or stranded on beaches.[81] There has to be something special about the sperm whale to achieve such dominance in such a challenging niche.

Look at a sperm whale, or more practically, a picture of one—they are hard to see in their natural environment. Special? Not everyone's cup of tea but certainly peculiar, very peculiar.[82] For nineteenth-century whalers, the whale they tried to catch was usually a sperm; for twenty-first-century children, the whale that they try to draw is usually a sperm. But sperms are nothing like other whales whether in their way of life, or nature, or looks. A long battleship-gray body is covered in huge wrinkles except at the front of the animal where a huge smooth nose swells out, rather phallic. This nose contains spermaceti, a substance that may look and feel like semen—hence the whale's name—but is actually oil. For many human purposes, such as the lubrication of precision machinery or making candles, spermaceti is the finest grade of oil known. It is also fine for whale purposes. The spermaceti is an integral part of the most powerful sonar in the natural world, the sperm whale's nose.[83] This sonar is key to the sperm whale catching all those squid, and it dominates the physical presence of the animal, making up a large part of its body. But there is more strangeness about the sperm whale: one blowhole on the left of the snout; a long, thin lower jaw studded with two rows of huge conical teeth that fit into sockets in the upper jaw; a crenulated ridge leading from the low dorsal fin to the flukes or tail; and, inside the whale, the largest brain on Earth.[84]

The contrasts between adult male and female sperm whales are so dramatic that sometimes they almost seem to be different species.[85] Female sperm whales, ten to eleven meters long and maybe fifteen metric tons, are large animals, but males are much larger: sixteen to seventeen meters and perhaps forty-five metric tons. The females have big heads, one-quarter of their bodies, but in the males, huge anyway, the spermaceti organ is even more prominent, a third of the entire body, and it appears even more swollen. The females are not particularly graceful creatures but they seem to move quite easily through the water whereas the males normally appear to have the nimbleness of a supertanker, although they can show remarkable dexterity and swiftness during their occasional fights.[86] But the most extraordinary difference between adult male and adult female sperm whales is in their habitat. The females, together with the dependent young and juveniles, live largely in the tropics and subtropics. Males, especially the largest males, principally use polar waters, sometimes near the Arctic or Antarctic ice edges. Happily for the continuance of the species, the big males, from their late twenties onward, make occasional trips back to the warm waters to breed.[87]

About half of a sperm whale's life is spent deep under water—usually three hundred to twelve hundred meters beneath the surface—in the search for squid. While foraging, the spermaceti organ is banging away: click-click-click-click- . . . at intervals of about half to one second. A sperm whale may make about a billion clicks through its lifetime.[88] The regular series is interspersed with "creaks" in which the clicks are much more closely spaced, and then accelerate becoming even more closely spaced still, sounding rather like a creaking door.[89] We assume that this is when the whale has something, often a squid, in its acoustic sights. Clever tags placed on diving sperm whales that record sounds and movements show creaking whales accelerating, twisting, and maneuvering underwater.[90] Often the creaks are followed by pauses; presumably the sperm whale is eating. Dives last about forty minutes, although sometimes the whales are under water for over an hour.[91] Between dives the whales come to the surface. Most of these surface intervals are about eight minutes long. During most of their visits to the surface, the whales do not seem to do much other than breathe, and they are alone or in pairs. But once or so per day they linger at the surface in larger groups, sometimes just for fifteen minutes, sometimes for hours at a time.

During these periods they socialize, sometimes lying quietly beside one another, at others times nuzzling each other, charging around, or breaching from the water.[92]

So what is community to a sperm whale? Well it depends on the sex and age of the whale, as well as where it lives. Let's start with a female in the North Atlantic, say, off the Caribbean island of Dominica, where we have been part of a team studying the social lives of sperms led by our colleague Shane Gero.[93] The adult female Mysterio is a member of the family unit that Gero calls the Group of Seven.[94] The Group of Seven live and move together, sometimes spreading out over a few kilometers of ocean as they make their long foraging dives for the deep squid, at others tightly clustered near the surface (plate 14). There are preferred relationships among the Group of Seven. Mysterio often babysits Pinchy's calf Tweak when Pinchy is making long dives—the calves cannot dive as long or as deep as their mothers—and the favor is returned: Pinchy accompanies Mysterio's young calf, Enigma, when Mysterio is feeding. The relationships evolve. In the first year of Shane's study (2005), Mysterio was the principal babysitter for another calf, Thumb—the progeny of Fingers. After Thumb died and Enigma was born in 2006, Fingers switched roles with her niece Mysterio, mother becoming principal babysitter, babysitter becoming mother. And in 2008, once Tweak was born, the two mothers, Mysterio and Pinchy, reciprocated caring for each others' calves, and Fingers, no longer a primary babysitter, was left on the periphery of the group.[95] Despite all the soap opera intricacies, it is clear that the Group of Seven is the primary community for its members, that this community is vital for them, and that relationships within the Group of Seven are structured by care for calves.

The Group of Seven is fairly typical of the family units that we study off Dominica. Their members are all females or young animals of both sexes. Each contains about eight individuals. The members of each unit are kin-related through the female line, so these social units are matrilineal. We generally see the units alone, but on occasion they associate with other units. There are preferences. The Group of Seven are particularly friendly with the Utensils, a small unit with two adult females, Knife and Fork, Canopener (an adolescent female), and Spoon (Fork's young son).[96] When they are grouped, the two units are well integrated; Canopener, for instance, will babysit and play with Tweak and Enigma, the Group of Seven's calves.

The different social units off Dominica have characteristic ways of life. The Group of Seven spend a disproportionate amount of time in the inshore Dominican waters and are particularly tolerant of whale watchers and other boats. Unlike some other units, the females of the Group of Seven do not

6.2. The authors' twelve-meter research boat follows a family unit (Unit N) of sperm whales off Dominica. Photograph courtesy of WingsOfCare.org/Jake Levenson.

suckle each others' calves, but they make an unusual "1+3D" coda vocalization (we will discuss codas a little later) that is very rarely heard from the other Dominican units.[97] Other units do things differently. Members of Unit V, "Vive la France," make another version of the "1+3" coda rarely heard from other units and are frequently seen off the neighboring French island of Guadeloupe and, more rarely, off Dominica. The females of Unit J, Jocasta's Unit, make a particular practice of nursing each others' calves but do not seem to have a distinctive coda variety.

We also study sperm whales in the Pacific. Here things are different. The units are a little larger than in the Atlantic, maybe eleven females and their young, and most units contain more than one matriline.[98] However, like the Atlantic units, they seem to have characteristic behavior, such as movement patterns.[99] But the Pacific units are much harder to get to know because they are not often by themselves. Compared with the Atlantic, the groups that we see and follow in the Pacific are much larger, typically containing two to three units in temporary association and so perhaps twenty to thirty sperm whales, in contrast to the single-unit groups of the Atlantic. As off Dominica, the associations among the different units are not random, but the nonrandomness has a feature that is particularly significant, as well as reminiscent of the resident killer whales of the eastern North Pacific: vocal clans.

Unlike the killers, who use clicks for echolocation and whistles or pulsed calls for communication, the sperms use clicks for just about everything. Maybe this is because the anatomical structures that the other toothed whales use to make whistles and pulsed calls were sacrificed in the evolution of the world's greatest sonar system, the spermaceti organ. Maybe, once one has evolved the world's greatest sonar system, it makes sense to use it for all sound production. Maybe a bit of both. So the spermaceti organ that goes "click-click-click- . . ." when hunting for food, and crams the clicks together into a creak when potential prey is found, can do other things with clicks. One of them is the "coda," a Morse code–like pattern of clicks lasting a second or so, such as "click-click-click-[pause]-click."[100] The sperm whale apparently has ways of adjusting the spermaceti organ to make the different types of clicks, so, for instance, the hunting or "usual" clicks are more directional and powerful than the coda clicks.[101] There are other types of communicative click in addition to the coda used by sperm whales, most notably the powerful "slow click" or "clang" of the breeding males. These eerie, slow and very loud clicks are sometimes audible through the hull of our research vessel (oddly, usually best heard in the bathroom!) and could well be the source of the old sailor's legend of Davy Jones, the ill-fated seaman who could be heard knocking to be let out of his watery grave, Davy Jones's locker. It is easy to see how superstitious sailors in the bowels of a quiet wooden sailing ship would have been affected by the slightly creepy quality of a distant slow clicker. It is the coda however that has attracted most attention from scientists.[102] "Coda" is not a particularly appropriate name; it derives from their first being heard at the end of long sequences of echolocation usual clicks, like codas in human music.[103] However, sperm whale codas are not usually made at the end of other vocalizations, instead being produced in a variety of contexts. Mature males seem to make codas rarely, and small calves "babble", so when we discuss coda repertoires, we are basically talking about the sounds of females and immature animals.[104]

Whales make most of their codas when they are at or near the surface, or at the start of their long foraging dives. The whales put the codas into sequences with about three to five seconds between the successive codas made by a particular whale. However, the sequences are often interleaved with, and overlapped on, those of other whales, so forming a kind of duet—and sometimes a chorus. The whales involved can be right beside one another, or a few hundred meters apart. We think that a major function of codas, perhaps *the* major function, is to label and reinforce social bonds.[105] The adult members of a social unit all have basically the same repertoire of codas.[106] However, as we described for the well-known units off Dominica, there are differences between the coda repertoires of units, and these are particularly interesting, especially as the pictures in the Atlantic and Pacific are so different. Ricardo Antunes, a PhD student supervised by Luke, looked at the types of codas made by sperm whales in a variety of locations around the North Atlantic. He found that social units in one area, such as the Azores, have similar coda repertoires, somewhat different from those in another area, such as off Dominica or in the Gulf of Mexico.[107] This might seem entirely unsurprising. It is the general pattern found with human dialects or the songs of many birds. But Antunes was following Luke's own PhD research looking at sperm whale coda dialects in the Pacific using coda recordings that we had made between 1985 and 2000. And Luke had found something very different.[108]

We had our best data from the waters off the Galápagos Islands where we knew many units. While many of the Galápagos units had similar coda repertoires, others' were completely different. These distinctions defined the clans. Mostly, the idea of science moving forward as a series of Archimedes-type "eureka" moments is a myth, but we distinctly remember the afternoon Hal came to see Luke with a scribbled piece of paper in his hands, breathing hard because he had run up the stairs between our offices. After Luke mentioned that one of the first analyses he planned was to look at the specific coda types produced by each of our identified units, Hal had realized he could do this with some data he had already at hand. Luckily for all, he wasn't in the bath like Archimedes when he did. He came running because he found a really striking pattern. Social units of sperm whales off the Galápagos could themselves be grouped together according to the kinds of codas they made. Some groups preferred to make codas that all ended with the "+1" motif—a pause followed by one click—while other groups preferred codas with a regular pattern. Still another appeared to prefer coda types that began with two rapid clicks followed by a slower pattern. We termed this higher level of population structure apparently defined by vocal repertoire "vocal clans," deliberately echoing the terminology used for dialect groups in killer whale populations, although it turns out there are important differences between the two systems, not least in the rather looser relationship between "clan" in the genealogical sense and vocal repertoire evident in the sperm whale case. We eventually documented five distinct vocal clans present in the South Pacific.

The two principal Galápagos clans are the Regular clan with codas such as "click-click-click-click" or "click-click-click-click-click" and the Plus-One clan who go "click-click-click-[pause]-click" or "click-click-click-click-[pause]-click." There were also a very few units from the Short clan: "click-click-click" or "click-[pause]-click-click." When we had recorded them years apart, units maintained the same repertoire.[109] We then looked at the social data. We found that Regular units only grouped with Regular units, and Plus-One units with Plus-One units, even though both could use the same waters.[110] Thus, because each group contained units from only one clan, we could refer to "Regular groups" and "Plus-One groups."

Although swimming the same Galápagos waters, the clans had characteristic ways of using them. The Regular groups stayed close to the islands, the Plus-One groups about ten kilometers farther from land.[111] Although they moved through the water at similar speeds, the tracks of the Regular clan groups wiggled much more, so that after twelve hours they were about fifteen to twenty kilometers from their start position, while the Plus-One clan groups, moving in straighter lines, had gone thirty to thirty-five kilometers.[112] These were pretty startling differences, and there was more.

Think of a group of twenty to thirty sperm whales, each diving for forty minutes or so and then coming to the surface for about eight minutes at a time. Sometimes they do this in synch, or nearly so—all up, then all down. At other times the dives are staggered so there are usually one or a few whales at the surface while the others are foraging beneath. Off the Galápagos, groups with calves are less synchronous, more staggered, than those without calves, which we interpret as active babysitting.[113] Even when accounting for the presence or absence of calves and group size, the Regular clan groups were much more synchronous than those of the Plus-One clan.[114] A few caveats are needed here. First, the data on synchrony of the different clans were from studies in one year only, 1987, and, second, we looked at a number of other measures of sperm whale behavior that did not show any statistically significant difference between the clans.[115] However, our measures of sperm whale behavior are necessarily crude, and the differences between the clans in movements and distributions were pretty remarkable. We wondered if they had an impact on how the whales managed to exploit their environment.

We cannot see the sperm whales feeding far beneath the surface, but we can see their defecations: brown patches sometimes appear behind the whales as they start their deep dives. So, as we sail along behind a diving sperm whale we first photograph the tail to identify the individual, and then look in the slick where it dove, searching for a brown cloud. By making the reasonable assumption that what came out must have gone in, the proportion of dives with brown patches, the defecation rate, is our measure of the feeding success of the sperm whales. We had good data from both main Galápagos clans for two study years. In 1989 the Regular groups had about twice the feeding success—twice the defecation rate—as the Plus-One groups. In contrast, 1987 was a warm El Niño year, which was bad news for almost all the marine life around the Galápagos, including sperm whales. In these hot-water conditions the relative success of the clans was reversed. Only about 2 percent of the Plus-One dives were accompanied by brown patches, but it was worse for the Regular groups: on most of the days in 1987 that we spent following them, we saw no defecations at all.[116] The clans' differences in behavior seem to translate into differences in their reproductive rates. For instance, roughly half of the Regular groups had small calves, while all the Plus-One groups had small calves.[117]

As obsessive sailors we have been good at finding research questions requiring extensive voyages to intriguing places. So we sailed to other parts of the South Pacific, recording the codas of the sperm whales. We found the same pattern of overlapping clans and could roughly trace their spread. While we only heard the Plus-One clan off the Galápagos and in the neighboring waters off mainland Ecuador, we recorded the Regular coda repertoires in Chilean waters.[118] The Short clan, so rare off the Galápagos, was heard almost everywhere else, including off northern New Zealand, seven thousand kilometers away, right across the Pacific. Off northern Chile, where we had spent nearly a full year studying the sperm whale groups, we found differences between the clans in movements and feeding success like those we had traced off the Galápagos.[119]

So what are these clans? Are they incipient species or subspecies like the killer whale ecotypes? No, while there are some differences between clans in the maternally inherited mitochondrial genes indicating that females generally stay within their natal clan, occasionally females transfer between clans, and this leads to genetic patterns that cannot alone explain the dialects.[120] In our preliminary analyses of the nuclear genes that are inherited from both parents, there was no hint at all of differences between clans, presumably because males often mate outside their natal clan.[121] There is no evidence of any morphological variation as between killer whale ecotypes. The only feasible explanation for the sperm whale clans is culture: the young sperm whales learn clan-specific behavior—dialects, movement strategies, and habitat use—from their mothers and other members of their unit and clan, and then, especially if they are female, these norms govern much of their way of life. For those of us studying nonhuman culture the clans of sperm whales in the Pacific are a great gift. They are not genetically distinct, they use the same areas, and they are made up of units that are themselves made up of unrelated animals. Hence, without some pretty contorted reasoning, differences in the behavior of clans at any location and time cannot be ascribed solely to genetics or ecology, so we deduce a significant role for culture.

The significance of the clans for the Pacific sperm whales is reinforced by Ricardo Antunes's discovery that there don't seem to be geographically overlapping clans in the North Atlantic. Remember, he found that the North Atlantic coda dialects are geographically based, with different repertoires in different ocean areas. However, the dialects of the sperm whales in these areas, which are mostly thousands of kilometers apart, are less distinct than those of the Galápagos clans that use the same waters.[122] It appears important for the Pacific clans to clearly mark their differences. This is reminiscent of the killer whale pods actively maintaining differences between the structures of their vocalizations.[123] As we discussed in chapter 2, and will again in chapter 8, the symbolic marking of cultural differences is an important attribute of human cultures.

You may be wondering, why no clans in the Atlantic? Or, putting it another way, why clans in the Pacific? We are not sure, but here is a scenario that makes sense to us. First, there seem to be almost no differences in nuclear genes between Atlantic and Pacific sperm whales, presumably because males quite often are born in one ocean and then breed in the other.[124] Second, maybe you recall that the North Atlantic social units rarely group with each other, while the Pacific units almost invariably do. This means that groups typically contain about ten animals in the Atlantic and thirty in the Pacific.[125] Ricardo Antunes has suggested that because grouping is so much more frequent and presumably important in the Pacific, the whales have developed the clan structure to manage these associations between units: we group with these units, but not with those.[126]

This brings up a more fundamental question, why is the grouping of units so much more important in the Pacific? Well, it could be because the oceanography is different in the two oceans, and this leads to differences in the availability of squid resources, which in turn affects social behavior. However, female sperm whales consistently form large groups in very different types of habitat in the Pacific (such as the enclosed waters of the Sea of Cortez and the wide oceanic waters of the eastern tropical Pacific), and small groups in different Atlantic areas (such as the Sargasso Sea, Dominica, and the Azores).[127] Another potential factor is the different histories of modern whaling in the two areas. Female sperms in the western North Atlantic were almost completely spared from harpoon guns in the twentieth century. Those in the southeast Pacific were targeted between about 1950 and 1982 by Japanese whalers working from Peru and Chile who were not even bound by the weak restrictions of the International Whaling Commission, as well as by pirate whalers, such as Aristotle Onassis's *Olympic Challenger*, which followed no regulations at all. Perhaps this whaling so weakened the integrity of the Pacific units they resorted to forming large groups. However, if social destruction by whalers caused clans in the Pacific, it did so remarkably quickly. Instead, we think the most plausible explanation for the different social structures in the two oceans is the subject of the first part of this chapter, the killer whale.[128] Killer whales were probably the only serious threat to a healthy sperm whale before humans started whaling.[129] In the scientific literature, there are six descriptions of killer whales attacking sperm whales in the Pacific, in at least one of which sperm whales were killed, but there are none in the Atlantic even though scientists have spent more time watching sperm whales in the Atlantic.[130] In the Pacific, sperm whales usually react quickly to the arrival of killer whales by forming a tight group at the surface, sometimes facing outward at the killers and sometimes putting their heads together, with their tails radiating outward like a wagon wheel.[131] However, in the Atlantic we have seen killer whales cruise right through a group of foraging sperms without any obvious reaction by either species. Thus we think the differences between the social behavior of Atlantic and Pacific sperms may result from the cultural conservatism of that other great toothed whale. Sperms are on the menu of some killers in the Pacific but not in the Atlantic. So the Pacific sperms form large groups and overlapping clans, and the Atlantic sperms don't.

The clans of the sperm whales in the Pacific are extraordinary. They stretch thousands of kilometers, contain thousands of members, and, most unusually, share the same habitats. There is nothing much like them in any animal other than *Homo sapiens*, and even among us, similar cultural structures are rare. For a female sperm whale in the Pacific, clan determines her vocal repertoire, her movements, her feeding success, and even her probability of giving birth. The population of sperm whales that uses the Pacific is much more clearly structured by clans than it is by geography.[132]

The sizes of clans are particularly noteworthy. In multitier, or multilevel, societies, like those of humans, elephants, sperm whales, and killer whales, in which one level of social structure is embedded in another, scientists have noted a general rule: the size of the groups at each level is about three to four times bigger than the size of its components. So a modern human is a part of a "support clique" with about five members, which is part of a "sympathy group" of about fourteen people, which is part of a "band" of about forty, and then, going on upward, we have "communities" of about one hundred thirty, "mega-bands" of about five to six hundred, and "large tribes" of around seventeen hundred.[133] Note the factor of approximately three between the sizes of consecutive levels. There is also a factor of about three to four between different levels of military structure: squads within platoons, platoons within companies, companies within battalions, battalions within regiments, regiments within divisions, and divisions within corps. The hierarchically embedded structures of human hunter-gatherer societies show a similar ratio between the sizes of successive levels, as do those of elephants and baboons.[134] For resident killer whales the mean ratio, going from matrilineal units to pods to clans to communities is 3.4.[135] It is not clear why we get these fairly universal scaling ratios; perhaps it has something to do with how mammalian brains work, or how an individual can apportion time available to different types of social relationship.[136] Now look at the sperm whale in the Pacific in terms of these rules. Social units of ten or so are formed from about three matrilines, and groups contain about three units (although groups are transitory, lasting only a few days, so perhaps they should not be included as an element in this analysis, in which case our point is even stronger). So far, so good. But going from groups, with about thirty members, up one level to clans, with very roughly ten thousand members each, the sperm whales have a ratio of around three hundred groups per clan.[137] What is happening? Perhaps we are missing four levels of sperm whale social structure, with about a hundred, three hundred, a thousand, and three thousand individuals, respectively, but we think not. We suspect there is something different about sperm whale clans; they are more than just a level of social structure, they are a cultural identity. So just as a modern human has a vital ethnicity, usually shared with millions, on top of his or her "large tribe" of a few thousand—that might be equivalent to a village, neighborhood, or high school—a female sperm whale in the Pacific has her clan. This leads to "us"—my clan, or nation—and "them"—members of other clans or nations. Outside humans, large-scale cooperation is rare, basically the social insects—ants, bees, and termites—as well as that strange subterranean mammal, the naked mole rat. But in all these cases the cooperators are related. Peter Richerson and Robert Boyd conclude that humans are unique in cooperating on a large scale beyond the confines of their kin.[138] The Pacific sperm whale clans suggest that this might not be so.

What about the male sperm whale? He leaves his mother's unit at roughly age ten. Between 2005 and 2011, Shane Gero has watched a young male, Scar, gradually disassociating himself from the Group of Seven.[139] We expect that he will soon be completely gone, as he heads gradually poleward from Dominica. Adolescent male sperm whales form temporary groups in temperate waters, but as they grow older, moving to colder and colder climes, they grow more solitary. However, these large adult males do sometimes associate with one another, and occasionally will strand on beaches together, seemingly unprovoked.[140] Communal suicide looks pretty social to us, so clearly the big males must get something out of being together. Novel behavior patterns—such as stealing fish from longlines in the Gulf of Alaska or entering shallow waters (forty to seventy meters deep) off New York—move through populations of male sperm whales.[141] We suspect that horizontal social learning likely plays a large role. Thus, while male sperm whales do not have social structures as obvious, or probably as important, as those of the females, they are significant to one another. The same may be true of the adult males using the same tropical breeding ground at the same time, although here competition is likely to be stronger.[142] The powerful slow clicks of the breeding males—which sound rather like a jailhouse door being slammed every seven seconds—may be primarily intended for the ears of other males.[143]

So there seem to be three kinds of sperm whale community that form reservoirs for social learning. Fundamentally, there is the social unit, the primary social environment of female and young sperm whales. Then there is the set of other units with whom the females group and associate: in the Atlantic, the other units that use the same range; in the Pacific, the other members of their clan, many of whom they may never meet. Finally, there are the communities of male sperm whales, both on their high-latitude feeding grounds and when breeding in the tropics—communities we know much less about.

Between us, we have been studying sperm whales for a total of nearly fifty years. What is the nature of this strange creature that we follow around the ocean? What is it like to be a sperm whale? We have not got far, especially for the males. But we have learned a little about the lives of females. They are constantly on the move with ranges that span hundreds and often thousands of kilometers. We think this nomadism shapes the character of female sperm whales. Almost continually in unfamiliar surroundings, they are nervous creatures, "timid and inoffensive" in the words of the nineteenth-century whaleship surgeon, Thomas Beale.[144] Their stable points of reference are each other, the members of the social unit with whom they travel. Female sperm whales are intensely communal: in addition to moving together, they socialize often and sometimes vigorously, they babysit and suckle each others' calves, and they defend themselves communally. "We" and "us" may be more important than "I" and "me" for female sperm whales. And "we" have a characteristic set of behavior, "our" culture, the way we do things.

Blackfish, Potheads, Mealy Mouthed Whales, and the Canaries of the Sea

In this chapter we focus on species with matrilineal social systems. A young individual in these societies naturally spends much time with its mother when suckling; this period is the prime opportunity to learn socially. If the period is extended to a lifetime and an individual is not only always accompanied by its mother but also by other close female relatives, all learning from one another, then matrilines can quickly develop characteristic cultures. If the ranges of the matrilines overlap, as is usual in cetaceans, then hierarchically structured social systems based on matrilines, with characteristic cultures at the different levels, may emerge, as we have described for resident killer whales in the northeastern Pacific and sperm whales in the eastern tropical Pacific.

So, if you are anywhere near as enthralled as us by the killer whales and sperm whales and all they do, you must be wondering what goes on in the other matrilineal species. We know rather little about the social systems of odontocetes other than bottlenose dolphins, killer whales, and sperm whales. With possible exceptions of the Arctic odontocetes, the narwhal and beluga, all the other good candidate matrilineal species are "blackfish," technically members of the subfamily Globicephalinae: the long-finned pilot whale (sometimes called the pothead), short-finned pilot whale, false killer whale, pygmy killer whale, and melon-headed whale (or mealy mouthed whale).[145] These are small- to medium-sized toothed whales, with adults ranging from two to seven meters long, that are mostly shiny black, have bulbous foreheads, and are highly social. We don't even know whether the false killer whale, the pygmy killer whale, and the melon-headed whale have matrilineal societies, but there are indications that way.[146]

Best known are the two pilot whale species.[147] They swim in all oceans—

the short-finned pilot whale in more tropical waters, the long-finned in the colder waters of the Southern Oceans and North Atlantic—and are often numerous. They have been hunted for centuries; and off Japan and the Faeroe Islands they still are. Japanese and Faeroese fishers drive whole groups of pilot whales into the shore and then slaughter them. The whales stick closely together as they flee from the noisy fishers' boats, which is why the drive hunt works. It is, we assume, a similar mechanism to that which leads the animals to strand together. The pilot whalers sometimes euphemistically call their technique "assisted stranding." Toothed whales sometimes strand in considerable numbers together on beaches. Most are healthy when they strand but most usually die. Mass stranding is one of the great puzzles of the Cetacea. It is particularly characteristic of matrilineal species. Killer whales mass strand, but the most common mass stranders are the sperm whale, false killer whale, and, above all, the pilot whale.[148] For instance, of the 816 animals recorded in mass strandings along the British coasts between 1913 and 1966, 53 percent were long-finned pilot whales.[149]

Observers of a stranding, whether natural or "assisted," naturally assume that the dead and dying pilot whales huddled on the beach form a natural social grouping: a pod or a school or, using our terminology, a community. At a time when DNA analysis was in its infancy, Bill Amos and his colleagues made several comprehensive analyses of samples of the animals killed in the "grinds," the groups of pilot whales driven ashore by the Faeroese whalers.[150] They found that the grinds contained several generations of matrilineally related males and females and that, while the mothers of nearly all of the younger animals were also captured in the same grind, the fathers never were. This fitted with the inclination to see the animals killed together in a grind as forming a primary element of pilot whale society and corresponded with the "momma's boys" social system of the resident killer whales: both sexes staying grouped with mother but finding nonrelatives for mates when groups interacted.[151] However, a major difference between the pilot and killer whales is in the sizes of the groups. Matrilineal units of resident killer whales contain about two to ten members, pods about ten to twenty. But the Faeroese grinds averaged a hundred fifty long-finned pilot whales.[152] The picture became more anomalous with the results of studies of photo-identified living pilot whales. Off Cape Breton Island, Canada, the long-finned pilot whales live in permanent social units of about eleven to twelve animals, and these aggregate into groups containing roughly two units, very like the Pacific sperm whales.[153] A study in the Straits of Gibraltar, on the other side of the North Atlantic, came up with quite similar results: groups of about fifteen animals, consisting of permanent "line units" with two to nine animals each and maybe a few more unidentified animals.[154] Unless the pilot whales do something very different off the Faeroes, the implication is that the grinds contain a number of social units and may constitute higher levels of social structure, so far unrecognized off Cape Breton and in the Straits of Gibraltar. The picture may be similar for short-finned pilot whales. Living animals off Tenerife in the Canary Islands form groups of about twelve animals, whereas the mean size of schools driven ashore off Japan was about thirty.[155] So, while the details of pilot whale social structure are a bit murky and may vary between the species and locations, it looks as though there are primary matrilineal social units of ten or so animals, which themselves associate to form larger structures.

We have seen how, in killer and sperm whales, such a social structure is a scaffold for characteristic cultural differences in behavior. What about pilot whales? Apart from a few hints, we don't know. In the Strait of Gibraltar, the different social units seem to have distinct diets, a pattern that may well have a cultural basis.[156] Leah Nemiroff studied the vocalizations of the Cape Breton pilot whales and found that the repertoires of pulsed calls, but not of whistles, seemed to vary among the different social units.[157] This fits with the killer whale pattern, where pulsed calls are, at least for us and probably the whales themselves, the primary cultural marker of group membership.

Robin Baird and his colleagues have used photo identification and satellite tags to track blackfish off Hawaii. There are three clusters of groups of false killer whales that use Hawaiian waters, with each of the clusters using the habitat differently.[158] The population and social structure of the Hawaiian pygmy killer whales and short-finned pilot whales seems to include characteristics rather like the distinctions between northern and southern resident killer whale *communities*.[159] We suspect that these results are the first visible tip of an iceberg of blackfish culture. But for now, we don't know how large that iceberg may be. It would be lovely to find out. While killer whales are powerful, conformist, conservative creatures of habit, and sperm whales seem mild, timid, giants of the deep ocean, the pilot whales appear confident, lively, full of fun and exploration. For instance, they seem to enjoy harassing sperm whales, charging around the sluggish sperms, nudging and nipping, to what effect we have no idea.[160] What cultures may the pilot whales and their blackfish relatives have developed?

Finally, we will make brief mention here of a species that we suspect has cultural riches yet to be discovered—the beluga, or white whale. This species has a relatively poorly known social structure, but one that has been described as "matrifocal," that is, loosely structured along maternal lines.[161] This animal has long been commonly known as the sea canary for its distinctive, high-pitched, and highly variable vocal behavior. In fact, we've recently learned that they are very capable vocal learners, from a report of a captive beluga called NOC who was recorded producing recognizable, though arguably not very passable, imitations of human speech.[162] There is also evidence from studies of genetic population structuring that belugas learn from both their mothers and her relatives how to navigate the migration routes of the Canadian Arctic in and around the complex geography of Hudson Bay.[163] We know very little about beluga society, and we include these small insights largely to highlight an area where we think significant discoveries are still to be made.

Matrilineal Lives, Matrilineal Cultures

Much as most of us love our mothers, we probably would not fancy spending all our time with her *and* all our sisters, various aunts, and a grandmother. For the matrilineal whales though, mother is not just the focus of life during suckling. She, and the social group within which she is central, continues to frame not only the social world but also the world in general. There are advantages to a matrilineal social system. Cooperation is likely to be both simpler to achieve and more effective among animals who have known each other their whole lives and can expect to be together until death. Just as cooperation is heightened by long-term relationships among close kin, competition is dampened, although there must often be competition for food, especially when it becomes scarce. The matrilineal whales have developed ways to signify the importance of their communities, whether they are the basal matrilineal units or the larger structures built on them. They have dialects and greeting ceremonies. Killer whales routinely share food; false killer whales pass food around.[164] These are the pillars of their cultures.

7 | HOW DO THEY DO IT?

We have implicitly assumed that much, or perhaps all, of the behavior of whales and dolphins described in the three previous chapters is the result of social learning. But is it? Are these animals actually able to learn from each other? If yes, then suggesting a role for culture in sponging dolphins or singing fin whales comes naturally. But if we look specifically for social learning and do not find it, then the whole "cultural whale" hypothesis promoted by those of us who study the animals in the wild looks pretty ropy.

The study of social learning is another facet of the study of culture and is usually the purview of experimental psychologists. Experiments are the gold standard of behavioral science. Conditions are changed in a carefully controlled manner, and the behavior changes, or it does not. So when studying social learning, one standard protocol is to compare when or if an animal adopts some form of behavior depending on whether it is, or is not, in the presence of another animal performing the behavior. If it does better with a model, then social learning is supported; if there is no difference in the learning rates, it is not.

To control conditions in this way almost always requires captivity for the experimental subjects. This brings problems. It can be argued that captive conditions for dolphins and small whales are so far from the natural environment of the animals that any results obtained under them are of little value—the ecological validity problem.[1] Furthermore, captive cetaceans are typically engaged in extensive training programs that can affect the ways in which they learn. For example, they may pay much more attention to a trainer than to other dolphins. It is incredibly easy to get a negative result from all kinds of captive animals in behavioral tests, not just cetaceans. Even if the experiment works, what does it mean? That one or a few dolphins can or cannot perform a particular task in a tank does not necessarily imply that they do or do not perform it in the wild. In addition to these scientific issues, there is a hot debate as to whether keeping whales and dolphins in captivity is humane or ethical.[2] Finally, whatever one's views on those issues, there is the problem that only the smaller species can be kept in captivity and so for species like humpback, bowhead, and sperm whales these kinds of data will never be available. (Stranded sperm whale calves have been recovered and kept for short periods, but they inevitably die.)[3]

However, if the animals do succeed in some task, then we could consider the results as showing us the minimum capabilities of their species. Such studies, if well done, do inform us about the general abilities of the species, which can help us interpret what we see and hear in the wild. The studies often appear motivated by a desire to see how dolphins measure up against humans—they start with something humans can do and look for whether nonhumans can do it too. This can be seen as an uninspiring and self-obsessed approach. Given the kinds of behavior we see in the wild, the main utility for us in these studies is to understand how well cetaceans learn from each other in experimentally controlled conditions. As we see it, the evidence suggests that they are adept social learners. In this chapter we describe these studies in some detail. We will also leave the lab to talk about intriguing observations from the wild that suggest some toothed whales go out of their way to teach hunting skills to their young.

Imitation—Do as I Do

Of all the forms of social learning that we introduced in chapter 2, imitation has been the primary focus of experimental psychologists studying nonhuman learning. Imitation is the learning of a new behavior pattern from seeing it done by another.[4] One of the reasons for this focus is the belief that, of all the forms of social learning, only imitation leads to cumulative cultural evolution and thus the creation of truly complex cultural products like jet aircraft, symphonies, or languages.[5] The preeminence of imitation among social learning mechanisms is debated outside experimental psychology, but that does not change the nature of the data: almost all experimental studies of social learning in dolphins refer to imitation.[6]

There is a good deal of anecdotal, and some experimental, evidence that bottlenose dolphins in tanks imitate both humans and each other. They imitate sounds and physical activities. Trainers use this ability: they shape the behavior of one animal to perform the desired display and then others imitate it.[7] Furthermore, dolphins can be taught the concept of imitation, understanding the "Simon says" game, so that if they are given a distinctive "copy" command, they replicate the subsequent action of the trainer, whatever it may be.

Their apparent intelligence and the relative ease with which they adjust to captivity means that bottlenose dolphins have long been the focus of scientific interest in cetacean cognition. Their ability to deal with captivity and the readiness with which they can be trained is illustrated by the fact that they have been drafted, for better or worse, into the U.S. military, which maintains an active-service marine mammal unit. These dolphins are trained to provide port security (primarily to detect underwater mines) and have been deployed in multiple theaters, most recently the 2003 invasion of Iraq.[8]

One of the first scientific reports of dolphins copying each other was published by Colin Tayler and Graham Saayman in 1972, in a paper that has become famous as the source of some of the most retold observations of dolphin behavior.[9] They did not actually perform any experiments but, instead, just observed the behavior of two female and one male bottlenose dolphins, named, for reasons we cannot even guess, Haig, Lady Dimple, and Daan. Tayler and Saayman watched and recorded the dolphins' interactions both with each other and with other things that entered their tank, in the Port Elizabeth Oceanarium, South Africa. One of these things was a human diver, whose job it was to clean algae from the viewing ports built into the side of the tank so that the paying visitors could see clearly. We don't know how long Daan was able to watch the diver cleaning, but it was described as "repeatedly." Daan was then observed using a seagull feather he had recovered to stroke the viewing glass, apparently copying the cleaning (needless to say, this kind of housekeeping has never been observed in wild dolphins). He then apparently became quite taken with this. He went on to use a variety of objects to fulfill his role as everyone's favorite tank mate, including stones, paper, and even the fish he had been given to eat. He remained quite faithful to his human "demonstrator" though, even mimicking the human divers' technique of holding onto steel bars beside the window to steady themselves, by placing a flipper in the same spot. He also became quite possessive of the viewing port and would aggressively prevent divers, and the other dolphins, from approaching it for a period of over fifty days. During this time though, he kept the window quite clean.

One of the females, Lady Dimple, had a calf, Dolly. In a short anecdote that has become part of dolphin imitation folklore, Tayler and Saayman describe how, at the end of one observation session, while Dolly was apparently investigating them through the viewing port, someone unnamed blew cigarette smoke at the window in her direction (the culture of smoking on the job has since declined among scientists). The observer was "astonished" by what happened next, and we can only imagine *how* astonished they must have been to allow that word to make it into the dry science literature. Dolly swam to her mother, from whom she was still nursing, and returned with a mouthful of milk and spat it back toward the window, producing an uncanny replication of a cloud of smoke. This apparently went on to become a regular trick. Dolly's mother was at the time being trained to perform in public shows, and it is telling that the calf was seen to begin copying her mother's tricks, so much so that with minimal positive reinforcement the calf became part of the performance team. This observation in captivity resonates strongly with the patterns of mother-calf similarity in feeding tactics in the wild.

The other female, Haig, had an imitation trick of her own. Another chore that had to be performed on their outdoor tank was cleaning the bottom—a diver used a kind of underwater vacuum cleaner to scrape off and suck out thickly growing seaweed. This vacuum cleaner was sometimes left overnight in the tank. Haig was observed to pick up the cleaner one morning between her two pectoral fins and rock it back and forth across the bottom of the tank. Intriguingly, when she succeeded in breaking out some seaweed, she promptly ate it. Plant eating has never, to our knowledge, been observed from dolphins in the wild, so we can only speculate why Haig found it so appetizing, but she did. After the cleaner had been removed from the pool for its own safety, Haig found a broken piece of the tiles used to line the pool and began scraping off more seaweed with it, sometimes breaking off to eat the fruits of her labor. Not only that, but then Lady Dimple picked up the technique, and soon both dolphins were to be seen working together to remove seaweed with broken tiles. Again, there is powerful resonance with reports from the wild of dolphin feeding specializations that scientists suspect are learned from each other.

These are incredible observations, no less vivid for the passage of time, and it is not surprising that they have captivated scientists and nonscientists alike. In a way, though, these observations are intruders in this chapter about experiments. They are observations only, not the results of controlled experiments, as those less easily impressed will quickly point out. Carefully observed and described anecdotes like these are a perfectly valid form of scientific information—after all, they really did happen, and an observer really did see them—but there is controversy over what they might mean. This is largely because of the way that we humans are so easily fooled by our own eyes, let alone anyone else's. It is only through careful experimentation, the argument goes, that we were able to see through Clever Hans, the horse that could apparently perform arithmetic in nineteenth-century Germany, and conclude that while he could do no such thing, he could pick up very well on subtle and inadvertent cues from his trainer.[10] We don't see how any kind of Clever Hans effect could account for the variety of observations reported, and in some cases recorded on video, by Tayler and Saayman, but we would be the first to admit that our eyes are as easily fooled as anyone else's.

One of the most important experimental research programs into bottlenose dolphin imitation was led by Louis Herman at the Kewalo Basin Marine Mammal Laboratory in Honolulu. The project began in 1979, when two female dolphins were brought into captivity, at two years of age. The facility was closed in 2004 following the death of the last dolphin at the laboratory.[11] During that time the dolphins took part in many illuminating studies of their cognitive characteristics and sensory abilities. The research was not limited to imitation and mimicry, but it is that aspect that we focus on here and, in particular, two studies by Master's degree students at the University of Hawai'i, Stacy Braslau-Schneck and Mark Xitco.[12] Both studies took advantage of the way the dolphins at Kewalo were trained to produce characteristic moves, such as slapping the water surface with the tail or performing an underwater barrel roll, in response to gestures made by their handlers.

The first study, by Xitco, used a setup in which each dolphin had its own trainer, and the two trainers were separated by a poolside partition that prevented each dolphin from seeing the other's trainer. During trials, one dolphin would be the demonstrator and the other the imitator. First, the researchers trained the dolphins to recognize a specific command for "mimic," the equivalent of "Simon says" (we use the term "mimicry" to refer to copying a behavior that you already know after seeing another do it, rather than a new behavior, as in imitation). To do this, they commanded the demonstrator dolphin to do a specific act. The mimic dolphin, which had not seen the command to the demonstrator but had seen the act, was then given the "mimic" hand signal, followed by the command for the same act that the demonstrator dolphin had just performed. After a little of this, they conducted probe trials where only the mimic command was given to the imitator following the demonstrator's act. The experimenters set a success criterion—the dolphin given the mimic command had to produce the correct act (i.e., the one performed by the "demonstrator") on 85 percent of the twenty-four to forty-eight trials in two consecutive training sessions. The frequency of probe trials was ramped up across sessions every time the dolphins reached the success level, until they were only receiving trials with just the mimic command. The two dolphins in the experiment reached the success criterion when receiving just the mimic command after seventeen and twenty-six of these training sessions, respectively. The researchers then carried out transfer tests to see whether the notion of mimic was generalized. They ordered the demonstrator dolphins to perform acts that, although known to both dolphins from their on-going training, had not at that point been used in the imitation study—again, the dolphins rapidly reached the success level. Xitco therefore showed it was possible to train a dolphin to mimic on command.

However, this is mimicry, not imitation of something new—the training behaviors were already known to both dolphins, and each had been trained to the same gestural command for them. For the experimental test, Xitco then went on to train each dolphin, on their own, to perform three new tricks (different ones for the different dolphins) and then introduced these into the mimicry sessions. The demonstrator was commanded to perform a behavior that the dolphin who would be asked to mimic had never seen before—for example, ringing a bell or putting a ring on a stick, so not things you would expect to be in a dolphin's natural repertoire. When tested, one dolphin, Phoenix, copied two of the three new behaviors she was exposed to after the second demonstration, while the other, Ake, copied one of the three after the third demonstration. The dolphins were not perfect, but clearly, on occasion, they could copy new behaviors—they could imitate, and they could do it pretty quickly after seeing a demonstration. Xitco also made the task a bit harder by separating the demonstration and the subsequent command to mimic in time. After eighty seconds, performance had degraded to 59 percent correct, which is still much higher than chance, but suggests that there is something about the dolphin's abilities that is focused on immediate synchrony of behavior, which jives with the field observations in chapter 5 suggesting that behaving in synchrony—doing the same thing at the same time—is an important social signal in dolphin societies, as highlighted by Louis Herman himself: "The unquestionable extensive imitative abilities of the dolphin may derive from the naturally occurring highly synchronous or closely coordinated natural behaviors often seen among pairs or groups of dolphins. Synchrony may function to assist in applied tasks such as foraging and prey capture, but may also be an expression of social affiliation."[13]

The second study, by Braslau-Schneck, was essentially a training program designed to build up to a situation in which the dolphins would understand and respond correctly to a command to choose a behavior themselves and then perform it together. It was partly designed as a follow up to Xitco's work to check that the imitator dolphin was not able to somehow see the trainer's commands to the demonstrator dolphin and take that as the cue for their behavior rather than the behavior of the demonstrator dolphin itself. Asking the dolphins themselves to select an act to perform removed this cue entirely—they would only be able to match by working together. The first stage involved training the dolphins to a gestural command, called "tandem," followed by a more conventional command to perform some behavior, such as an aerial somersault, so that, when the pair of commands were given to two dolphins, they would do the right thing and do it together. The dolphins mastered this so well that by the end of training they could be commanded to perform not just single tricks but also sequences of acts together. The dolphins were also trained to a command called "create." When this was given, the dolphins were rewarded for performing any trick they wanted, provided it was new to the session. In the second stage of the study, these two commands were combined, "tandem+create," meaning do something new and do it together. To give this combined command, the trainers would tap two index fingers together ("tandem") and then throw their arms up in the air ("create"). The training of these commands took place over several years, with the dolphins' trainers incorporating the tandem+create game into general training sessions. Braslau-Schneck then quantified how good the dolphins were at the game. During her study sessions, the dolphins performed no less than seventy-nine different behaviors together but always with a delay of about ten seconds, during which they would swim around the pool together. Were they planning? How would we ever know? Even though the sessions were videotaped, in many the experimenters could not detect whether one dolphin apparently led the other, although in forty-four of the sessions they could detect one dolphin marginally in the lead. The lead role was shared between the two dolphins.

The research at Kewalo Basin would appear to suggest that the case for imitation in dolphins is open and shut, but it is not the end of the story. In another study published in 1994, Gordon Bauer and Christine Johnson attempted to replicate the results of Herman's students, using dolphins held at the EPCOT Center in Florida.[14] Their subjects were two male bottlenose, Toby and Bob, both around twelve years of age at the time, who had been captured from the wild around ten years before. Instead of introducing a "tandem" command given by a single trainer, these dolphins were trained from the outset by two different trainers as in the setup of Xitco's experiments. In the first training sessions they were given the same command, so that they produced some trick from their repertoire, like tail slapping or water squirting, simultaneously. Once that was mastered, the protocol changed so that one of the dolphins would be given the signal two seconds later than the first. Then the researchers introduced a new signal, which they wished the dolphins to interpret as "mimic," before giving the second dolphin its behavioral command. To check that the second dolphin was indeed responding only to its own trainer (rather than somehow being cued in to follow the other dolphin's trainer), it would occasionally be given a command to perform a different behavior to that given the first. In these cases the second dolphin always did what it had been commanded by its own trainer. Then the researchers started giving the second dolphin just the "mimic" command, and kept going until both dolphins were able to successfully mimic on at least 75 percent of their trials in a given session. It is here that the first differences to the Kewalo Basin results show up. Toby and Bob were much, much harder to train to this level of success at mimicking, with Toby taking "many hundreds" and Bob over a thousand individual trials. Not only were they apparently slower to pick up this command, there was also a marked difference between the two dolphins—Bob was much less into it.

Having reached 75 percent successful mimicry of the training behaviors, the dolphins were classed as successful mimics, and the testing began with a different set of tricks than had been used up to that point. Of these new moves, nine were familiar in the sense they had been taught to both dolphins before, and each dolphin was also taught two tricks on their own, to be used as novel behaviors in the final crucial test. Toby mimicked four of the nine familiar behaviors (one of which was not actually in the planned set but was performed by Bob "by mistake" and copied the very first time by Toby), but Bob didn't mimic the familiar behaviors once. When it came to the novel acts, neither dolphin successfully imitated any of them to the satisfaction of the researchers. The researchers were not easily satisfied and, in fairness, did note that the dolphins were not total failures. For example, one of the new tricks involved touching something with the tail. When asked to mimic, the second dolphin "swam parallel to the model, assuming the same lateral posture but located several feet away from the bumpers" that they were required to touch.[15] The dolphin reproduced the exact body movements of the model but failed to replicate the end result—touching the bumper with its tail. Similarly, another of the novel tricks was tailwalking—the same thing Billie appeared to learn without any training, as we described in chapter 5. Here, Bauer and Johnson noted that Toby, when asked to mimic this, "spyhopped with vigorous fluke movements, but did not move forward as did the model," and so missed out on a fish for the sake of a few feet of forward movement, despite having got the idea of forcing himself vertically out of the water. As these tests were cut off after ten trials, we will never know if Toby would have eventually cottoned on that he had to touch the bumper with his tail or move forward a bit, in order to get his fish. Nonetheless, the fact remains that these dolphins did not imitate nearly as impressively as those at Kewalo Basin.

There are many reasons why an experiment can fail to reproduce a finding in animal behavior. One of them is that the original finding was bogus.

However, given that in both experiments the dolphins could be trained to mimic and that, in the second, some of the actions attempted by the mimicking dolphin were pretty close to what they had been asked to copy, this seems unlikely to us. There are a number of other differences between the two studies that are equally plausible explanations. Perhaps most obviously, the Kewalo Basin subjects were female, while the EPCOT ones were male. We don't know the history between the EPCOT males but given the high tensions that can surround alliances between male dolphins in the wild, and the suggestions that behaving in synchrony with another male is an important social signal, the researchers could have unwittingly been asking the dolphins to do something entirely against their social instincts at the time. We do know that Bob was dominant over Toby.[16] This might therefore explain why Bob seemed much less willing to copy Toby than vice versa: that is, if being the leader in a bout of synchrony is a signal of dominance, that could account for why Bob was just not willing to concede. Again, we don't know enough about the relations between the two before the study began to figure out if this is a credible explanation.

Another difference is that in Kewalo Basin the dolphins took part in years of cognition studies before their imitation trials, and taking part in these studies was virtually their entire job description. At EPCOT, the dolphins had taken part in "several" learning studies beforehand and, in addition, had roles in the Disney tourist attraction—the effects of which on their performance in these trials is unknown. The idea that being used over and over in experiments, involving lots of contact with humans as in these studies, can affect an animal's performance has also been raised as an issue in studies of chimpanzee learning, where it has been given a name from the theory of child development, "enculturation."[17] This is the idea that one of the factors that makes humans so smart is that they are raised by other humans and are thereby exposed to endless opportunities for stimulation and learning—they are "enculturated."[18] Some chimp researchers argue that similar things can happen to animals that are exposed again and again to interaction with humans, so that their apparent cognitive performance is ratcheted up to levels that are not seen in the wild. This might seem a pretty egotistical stance, but so-called feral children, raised by animals or in isolation with little or no contact to other humans, show remarkable deficits in social skills, including social learning.[19] So perhaps it is the case that the Kewalo Basin dolphins were enculturated by their prolonged experience and interaction with their human trainers, and they are in fact "super dolphins" in terms of their brain power. Perhaps. The counterargument is that you can try to enculturate other animals—cats, for example—as much as you like, but they'll never do what Louis Herman's dolphins did, even after millions of trials, so something concrete about the underlying abilities of a species is revealed by these kinds of experiments.

It is clear that a bottlenose dolphin can readily mimic others producing a behavior it already knows. It may surprise you to learn they do not always need to even see the other dolphin in order to copy it, but this is just what a series of experiments led by Kelly Jaakkola at the Dolphin Research Center in Florida shows. In her first study, Jaakkola worked with a seven-year-old male dolphin called Tanner, who had been born in captivity (and so could be classed as enculturated).[20] Tanner was trained to recognize a "mimic" command just like the dolphins in the studies we already described, but the twist in this study was that in half of the test trials the scientists would place eyecups over Tanner's eyes, effectively blindfolding him. Some of the behaviors, all of which were already known to him, were purely vocal—producing a raspberry sound, for example—and perhaps unsurprisingly his performance at copying these behaviors from other dolphins was unaffected by the blindfolding. However, most of the behaviors in the test did not involve producing vocalizations, and in these, Tanner continued to perform well above chance levels in matching behaviors when blindfolded. Sure, his performance dropped, from 61 percent correct matches when sighted to 41 percent when blindfolded, but with twenty such behaviors to choose from, getting 41 percent right is still many times more than could be expected by chance. How could Tanner achieve this? One possibility is echolocation, and indeed Tanner did echolocate more—about three times as much, in fact—when blindfolded, so perhaps he was able to identify the behavior by tracking the demonstrating dolphin with his clicks. Jaakkola doesn't think this explains it all, though, because she found no relationship between whether Tanner echolocated or not during a given trial and the accuracy of his copying. Although these behaviors did not involve producing a vocalization, there were sounds, such as varying patterns of splashes and occasionally jaw claps, which could give cues to an attentive listener. Jaakkola demonstrated that this was plausible by having blindfolded human trainers try and identify which behavior a dolphin was performing. The humans got it right about 55 percent of the time, so this could explain Tanner's performance. However, a follow-up study by Jaakkola and her colleagues suggests things are a bit more complicated than that.[21] In this experiment, a human was tasked to perform, as best he or she could, one of the subset of test behaviors that a human could reproduce reasonably well, and Tanner was asked to copy them. The rationale was that if Tanner was just using the characteristic sounds of the behavior, he would be flummoxed in the human-model trials because the behavior produced by a human will sound different from the same thing done by another dolphin. Again the scientists recorded Tanner's echolocation activity during the trials. The results? When asked to copy a human, Tanner performed just as well as when asked to copy another dolphin, but he echolocated around 50 percent more often. So it seems that Tanner was able to switch flexibly between strategies to achieve his goal of identifying which behavior he should produce—passive acoustic cues when he could, and echolocation when he had to. This tells us something about how the dolphin viewed what he was being asked to do. In Jaakkola's words, "This active strategy-switching precludes the possibility of automatic imitation via response facilitation, and rather paints a picture of a dolphin engaged in imitation as an intentional, problem-solving process."[22]

So what are we left with? Certainly, dolphins can be trained to copy each other's behaviors if they already know the particular trick—even when they cannot see! The only theoretically feasible counterargument we have heard (verbally, or off the record) is that some unknown third stimulus was causing both dolphins to produce the same behavior at the same time. To us, this is a perfect illustration of how absurdly far some are prepared to go in their skepticism—to suggest that perhaps an odd sound from a water pump in a tank would cause two dolphins to spin on their axis together, in the same direction, rather than concede they can copy each other. At least two dolphins have been shown to imitate behaviors that they've never been asked to do before. Others have failed when asked to precisely reproduce a new trick, even though they have often done a pretty good job of copying the body movements involved. Animal behavior is complex, and its motivations are often opaque, so it is far, far easier to design an experiment that fails to show that an animal is capable of something than vice versa. Once one member of a species shows a capability, then this shows that it is attainable. The current balance of evidence therefore supports the notion that dolphins are capable of imitating each others' body movements, even when the behavior is novel—so-called true imitation. We are not alone in this view—Andrew Whiten, a leading expert on imitation in children and other primates, has noted that "dolphins may ape humans more clearly than apes."[23]

Although most of the experimental evidence for imitation comes from bottlenose dolphins, there has also been a significant study of killer whales that demonstrates genuinely impressive learning skills.[24] José Abramson and his colleagues (including Josep Call, a leading light among researchers studying primate learning) took as one of their starting points the claims we have collated in chapter 6 about the behavior of wild killer whales, namely, that much of the variation we can see is likely the result of cultural transmission. They wanted to see how readily killer whales could imitate each other when "asked" by their trainers, even if they had never seen the behaviors before. They began with a single trainer and a pair of whales, by giving a command to perform a particular act to just one of the whales (e.g., "slap the water with your pectoral fin") and then using pointing and eye gaze to direct the second whale's attention to the first's performance of the trick. The whales got a pep talk, a pat or a rub, and a fresh fish every time they did the same thing as they saw the other doing. They caught on quick. "Surprisingly," the scientists wrote, "the subject reproduced the demonstrator's action, generalizing the 'game' in a single trial."[25] Encouraged, they then moved to using two trainers. The first gave a "demonstrator" whale the command to produce a certain act. The other focused on the "imitator" whale and did two things—pointed to the "demonstrator" and gave to the "imitator" whale a single gesture that the scientists made up to use as a general "copy" command. The whales didn't miss a trick. Thumbing their thesaurus, the scientists wrote that "Unexpectedly, as every tested subject copied the demonstrator's behavior in 70% of the trails from the very beginning, individuals only received one or two training sessions."[26] In other words, the whales caught on so quickly that the scientists did not need to spend weeks training their subjects to the new task, usually a necessity (that can sometimes be swept under the carpet) in studies of nonhuman brainpower.

Impressive, but this is not learning a new behavior—all these acts were well known from the many shows these animals would give. So then the scientists had each of the three whales trained on some new behaviors, on their own, as in the dolphin studies. The whales had to then reproduce their new trick in the same two-trainer setup, with the whales able to each see only one trainer, in front of another whale that had never seen it before. The results were beautifully simple: "All three subjects copied correctly 100% of untrained behaviors."[27] Apparently now used to these animals' abilities, the scientists proffered no expression of surprise at this result. This simple and elegant study leaves little doubt that killer whales are perfectly capable of learning new behavior by copying each other and fits perfectly with the observations field biologists have been making for decades.

During the study, the researchers observed something so telling that they were moved to report it as an anecdote. One of the whales in the study had a calf that, at two months old, was seen accurately matching, within a few seconds, two of the behaviors that her mother had been trained to produce and also made a pretty good attempt at a third. The calf had no training and was never rewarded—the copying was spontaneous. It is easy to see how such copying would be pretty useful for coordinating hunting and could lead, for example, to the transmission across generations of complex patterns like hunting by intentional beaching. The one note of caution we could sound about how eye-opening this study was is that the animals concerned were undoubtedly heavily enculturated—they were born in captivity and spent their whole lives being trained to do things in synchrony for their shows. Whether anyone attempts to replicate this study, like the dolphin one, and what the outcome of the attempt is, we will have to wait and see. We suspect that most scientists will be convinced.

Vocal Learning—Say as I Say

Vocal learning—the ability to learn to produce a sound just from hearing it—is a form of social learning when another individual is the source of the sound being learned.[28] Dolphins readily reproduce sounds that are played to them and appear to use their vocal learning abilities in several ways. The ability to copy novel physical behavior—imitation—has long been a focus of interest for comparative psychologists, but that field has shown relatively little interest in the ability to copy new sounds. This puzzles us. Our ability to copy sounds that we hear is the foundation of how we learn language, a major part of human culture, so it is not clear to us why comparative psychologists have not shown equal interest in comparing vocal learning across species. It may be because in our primate cousins, our closest evolutionary relatives and the typical research focus of comparative psychologists, there is virtually no convincing evidence for vocal learning at anything like the levels that are routine in humans.[29]

Other arguments on the lesser significance of vocal learning center on the concept of "equimodality," a fancy way of saying that when you hear a sound made by someone else, you perceive it using the same hearing kit with which you perceive sounds that you make. In contrast, to copy actions requires "visual-tactile cross-modal performance," which is a fancy way of saying that seeing an action performed uses a different set of sensory systems to those you use to sense what actions your own body is performing.[30] Supposedly, the argument goes, this is harder. We find this unconvincing for two reasons. First, if you've ever listened to a recording of your own voice, you'll know that what you hear yourself producing and what other people hear are not always so similar—among other things, we typically sound lower in pitch to ourselves than to others because lower pitches are transmitted better to our ears via the tissues and bones separating our larynxes and inner ears. Second, the discovery of "mirror neurons," nerve cells in the brain that fire both when an action is seen and when it is performed, fundamentally calls into question the assumed separation of the sensory systems used to observe your own and others' actions.[31]

Behavioral biologists, in contrast, have long had a healthy interest in vocal learning, prompted largely by a fascination with one of the most captivating of animal displays, birdsong. There is a long tradition of studying how birds learn their songs and calls from each other.[32] More recently there has also been interest in the vocal learning capabilities of bats, many species of which, like toothed whales, have evolved echolocation abilities.[33] The complex patterns of vocal behavior shown by whales and dolphins in the wild all suggest that these animals represent another independent evolutionary peak in the evolution of vocal learning capabilities. Many scientists who spend a lot of time with wild whales and dolphins have personal anecdotes of vocal mimicry. We have our own, from the fieldwork of Master's degree student Andrea Ottensmeyer, who studied long-finned pilot whales from whale-watching boats. An apparently excited pilot whale produced awfully convincing copies of the whistle used by a member of the boat's crew to call his dog, after the crew member began directing it at the whales when they approached the boat. These observations rarely appear in the scientific literature, though, because of the difficulty of pinning down just what was going on from rare chance observations.

In short, documented direct evidence of vocal learning is surprisingly rare across the range of whale and dolphin species, and the best of it comes again from the bottlenose dolphin. The key demonstration came again from the Kewalo Basin captive facility, by the simple expedient of playing sounds through an underwater speaker and rewarding dolphins when they produced a good copy of the sounds.[34] Recall that dolphins in this facility were trained to a "mimic" command. By associating this gesture with a short sound, quickly followed by a test sound, one dolphin, Ake, was trained to produce convincing copies of nine different sound types. None of the test sounds were heard from the dolphins themselves prior to the experiment and three of them were extremely artificial, with, for example, abrupt and discontinuous pitch shifts. This study was somewhat corroborated a decade later by Brenda McCowan and Diana Reiss, who introduced an underwater keyboard to their captive dolphins as part of a communication experiment.[35] The device made sounds underwater when keys were pressed, and when the dolphins started pressing the keys, they quickly began making copies, again very convincing ones, of the sounds produced by the keyboards. The reason for our caveat is that the sounds produced by the keyboard were sounds that were very close to whistles already made by dolphins in the tank—this is because the primary aim was to see if the dolphins could make a link between the sounds and particular images represented on the keyboards, rather than to demonstrate vocal learning per se. So while these results show the dolphins can readily match sounds that they hear, the evidence that they can learn something completely novel is restricted to those sounds from the Kewalo Basin study that were clearly unlike any naturally occurring dolphin vocalization.

These examples all involve specific training in which the dolphins were rewarded for saying the right thing. Many observations of mimicry in the wild appear to be spontaneous. One example of spontaneous vocal learning that has been published, because it originates from a captive animal and thus was investigated in some depth, is a report by Sam Ridgway and colleagues with the self-explanatory title "Spontaneous Human Speech Mimicry by a Cetacean."[36] This did not concern a dolphin, but a captive beluga, a species with a justified reputation for vocal exuberance, nicknamed by seafarers "the canary of the seas." The researchers were first alerted to the mimicry by a human diver surfacing next to the beluga's enclosure asking who had told him to get out. Nobody had. The source of the sound the diver had interpreted as "out" was identified as a maturing beluga called NOC, and subsequent investigation showed that this whale had taken to producing sounds that closely matched the rhythm and pitch of human speech, although few would recognize the "words." The pitch, in particular, seemed to cause the whale some effort to produce, probably because it was "several octaves lower than the whale's usual sounds," as NOC's internal air sacs, used to produce the sounds, would inflate so much that they could be seen on the surface of the whale's head, which is never seen in "normal" sound production. These observations are very similar to a much earlier study by John Lilly in which he trained bottlenose dolphins to reproduce the pitch and rhythm of human speech for a reward.[37] The beluga, however, was not being rewarded. The spontaneity is intriguing, though, because it speaks to a naturally occurring drive to copy sounds and shows how the animals put a lot of effort into driving their vocal systems outside their normal envelope to achieve it. How do these animals use these abilities in the wild?

We certainly don't know most of what's going on, but one piece of the picture we do know something about is the bottlenose dolphin's signature whistle, which we briefly discussed in chapter 5. We will talk more about those whistles here because there is excellent evidence that learning is important in their development, and thus they provide a window into understanding how vocal learning benefits dolphins in the wild. A young dolphin is not born with a signature whistle; it develops over time. When Jennifer Miksis-Olds compared the whistles of wild dolphins with dolphins raised in captivity, she found that the captive dolphins had apparently been influenced by the sounds they had been hearing from their trainers, who used whistles to give commands.[38] The captive dolphins had signature whistles that were demonstrably more similar to the flat-pitched and short whistles made by the human trainers than those of the wild dolphins. Male dolphin calves appear to develop signature whistles that are notably similar to those of their mothers, whereas females, for reasons we do not understand, develop whistles that are more like other members of their community rather than like those of their own mother.[39] The learning does not stop there, though, as bottlenose dolphins appear also to learn the signatures of other members of their community. They can often and readily be heard making each other's signatures, both in captivity and in the wild.[40] When adult male dolphins form alliances in the wild, their signatures become more similar to those of their alliance partners, perhaps to solidify the cooperative bond between them and perhaps to signal to other male competitors that their alliance is strong, having a similar role to the synchrony in their physical behavior.[41] The signature whistle is a cornerstone of dolphin communication, then, and its development is strongly influenced by social learning. Signature whistles are only a part of the dolphins' communication, so we can ask how much other known aspects of dolphin communication are influenced by social learning. These questions are going to be important in developing our understanding of dolphin cultures. We should not lose sight of the fact that all this effort has told us only part of the story of part of the communication system of only one of the tens of dolphin species in the oceans. So much remains to be understood.

What about killer whales? We've seen in chapter 6 how they have such complex variation in their shared vocal repertoires, and even how we can track the way their calls are changing over time but not diverging between groups that spend a lot of time together. They also engage in the same kind of call matching in the wild that dolphins do.[42] They are regularly held in captivity but have apparently never been subject to the same kinds of vocal learning experiments as bottlenose dolphins. However, there are two reports that pretty much seal the case made so strongly by observations in the wild. The first originates in captivity and results from the practice of moving killer whales between captive facilities. David Bain describes in his doctoral thesis how a young killer whale captured off Iceland was subsequently housed with an adult that had been captured off British Columbia.[43] The respective origins of the whales make it impossible they could have heard each other's repertoires before captivity, yet the younger whale went on to reproduce the calls of the older British Columbia whale (but not vice versa, showing perhaps how respect for elders leads to cultural conservatism in wild killer whales). The second report derives from a "natural experiment," in which two young killer whales became separated from their families, for unknown reasons.[44] One of these, Luna, was then recorded producing very faithful imitations of sea lion barks. This calf, isolated from other killer whales, was repeatedly observed in "close association" with what were probably some quite nervous sea lions, so they are credible acoustic models. These studies demonstrate an ability to acquire new sounds, a hallmark of vocal learning.

Thus learning sounds from each other is a big part of whale and dolphin cultures. Wherever we bother to look with any real effort, we again and again find evidence that the ability to do this is widely spread among the cetaceans. We have discussed the songs of the baleen whales already, but they bear mentioning again here. There is just no mechanism other than the kind of vocal learning that we have such strong evidence for in dolphins and killer whales that can explain the extraordinary way that the songs of humpback, bowhead, and blue whales change across populations and across years. Both the toothed and the baleen whales have evolved this ability, and we think this speaks powerfully to the ubiquitous importance of learning from others in these societies. The sharing of vocal patterns may well have been a key foundation in the evolution of cetacean cultures.

Teaching—Do as I Say

Our discussion of social learning up to this point has focused on the learner—an individual learning something by watching what another, the "model" or "demonstrator," is up to. The "watched" individual in this situation is just going about his or her business, whatever he or she would do regardless of being watched or not, and there is no implication either way that the demonstrator cares about the learner's learning. In our societies, though, we regularly encounter situations where this is not the case—where the demonstrator really does care whether the learner gets it and goes out of her way to explain, demonstrate, and guide others toward a new skill or piece of knowledge. Humans teach. We actively pass on cultural knowledge, informally to those around us and also formally in cultural transmission institutions—schools, colleges, and universities—and the process plays a large role in the maintenance of cultural knowledge within our populations. Perhaps it is key to the transmission of cultural knowledge—and one of the things that sets human culture apart. Some psychologists have suggested that its prevalence has become evolutionarily embedded in our development in ways we can observe through watching infants follow their mother's gaze while they are still incapable of doing much else. If the first adults you encounter are likely enough to be both sources of highly relevant information and have specific interests in making sure you get that knowledge, then this makes evolutionary sense.[45]

Teaching has traditionally been seen as a more complex transmission process than imitation because of its implications for "theory of mind." "Theory of mind" is scientific shorthand for the idea that we recognize, as a key part of our social world, that other individuals have minds that, although broadly similar to ours, will view the world from their own perspective. We can conceive that other humans do not know everything we do (and vice versa) and can have quite different thoughts, feelings, and desires from us. We can construct theories about what other's minds may contain in terms of knowledge, desires, and attitudes—hence "theory of mind." From this perspective, you have to recognize others as individuals with potentially different knowledge states from your own, before "intentional" teaching can occur.[46] The teacher, unlike the imitative model, recognizes that the learner has different knowledge from herself and performs behavior with the conscious intention of altering the knowledge state of the student. A good teacher monitors her student's progress and adjusts her subsequent behavior based on feedback about the current progress of the learner. This need to understand and react to a perspective and state of knowledge different from your own means that teaching has long been considered a "higher" form of social learning. But is it unique to humans? Does teaching play any role in propagating knowledge in cetacean societies?

These are difficult questions. How would we recognize teaching in a nonhuman, when we can't ask the individuals in question what theories of mind they have about their potential students? If the answer demands that we access the internal representations that potential teachers have about what others around them know and do not know, as well as their intentions to change the knowledge of others, then we are stuck. For some, that is sufficient to preclude entirely the idea that nonhumans might teach.[47] For other, more evolutionarily minded scientists, this doesn't make sense, for similar reasons that they dismiss a concept of culture that requires access to internal representations.[48] Animals may or may not possess such representations, but they are inaccessible to us. As evolutionary biologists, we are interested in the functional outcome—that an individual learns something, or learns more quickly, because another individual goes out of its way to facilitate that learning. This view was first put forward in an influential paper by Tim Caro and Marc Hauser, who argued that the functional outcome can be achieved in multiple ways that require only a certain sensitivity to the behavior of others, rather than the full-blown theory of mind used in the form of teaching sometimes shown by humans.[49] This parallels our perspective on culture in general, which also focuses on the functional outcome—the nongenetic transfer of information—above specific mechanisms like imitation used in the form of cultural transmission shown by humans (this chapter notwithstanding!). To this end, Caro and Hauser proposed an operational definition by which we might recognize teaching without recourse to the internal representations of the individuals involved.[50] To qualify as teaching, an interaction must satisfy three conditions. First, the "teacher" must do something they would not do if the "pupil" was not around. Second, the teacher should not receive any immediate payoff for what he is doing, and finally, obviously, the pupil should learn something, or learn it more quickly, as a result of what its teacher is doing. This definition allows us to identify behaviors that function as teaching because they meet the operational definition. We should bear in mind, however, that this is not necessarily the same as intentional teaching involving theory of mind. We should also be aware that overreliance on the strict terms of the definition may distort our understanding because it might reject things that we would accept readily as teaching if a human did them.[51] Nonetheless, the Caro and Hauser approach has been very influential.

Using their definition, the distribution of strong evidence for nonhuman teaching is eye-opening. It is, by current reports, rare, although given the stringent set of conditions that have to be met, this at least partially reflects how much easier it is to study, and especially experiment with, some species than others.[52] The strongest evidence to meet Caro and Hauser's definition comes not from our closest primate relatives but from meerkats, which teach young how to handle dangerous scorpion prey; a group-living bird called the pied babbler, which teaches its young about food using a specific call to attract them to food patches; and perhaps most extraordinary, a species of ant that teaches nest mates about the location of food by leading them in a behavior called "tandem running."[53] These discoveries are not problematic for scientists with an evolutionary perspective on teaching that sees it as a cooperative behavior, likely to be most common in social settings (all three species are highly social) where complex, difficult-to-learn behavior is valuable not only for the student but also for members of its long-term social group, which includes the teacher.[54] Cetaceans might seem to fulfill these conditions. However, proving that teaching is occurring is difficult, especially in the wild, and even more especially in the ocean, where it is very hard to set up experimental situations.

There is perhaps no field site where this is truer than the inhospitable Crozet Islands, deep in the "Roaring Forties" of the southern Indian Ocean. We described in chapter 6 how killer whales there intentionally strand onto beaches in order to snatch elephant seal pups from breeding beaches. This behavior is also known from Punta Norte, Argentina, but Christophe Guinet and Jérome Bouvier, two French scientists with the levels of dedication needed to spend long periods observing the whales of the Crozets, documented the development of this behavior in two calves, unromantically dubbed A4 and A5, that they observed from shore over the course of three years.[55] Although they did not run any experiments, which would have been impossible, their observations were fascinating. The stranding hunting technique is profitable, perhaps essential to survival, but has to be done carefully, as getting it wrong means the whale is left high and dry and eventually, dead. The risk was graphically illustrated when Guinet and Bouvier found one of the calves, A4, permanently stranded and facing death. The researchers decided to intervene and push it back into the water.

These whales do not just strand when there is prey about—they can also be seen swimming up beaches with no seals on them, in groups containing calves. These beaching play sessions appear to be where young killer whales learn the skills needed to beach successfully, capture a meal, and refloat themselves. As killer whale calves spend nearly all their time with their mothers, and there was no pod in the study that did not have a calf, we cannot really know if this beaching play would still occur if there were no calves present. However, Guinet and Bouvier describe the mothers pushing their calves up the beach and then back down, and on occasions when they did do it at seal breeding sites, pushing them with their heads toward young seals. These are all behaviors that simply could not occur without the calves present. Do the adults get some immediate payoff themselves from swimming up a beach with no prey on it? This is hard to assess definitively—perhaps, for example, parasites are removed from the skin, so there is a function to swimming up the beach even with no hope of catching prey—but it seems unlikely that they do. At the Punta Norte site in Argentina, adults have been seen throwing already captured prey at calves.[56] Literally throwing away lunch is a pretty costly behavior, but it is unclear whether this actually contributes to the calves learning how tasty young pinnipeds are.

What is the evidence that mothers accompanying calves during beaching play sessions affects the rate the calves learn to catch prey by themselves?

We can consider here the differences between A4 and A5, the two calves Guinet and Bouvier followed. At the start of the study, A4 was four years old and A5 was three, and they regularly took part in beaching play sessions. Both calves were observed to strand alone for the first time at age five. A5 became a fully fledged beach hunter near the end of the observation period, catching its first "solo" seal pup at age six (learning this technique is a long-term investment). A4, however, although a year older than A5, was never seen making a catch on its own. What could be behind this difference? One possibility is that A5's mother was a better teacher. During Guinet and Bouvier's observations, when A5 stranded, its mother was always present. A4, however, had mom around only twice in thirty-five observed stranding attempts. A4's mother was apparently not that into beach hunting—she rarely took part in beaching play and was never observed hunting herself in this way. A5's supermom, however, closely supervised her calf's strandings. She was observed pushing the calf up the beach and stranding onshore in order to push the calf back into the water, accompanying the calf on unsuccessful hunting attempts and finally assisting in the first successful capture by pushing the calf toward the prey and helping the calf to return to the water following capture. In contrast, recall that it was A4 that the researchers rescued when it became properly stranded, suggesting a severe fitness cost for mothers who do not give their calves much attention when they practice beaching. It is tempting to conclude that the behavior of A5's mother seems to have enabled her calf to learn the hunting technique at least one year earlier than A4, who received very little "instruction," was never seen successfully hunting independently, and would have died but for the researchers' intervention. If the beaching play behavior results in a skill being learned more rapidly than it otherwise would, or less riskily, then it supports the case that the killer whales are teaching their calves the beach-hunting technique. Like meerkat scorpion hunting, it fits with the evolutionary perspective of when we might expect to see teaching—highly social animals learning a complex and dangerous foraging skill. Unlike the meerkat example, however, it is not backed up by experimental manipulation. For example, Alex Thornton and Katie McAuliffe, who carried out the meerkat study, showed how they could change how much adults disabled a scorpion before presenting it to a pup by playing back calls recorded from other pups of various ages.[57] Play an older pup's call, and the adult presented a relatively intact, and relatively dangerous, scorpion. In the killer whales, the case is hampered by the fact that only two calves were involved, so the difference in their learning rates could have been a result of other factors specific to those animals, such as A4 being impaired in some other way (its mother certainly did not seem too switched on). The evidence for killer whale teaching has been described as "compelling but not convincing" by Louis Herman and Adam Pack, and we think this is a pretty fair description.[58]

The clear warm waters of the Bahamas, quite a contrast to the Roaring Forties, make it much easier to observe underwater behavior in some detail. On the Little Bahama Bank, north of Grand Bahama, Atlantic spotted dolphins root out fish hiding in the sand on the sea bed. They use echolocation to find the fish and then plunge their snouts into the sand to rouse them out into the water column, where they are easily caught. Because of the dents this leaves in the sand, the technique has become known as crater feeding. Denise Herzing has amassed a video archive of this behavior, and her student Courtney Bender has used this archive to produce evidence that dolphins with calves change the way they hunt.[59] Since the archive spans thirteen years, they were able to compare females hunting when they did not have an accompanying calf to when they did. They found that when calves were around, dolphin mothers took around five times longer on average to catch their prey once it had been forced out of hiding, and made many more of what the researchers called "body-orienting movements," compared to when there was no calf around. These body-orienting movements are described as "exaggerated movements in the direction of the prey." These movements were of interest to the researchers because bottlenose dolphins in captivity have been shown both to understand human pointing as a way of referring to something and to use similar body orientation spontaneously to indicate the location of bait-filled jars to a human trainer.[60] This was interpreted as "pointing" because the dolphins did it to divers facing them twice as much as to divers with their backs turned and nearly ten times more than to divers that were swimming away. This implies the dolphins were sensitive to whether their pointing was going to produce a worthwhile effect in its intended target. Accordingly, in her study, Bender was examining whether the spotted dolphin mothers were "pointing" more at their prey once it had emerged from the sand, and it appeared they did. It does, then, seem that there is behavior that changes only when calves are present, and it also seems that the changes provide no immediate benefit to the mother; indeed, given that it delays the mother eating her prey, it is arguable costly. However, we have no evidence that the calves learn, or learn more quickly, because of their mother's actions. Short of interfering with their natural behavior in ways that would probably be ethically unacceptable, it is difficult to see how this could ever be shown experimentally. The best hope would perhaps be correlational evidence—for instance, if the calves of mothers who "pointed" more also foraged more effectively later on. But even this would be vulnerable to suggestions that there are some mother-calf pairs who are more active or intelligent than others for genetic reasons, which lead to both more pointing in the mother and better feeding in the calf with no direct link between the pointing and the feeding.

Therefore, while it might look like both killer whales and spotted dolphins teach, the evidence does not completely satisfy the full functional definition. If one were forced into a yes-or-no choice, one would have to say no, but we think that it is counterproductive and potentially misleading about the true extent of nonhuman teaching to approach the issue in this way. We agree with psychologists Dick Byrne and Lisa Rapaport, who suggest we should emulate Scottish law and have a third "not proven" category of evidence that does not simply shut us off from thinking about teaching in such cases.[61] There may well be more suggestive examples of teaching out there among the whales and dolphins, but they will be hard to see and study.

Beyond Imitation and Teaching

Other types of social learning seem less appealing to scientists and have received much less attention than imitation, and even teaching. But other types of social learning can nevertheless be important. For instance, young humpback whales and right whales likely learn their annual migration route from the warm-water breeding ground (where they were born) to the cold-water feeding grounds (where they spend the summer) simply by following their mothers.[62] This leads to whale populations stratified by feeding grounds. Similarly, beluga whales spend the summers in restricted but traditional feeding grounds, a migration probably learned from following mother.[63] Although the experimental evidence for imitation in captive dolphins is impressive, this does not mean that the spongers of Shark Bay learn their skills through imitating their mothers. It could be that following their mothers around, as is typical of a calf, leads to calves being exposed to the right learning environment to pick up the same skills. From our evolutionary perspective on culture, the result is largely the same—information about the use of sponges flows down the generations independently of the dolphin's DNA.

The stance that only controlled experiments in captivity can inform us of social learning is wrong. Observations of the natural behavior of the animals can indicate, sometimes clearly, that social learning is occurring. For example, we can only observe the changes in humpback song or bowhead song over space and time, but what we have observed cannot be reasonably explained by anything other than the whales learning from each other. Controlled experiments in the wild would remove some of the drawbacks of the captive studies, and in some ways they are the holy grail of social learning research—combining the ecological validity of a natural social group with the control of a carefully constructed experimental manipulation.[64] These kinds of experiments have been immensely informative in the species amenable to them, such as meerkats.[65] They have the potential to tell us more. But this experience has also shown that "controlling" environments in the wild is extremely difficult, and to make good sense of the results of such experiments we would need to know more of the natural behavior of the animals than we currently do, so the observational field data will always be essential.

How big is the gap between human and cetacean social learning? We believe that this is the wrong question. Recently the hierarchical edifice of social learning processes that experimental psychologists have spent decades building has come under attack from within. Psychologist Celia Heyes has made a provocative and, to us, really interesting argument about how we are asking the wrong questions.[66] For her, there is nothing really special about social learning—in humans and nonhumans alike it depends on a common neural framework for learning that is "adapted for the detection of predictive relationships in all natural domains." Moreover, it is the same learning framework, called associative learning, that we and other animals use for learning when there is no input from others. Any animal capable of learning in this way should in theory be capable of social learning if put in the right circumstances, and Heyes points in support to evidence of social learning in animals that typically live very solitary lives in the wild—octopus and tortoises.[67] From this perspective, social learning is the acquisition of knowledge based on information that arrives via a social channel. Social learning varies from other types of learning in that it is biased toward social information—learners are motivated to look to others for their information and may have evolved mechanisms for perceiving the relevance and intent of others' actions. Given, for example, the way synchrony appears to function as a signal in dolphin society, it is easy to understand their social learning abilities as reflecting this imperative to pay attention and match others' behavior. We will have to wait and see how Heyes's bold argument is accepted, but to us it represents a refreshing break from the ranking of nonhumans on a ladder topped by the gold standards of imitation and teaching, which has provoked a rather sterile research effort dedicated to listing which other species might or might not measure up. However, Heyes's integrated perspective on the mechanisms of social and asocial learning does not alter the population consequences of social learning when social learning is shared: culture. The consequences for evolution and ecology that we explore in chapters 10 and 11 are unchanged.

Also unchanged are the facts that we have outlined here. Bottlenose dolphins and killer whales, the only cetaceans comprehensively tested, have both been shown to be perfectly capable of copying each other and learning new behavior from each other. Both have also been shown capable of learning new sounds from each other. To us, the results of this research make it more likely that much of the behavioral complexity we observe in the wild also results from this kind of learning. Killer whales and dolphins, in this case spotted dolphins, both perform behaviors that look like teaching but that have not been experimentally demonstrated to be so. We would love to have similar investigations of sperm whales and humpbacks, for example, but the experiments are not happening any time soon. In the next chapters, we will ask how all the evidence we've presented supports the notion that whales and dolphins have something we might usefully call culture and go on to explore what this means for their lives and ours.

8 | IS THIS EVIDENCE FOR CULTURE?

We have been to sea. We have been surrounded, sometimes literally, by the behavior of whales and dolphins in its extraordinary richness. We, as two scientists, have been astounded, moved, and entranced, but we have generally had little idea of what was really going on from the perspective of the animals themselves. We, in the more general sense of the scientific community, despite decades of dedicated research and volumes of hard-won data, have achieved but a glimpse of the world of the whales and the dolphins. Nonetheless, we hope you agree that what has been uncovered is extraordinary. In this chapter we're going to drop anchor for a while and start to think about what it means for our exploration of culture. When we do this for real, resting at anchor on our small research vessels, both of us find the process is helped by one of the true pinnacles of human culture—whisky from the Scottish island of Islay. There is an initial burn that soon gives way to a wonderfully rich flavor. This works as a metaphor for this chapter, where the burn will involve facing up to the criticisms that have been made of the evidence we've outlined, and our interpretation of it, but the hope is that our notion of cetacean culture will be richer in flavor for it. The question we shall take aim at is whether it is justified, given the evidence at hand, for us to talk of cetaceans "having" culture in any meaningful sense of the word "culture." We think it does, but critics of whale and dolphin culture have two major bones of contention, which we introduced in chapter 2: first, that the scientific case that these behaviors really are learned from others is not convincing and, second, that in any case they fall so far short of what we call culture in humans that it makes no sense to call it culture.

Critiquing Animal Culture: "Where's the Beef?"

In plain terms, there is no case we are aware of where a cetacean behavior observed in the wild has been experimentally demonstrated to rely on cultural transmission.[1] This is the first bone of contention. If your perspective requires such evidence before we can even talk about culture in whales and dolphins, then you will find this chapter disappointing.

On this front, then, critics who accept something like our definition of culture but balk at the claims for it in cetaceans, generally question the evidence that social learning is behind the behavioral patterns that we observe. First, they sometimes suggest that the behavior patterns might be the result of ecological differences, with animals in the same environment learning the same behavior individually and without reference to each other or having their behavior shaped across space or time in response to changing conditions. If this were so, claims of culture would be spurious. Second, they suggest that genetic variation could be causal in some subtle way that is easily missed. Generally, this kind of criticism is a healthy component of the scientific process and what gives it its power. We might grumble that our critics are prepared to accept alternative explanations for observed behavior that we think are wildly implausible compared to the quite straightforward notion of cultural transmission, but we have to accept the rules of the game if we want to build scientific knowledge. Otherwise we could become priests.

The traditional "exclusion method" of ascribing culture makes these attacks potent. We discussed in chapter 2 how methods of excluding genetic or ecological causation are labeled "Handle with care!" Using exclusion, culture can only be invoked if ecological differences and genetic influences can be ruled out. This involves the logically thorny task of proving a negative and unfortunately reprises the nature/nurture argument. Sometimes, the proponents of animal culture, including ourselves, have not been sufficiently rigorous when considering these alternatives. Quite often, genetics and ecology cannot be excluded definitively. To our minds, the important question is whether social influence is necessary, even if not sufficient, for a behavior to develop fully.

The other criticism of the exclusion method is that it can easily underestimate the importance of culture. One of the reasons culture is thought to be so adaptive is precisely because it allows populations to adjust to varying ecological conditions. But using this method, any example where behavior follows variation in ecology is automatically "excluded." The method also excludes any behavior that is universal—any truly good cultural innovation in a reasonably connected population is likely to spread to everyone, so will lead to homogeneity, not variation.[2] Similarly, if the predominant mode of cultural transmission is vertical, then most culture is transmitted in parallel with genes from parents, and we might expect correlations to build up between genes and culture that would lead to another false exclusion.

Generally we can be more certain that a form of behavior is culture when the behavior of an entire community systematically changes within animals' lifetimes, thus ruling out genetics as a cause. The same applies when the behavior has to do with communication rather than resource exploitation (as this lessens the role of ecological variation), when the behavior is necessarily performed by several animals at once, and when the population contains clearly defined communities that behave differently but are not genetically distinct and whose ranges overlap. This does not mean that stable, individually performed foraging specializations in networked societies are not culture or that they are unimportant to the dolphins or whales themselves or to evolutionary or ecological processes. These may, in some respects, be the most significant of all cultures. We just have a hard time identifying them as culture because if they really help certain populations or communities to adapt to local ecological conditions, then inevitably the behavior is going to vary between communities in ways that are not independent from either genetic signatures or local ecology. If a certain foraging innovation allows a female and her descendants to take advantage of a prey source specific to their local habitat ecology, then that behavior will be linked both to her genetic lineage and to her ecological experience, so it will be more difficult to disentangle what causes that behavior to develop in individuals.

So what should we do? Instead of taking the all-or-nothing approach of the exclusion method, the goal should be to estimate how much of the variability in the behavior is caused by genetics, ecology, and culture, respectively. If social learning is a necessary condition, models incorporating social influence should be much more powerful at explaining these data. We need good multifaceted data to do this, data on behavior, genetics, and ecology, and how each varies with space, time, and social relationships. We have this for only a small minority of putative cetacean cultural behavior at the moment—the analysis of the spread of lobtail feeding by humpback whales in the Gulf of Maine described in chapter 4 is the best example. But what we can do here is consider the likelihood that culture is contributing to the behavioral patterns described in chapters 4–6, bringing in where appropriate the results of experiments from chapter 7. In some cases, few can doubt that culture, as we define it, is behind the behavior. In many instances, culture seems very likely to be the primary driver of the behavioral variation. Potential alternative mechanisms, while just about feasible, require some pretty tortuous conditions and coincidences to be present. For some of the behavior that we have listed, the role of social learning is not clear, so perhaps it may not be culture. We think that social learning is plausibly the cause, but it may not be, and we do not yet have the crucial data.

False Positives and False Negatives

In considering this evidence, it is important to understand the decision process in the context of how science works. Every time we decide to investigate whether something is or is not true in nature, we perform—implicitly or, ideally, explicitly—a balancing between the risks of two suboptimal outcomes. The hope is that we get the answer right, but there are two ways we can get it wrong. The first is a false positive, in this case ascribing culture where there is none, and the second a false negative, or failing to ascribe culture when it plays a significant role. Being conservative, scientists tend to focus on the risk of false positives and do not as often consider the consequences of false negatives. So what are the consequences? Are they more severe for one kind of error than the other? The experimental and comparative psychology perspective relies on a directive laid down by C. Lloyd Morgan, a nineteenth-century British psychologist: "In no case may we interpret an action as the outcome of the exercise of a higher psychical faculty, if it can be interpreted as the outcome of the exercise of one which stands lower in the psychological scale."[3] This is today known as Morgan's canon. It remains massively influential and is the constant refrain of those skeptical of claims for complicated behavioral phenomena—like culture—in nonhumans. However, the ease with which his ideas were subsequently, in his view, misused to assert that the simplest sufficient "faculty" should always be the principal explanation irrespective of the context, led Morgan to later write a caveat: "To this, however, it should be added, lest the range of the principle be misunderstood, that the canon by no means excludes the interpretation of a particular activity in terms of the higher processes if we already have independent evidence of the occurrences of these higher processes in the animal under observation."[4]

Morgan's caveat is much less repeated than his canon, which has led to an unfortunate, perhaps even damaging, misuse of Morgan's canon to emphasize discontinuity between psychological processes—like cultural transmission—in humans and other animals, contrary to the Darwinian principle of continuity across species.[5] It is possible to find scientists using the canon to emphasize one (less "sophisticated") explanation for behavior over another not because of any data favoring one or the other but simply from the authority of the canon itself. The problem is that this just replaces one bias—toward anthropomorphic or sophisticated explanations—with another—"simple" explanations, even if they are equally unsupported by the data at hand.[6] This doesn't seem particularly useful to us in thinking about cetacean culture, where, for example, we might emphasize Morgan's caveat rather than his canon when considering foraging specializations in dolphins—given that we have independent evidence of their social learning abilities, then it seems quite reasonable, even parsimoniously simple, to have social learning as a leading hypothesis to explain the observed variation in foraging techniques.

More fundamentally, when studying culture and its principal alternatives, ecological variation, and genetic causation, there is no consensus as to which direction we should lean—or, more formally, what is the null hypothesis (or simple explanation) and what is the alternative (or sophisticated explanation). In studies of humans, culture is usually taken as the null, and so the burden of proof lies with those testing hypotheses about genetic causation, while in nonhuman studies the direction of proof is always reversed and, thus, ascribing culture without reason would be the false positive. This is highly unsatisfactory. The approach is set up along the lines of addressing "the question of culture."[7] This implies there must be a yes or no answer. Set up as a dichotomous hypothesis-testing approach, where everything hinges on getting across a completely arbitrary "significance level" in statistical tests, the approach is almost guaranteed to give incomplete answers. This approach has been much discredited by statisticians themselves and by scientists working in applied areas who have to make real decisions about, for example, wildlife management.[8] We therefore favor conceptual frameworks that treat the potential causes of behavioral variation on the same basis and take a model comparison rather than binary hypothesis-testing approach. In other words, the starting point should be that genetic, ecological, and social influences may all play some sort of role and that researchers should try to understand the relative contributions of each by comparing how well statistical models in which the relative contributions of each are allowed to vary are able to explain what we see in nature.[9] This seems to us much more likely to bring useful insights.

In some contexts, there are reasons to think that the default assumptions should even be flipped. The Japanese whale scientist Toshio Kasuya (fig. 8.1), in a lecture accepting a lifetime achievement award from the Society for Marine Mammalogy, put the case that when it comes to conservation, we should consider cultural diversity as an element of biodiversity (we will return to the theme of conservation and this remarkable scientist in chaps. 11 and 12). He cautioned against equating absence of evidence with evidence of absence, writing: "Rather it will be safer for conservation purpose to assume that culture exists if a cetacean species exhibits a behavioral trait that suggests culture, or if it has life history, or social structure that is suitable to maintain a culture, such as the short-finned pilot whales and some

8.1. Japanese scientist Toshio Kasuya has made important discoveries about whale societies and exposed illegal practices of the Japanese whaling industry. He has eloquently highlighted the implications of whale and dolphin culture. Here he is searching for Irrawaddy dolphins on the Ayeyarwady River, Burma. Photograph courtesy of Toshio Kasuya.

other toothed whales."[10] The balance of evidence in any area of scientific uncertainty should always incorporate considerations of the risks that lead from the possibility of our getting things wrong. In the case of conservation then, a precautionary approach might be to seek to conserve behavioral diversity in order to mitigate the risk that, if it is cultural and it is lost, it might prove very difficult to regenerate.

Not all the evidence for cetacean culture is equal. Looking again at the evidence we have presented in the previous chapters, where we didn't include anything that we didn't think was at least plausibly culture, we can split it up into three categories based on how sure we can be that culture is actually the underlying explanation—definitely, likely, or plausibly.

Definitely Culture

The best evidence for culture—by our definition—in the cetaceans is the song of the humpback whale, and it comes not from experiment but careful observation. It is Exhibit A in the case against the dogmatic view that only experiments can inform us about nonhuman culture. Scientists have documented how the song changes in both evolutionary and revolutionary mode within the lifespan of individuals. There is no way even the most outlandish scenarios can explain this pattern with genetics alone. Even some of the more hard-nosed critics of animal culture do not dispute this.[11] There is no realistic way all the males in an ocean basin could sing the same song and have it evolve over months and years without them listening to each others' songs and adjusting their own accordingly. We can't be sure about the precise mechanisms involved—we think it's most likely down to vocal learning, where individuals acquire new songs and themes from hearing them sung by others. However, it is theoretically feasible that humpbacks are born with innate vocabularies of song units, and their production is triggered by hearing them produced by others. These are interesting questions but irrelevant to the central one at hand here. Irrespective of which mechanism is at play, the song changes rely on humpbacks hearing each other. Thus there is social learning, and the behavior is communal, so we have culture by our definition. The other baleen whale songs in which the form of the song used by all singing whales changes systematically over timescales much shorter than population turnover, those of bowheads and blue whales, although less well known, must also be culture by the same reasoning.

The key thing about humpback song is that it changes greatly within the lifespan of individuals. There are other examples where behavior changes rapidly, so by similar logic, arbitrary normative behavior spreading quickly through populations, such as the dead-salmon-pushing fad of the killer whales and the tailwalking that Billie the dolphin introduced into her population, must also be culture, albeit relatively ephemeral and so perhaps unlikely to have persistent effects. Similarly, the details of how lobtail feeding has spread through the population of humpback whales in the Gulf of Maine are incompatible with scenarios that do not involve some level of social learning, although changing ecological conditions probably made such behavior more advantageous and therefore also more likely to be uncovered through individual learning—modeling of the spread of lobtail feeding indicates that both processes played a role.[12]

Finally, we know enough now to be certain that the pulsed-call dialects of killer whales are cultural. We have evidence of how calls vary between pods, clans, and *communities*.[13] Scientists have tracked how specific calls accumulate gradual small changes over time and how these changes occur in parallel in pods that associate a lot with each other.[14] If you move a young killer whale into a tank with adults that use a different dialect, it acquires that dialect.[15] There is no room to doubt the cultural nature of killer whale communication.

Likely Culture

The other vocal and nonvocal behavior we have described in the large matrilineal whales are almost certainly culture. It is possible to devise scenarios that are theoretically feasible but unlikely in practice in which genetics or individual learning and environmental variation explain the characteristic movement patterns of sperm whales or the beach rubbing by one community of killer whales but not others. However, it is much, much more likely that the sperm and killer whales learn this behavior, probably mostly from their mothers.

We have described in chapter 6 the diversity of foraging strategies that killer whales use in different parts of the world, including herding herring, intentional stranding on beaches, and washing seals off ice floes. Sometimes different strategies co-occur in the same region, such as the fish and mammal eating ecotypes (the residents and transients) of the Northeastern Pacific. We can, though, detect genetic differences between these populations, which could lead to doubt about the role of culture and leads us to placing these behaviors in the "likely" category. However, we also know that, when tested, captive killer whales have proven impressively proficient at copying each other's behavior, even behaviors never seen before. Hence, remembering Morgan's caveat, we have independent evidence that killer whales are quite capable of cultural transmission, and this makes it very likely that this process has an important role in explaining behavioral variation in the wild.

When it comes to sperm whales, we again have strong evidence of different dialects and different habitat use between groups of whales that occupy the same area in the eastern tropical Pacific. Here, we have looked for genetic differences that could play a meaningful role in generating this variation in behavior when it comes through the female line, by analyzing DNA from naturally sloughed skin. We could not detect them. It seems implausible that this variation could be generated and maintained without cultural transmission playing a significant role. However, an element of doubt lies in the fact that we have only properly tested the genetics of the maternal lineages. It is theoretically feasible that sex-linked genes passed down the paternal line could produce these patterns if male sperm whales only mated with females from the same vocal clan they were born into, something we don't have the data to properly look at yet. It seems rather unlikely, however, since males leave both the tropics and the groups they were born into for ten years or more before returning to look for mating opportunities and, when on mating grounds, rove around between different groups of females. The presence of such paternal genes specific to clans would also completely contradict the results of a preliminary study that found absolutely no difference between clans in genes passed down through both parents.[16] It would be astonishing were such genes to be identified. Nonetheless, because we cannot rule them out, we must accept an element of doubt.

It is rather harder to pin down the foraging specializations of the bottlenose dolphins, from solitary pursuits like sponging to genuinely unique phenomena like cooperative fishing. For instance, we know now that the use of sponges is carried out in particular places by particular matrilines, so both environment and genetics could reasonably have a role (although this needn't exclude a role for culture). The technicalities of the genetic studies of sponging mean that a small element of doubt must remain. We know that in one part of Shark Bay, all the sponging dolphins bar one have the same mitochondrial genotype passed down the maternal line, whereas in another part of the bay, most of the spongers have a different genotype. In some cases the difference between the genotypes associated with sponging and not sponging are as small as a single nucleotide substitution. This is a bit like swapping a single letter in a word—although it is a small change, in the right place it can make a big difference. However, the parts of the mitochondrial DNA that have been studied—the "hypervariable regions"—are thought to be noncoding, in that they have no role in producing the respiratory enzymes vital for life. Where scientists have looked at areas of the mitochondrial DNA that do code, they find no evidence that specific sequences are associated with the sponging behavior.[17] This means the window in which noncultural explanations must operate is really quite small, with the only plausible genetic explanation relying on complex additive interactions between genes that somehow occur only in specific matrilines that are also different in different places. Nonetheless, the window exists, so we must retain a degree of uncertainty. Less certain, because they are less studied, the human-dolphin fishing cooperatives nonetheless seem very likely to us to have to contain some cultural element for them to have persisted in subsets of the local populations over multiple generations, although environmental conditions—the presence of fish and human fishers—are clearly important, and genetics might also have a role. Provisioning and begging also look pretty cultural. In both Shark Bay and Cockburn Bay, Western Australia, the pattern by which dolphins acquire the habit of obtaining fish from humans seems to follow the path of the dolphins' social network.

If we knew nothing else about dolphin learning abilities we might have placed these behaviors in a more tentative category. However, as with the killer whales, we can add to this the experimental results showing that dolphins can copy each other. Granted, none of the wild behaviors have been specifically tested in this way, so they can be individually criticized as unproven. What we do have, though, when the evidence is considered as a whole, is a combination of incredibly diverse behavior in the wild and evidence for sophisticated social learning abilities in captivity. On balance, the evidence in this picture tilts heavily toward a cultural explanation for dolphin behavioral diversity, although because they have been less studied, we will place some specific examples in the "plausible" category below.

We also place the songs of the minke and fin whales in the likely culture category. In the songs of these species we do not have the definitive "progression over time" evidence that seals the cultural argument for humpback, bowhead, and blue whale songs. But these songs do vary spatially, and while it is theoretically possible to come up with genetic or environmental scenarios that cause the spatial variation, they have little plausibility. The songs within particular regions are highly stereotyped, while those of different regions are quite distinct. Neither the whales' environment nor their genetic variation follows this pattern. Instead, they follow more gradual clines.[18]

Finally, another area where the evidence makes it very likely culture is playing a role is in the seasonal migration movements of a number of species. We know this because of photo identification and genetic studies that reveal patterns of matrilineal segregation across feeding grounds—whales consistently use their mothers' migration destinations, not those of their fathers. The simplest explanation is cultural transmission of migration routes from mother to calf. So we can put the migration behaviors of right and humpback whales in this category, too. We don't think the urge to migrate is itself necessarily culturally transmitted—this is easily explained with genetics alone—but the specific routes, as well as specific summer and winter grounds, are most likely learned by young during the first migrations of their lives as they follow their mothers. A similar process appears to occur in populations of beluga whales.

Plausibly Culture

Behavioral variation between communities or areas suggests where to look for culture but doesn't prove that the variation is explained by social learning. While we might strongly suspect the origin of some behavior is cultural, with an absence of other evidence we are limited in the strength of conclusions we can draw. Note that this is not the same as evidence of absence of culture. We think that the reason we have to consider these cases as merely plausible is that they have not been investigated to any great degree. We are not aware of any case in cetacean research where a scientist looking at how behavior develops has produced evidence that is not consistent with a large role for culture, but consistency is not confirmation. So in this section we place some behaviors that we have noted in previous chapters because their complexity and/or their cooperative nature suggest to us that culture is playing a role but about which we know little more than their initial description. These include the ways bottlenose dolphins process cuttlefish and engage in various cooperative feeding techniques like mud-ring feeding and working in groups to chase fish onto mud banks. We would also include in this category the various foraging techniques of humpback whales (apart from lobtail feeding where the case for culture is good), bottlenose dolphin dialects, and the migrations of dusky dolphins along the New Zealand coast, as well as those of Antarctic killer whales to Brazil and back.

A Whale without Culture?

One obvious way to assess the importance of culture in the life of an animal is to ask the reverse question—what happens when the opportunity to acquire culture is restricted? This can be done experimentally if you remove any potential sources of social learning by raising individuals in isolation. Another related approach is to change the nature of the culture available to acquire, by transferring young or eggs into a context in which the social information available is different from that offered by their natural parents, termed cross-fostering. If individuals raised in isolation do not develop the behavior being studied, then we might conclude that some kind of social learning is necessary. Such isolated rearing has been an important part of our understanding the role of cultural transmission in the development of birdsong.[19] Similarly, if cross-fostered individuals develop behavior more similar to their adoptive parents or social group than that of their biological parents or the social group they were born into, then again it seems that social learning has a big role to play. This approach has been used to great effect by the Norwegian scientist Tore Slagsvold to show that social learning plays an important role in how foraging develops in small forest birds.[20]

It is also obvious though that there are some problems with these approaches, which is why they haven't been as widely used in studies of mammals. Raising individuals in isolation requires hand rearing by humans. In some species, this is straightforward, but in others it is notoriously difficult to achieve. Moreover, rearing in these conditions could also deprive the young of a whole suite of learning opportunities, not just the social learning ones, because of the restrictions that are required to achieve a genuinely isolated upbringing. Cross-fostering can be very challenging because, unsurprisingly, natural selection has equipped many species with various ways to avoid parents giving care to young that do not carry their genes. In some cases these can be subverted—imprinting in birds, for example, where hatchlings that normally experience their parents as the only large moving thing in their immediate surroundings can be "fooled" into imprinting on humans, or anything else we might choose to confront them with immediately after hatching—but often they cannot. For cetaceans, these problems are so serious that the likelihood of failure and dire consequences for the animals involved almost entirely precludes us for ethical and logistical reasons from setting out to investigate cetacean culture in this way.

Circumstances sometimes conspire, however, to create situations that we would never think to deliberately engineer but from which we can learn a lot. For example, most people would consider it immoral to deliberately study humans in this way, by raising babies with robots, devoid of human contact, or by swapping newborns around in maternity units. Nevertheless, it sometimes happens that twins are separated at birth—fortunately less so nowadays—and we can learn a lot from studying their subsequent development.[21] There are also cases in history of human children being largely cut off from human contact for the early years of their lives—the so-called feral children.[22] There is for example the case of Victor, the "wild boy of Aveyron," a region in southern France, who was discovered wandering naked in woodland searching for acorns, in 1799. With an apparent age of eleven or twelve (obviously, impossible to know for sure), he was described as "a dirty, inarticulate creature who trotted and grunted like the beasts of the fields."[23] Victor was made a ward of the state. A young French psychologist, Jean Itard, just out of his doctoral studies, spent much of the next seven years working with and studying Victor in an attempt to teach him to live in human society. He eventually gave up. Victor never learned to speak more than a couple of words, and even then Itard was dubious whether he really understood their meaning. He lived out the rest of his life until his death in 1828 in a state we would today diagnose as having severe learning difficulties. Despite lacking language, or any other kind of human enculturation, he was nevertheless apparently capable of empathy, weeping alongside his primary caregiver following the death of her husband. This observation alone has profound implications for how we think about the relationships among culture, language, and emotions. Then there is the incredible but true story of the wolf children of Midnapore, India. Two girls were discovered, around aged eight, living alongside wolf cubs, apparently accepted as part of the brood by the wolf mother. Again, these children received intensive attention from their carers for the rest of their lives, which were tragically cut short by illness, with the oldest living to around seventeen. This was apparently more fruitful than the efforts expended on Victor, in that the longer-lived of the girls eventually learned approximations of around thirty words, but the wolf girls still fell far short in this and other social skills of the level needed to function properly in human society. The crucial point is that in these and other cases of genuine lack of human contact during their crucial early years, the children were profoundly incapable of functioning independently in human society, with no evidence they ever approached concepts of things like shared meanings or cultural identity. Their stories illustrate quite starkly just how vital social input is to the development of human behavior.

In cetaceans, there are examples where the various translocations of captive animals have created somewhat similar inadvertent experiments. One was the source of significant insight into how killer whales learn their group calls, when a young killer whale was moved into a tank with other older animals with a different dialect, described in chapter 6. A mirror image of the cases of Victor and the wolf girls arises from the story of a captive killer whale named Keiko, who starred in the 1993 blockbuster movie *Free Willy*, the sugary tale of a boy who rescues a killer whale from captivity.[24] In reality, after the wrap party, Keiko went back to his life in a captive facility in Mexico, where he was kept alone, with only a couple of bottlenose dolphins for company. In just one of the many onion-like layers of irony in this story, the name Keiko is Japanese, meaning "lucky" or "blessed child." Not so much. Keiko was captured in 1979, aged around two, from the waters around Iceland. He was kept first at an aquarium in Iceland and then sold to a Canadian aquarium before eventually moving to a facility in Mexico City in 1985. In Iceland and Canada there were apparently other killer whales in the tank with him, but in Mexico City there were none. Then he got his big break as Willy. Following the success of the movie, a campaign to free Keiko attracted several million dollars of funding, some from the Warner Bros. Studios themselves. The level of funding, tied to the fate of a single whale, was extraordinary in comparison to the budgets marine mammal scientists typically get to work with when studying entire populations—another level of irony. Keiko was moved to a temporary rehabilitation facility in the

United States before eventually being flown back to Iceland to a sea pen in the archipelago of Vestmannaeyjar just south of the mainland to start a training program leading up to his eventual release.

A major challenge was that Keiko had received all his care, contact, and food from humans for all his adult life, from which he would have to be weaned if he was ever to prosper in the wild. The training started by having Keiko follow his caretakers' boat on excursions from his home pen into areas frequented by the wild killer whales off Iceland, over the three summers of 2000–2002, with the hope that he would eventually become more integrated with them than with his caretakers. In 2002 he spent two distinct periods associating with wild killer whales, but as the team of caretakers noted, his interactions with them appeared distinctly awkward. He kept his distance and did not appear to join in with the foraging activities of the wild whales, who in turn apparently ignored him. During this time, Keiko's caretakers were able to take stomach samples from him on a couple of occasions, both of which showed no evidence he had been able to feed for himself. After a couple of weeks, things appeared to warm up a bit, socially, but then one of the wild whales took a step too far for Keiko:

> A brief physical interaction was witnessed on 30 July, when Keiko dove among foraging whales and surfaced in very close proximity to three adult males and at least two females or immature males. There was a splash from the tail of one of the wild whales, which was swimming ventral side up, with his head below Keiko, while he was at the surface. The splash was accompanied by a "startle" reaction from Keiko who swam to the tracking boat, while one of the female/juvenile whales surfaced after him.[25]

In other words, one of the wild whales appeared to initialize what for them would be a fairly typical social interaction, but poor Keiko got so freaked out he immediately ran off to seek the only source of security he knew—humans.

In the days following these interactions, Keiko swam away from Iceland. Over the next three weeks he made his way to Norway, some thirteen hundred kilometers distant. The scientists tracked him over this period with satellite transmitters attached to his dorsal fin, hoping this was going to be a positive development but also fearful of the risks, for example, of heading north and getting trapped in ice. We don't know whether he undertook this journey in the company of other killer whales, but when he was next sighted, just a few meters from shore at a place called Kristiansund, he was alone. His first recorded act in Norway was to follow a small pleasure boat into a nearby fjord, Skålvikfjorden, where the boat was based. Keiko then appeared to adopt Skålvikfjorden, a sheltered inshore fjord some ten kilometers from open water, as his new home, such that the foundation then constructed another open sea pen, like the one he had lived in while in Iceland. He received both food and care at this facility, with his caretakers making the decision to begin provisioning him again after locals started feeding him anyway (it is unclear whether he actually fed at all up to that point, since leaving Iceland). His caretakers also began taking him on excursions around the local waters, as they had in Iceland, but as far as his caretakers knew he made no more contact with any wild killer whales. He lived the rest of his life in this manner before he died in December 2003, apparently of pneumonia, at around age twenty-six. Danish scientist Malene Simon and her colleagues who worked with Keiko during this time concluded that the attempt to release him was a failure—he never became truly independent, he never successfully integrated with any social groups of his own kind, and he died still entirely dependent on humans for food.

Not all attempts to re-release killer whales have failed in this way. In 2002 a young killer whale, thought to have been around two years old at the time, was encountered swimming alone in Puget Sound, off the U.S. city of Seattle. The whale was photographed and its calls were recorded. It was subsequently identified as a calf designated A73, nicknamed Springer, who had been born into the A24 matriline of the A4 pod of the northern resident killer whale *community* in 2000.[26] Springer and her mother had, however, not been seen in Johnstone Strait, British Columbia, a core part of the northern resident range, at all in summer 2001 and had been feared dead. Her mother, A45, was never seen again so the likelihood is that Springer was orphaned sometime in 2001. The U.S. National Marine Fisheries Service decided to intervene and support an effort to reunite Springer with her group, back in Johnstone Strait, some four hundred kilometers north of where she was discovered. Springer was successfully captured and maintained at a sea pen facility in Puget Sound for around four weeks, with restricted human contact, fed live salmon delivered by a chute so that she could not see the person feeding her. She was then transported north to another sea pen at the northern end of Johnstone Strait, after researchers there used hydrophones to determine that members of the A4 pod were in the area. After arrival, Springer became visibly excited when the A4 pod calls were heard and responded to them with calls of her own; she was then released and rejoined the pod. She has continued to be sighted with the A4 pod.[27] In July 2013, Springer, now aged about thirteen, was seen with her first calf.[28] Springer had reintegrated successfully.

So why was Keiko different? Simon points to a number of guidelines for cetacean release outlined some years earlier by the American dolphin expert Randy Wells after he supervised the successful release of two bottlenose dolphins in Florida.[29] The attempt to release Keiko failed to meet the first three of these—specifically, that animals should be released together in some form of functional social unit, that released animals should be young, and that they should have been short-term captives in the first place. In short, Keiko was never a good candidate for successful reintroduction. The contrast with Springer is clear—Keiko was kept in captivity much longer, and the Iceland killer whales were never well enough known to him, or to scientists, to be able to release him into an appropriate social context. But why are those factors in particular so important? To us, this makes sense if we understand Keiko as a whale without culture, or at least missing very important parts of it. Captured at an early age, and spending his developmental and early adult years in captivity, he never acquired that "complex whole" of cultural knowledge crucial for a killer whale to survive in the waters off Iceland and to function effectively as a member of the local killer whale society (or perhaps he lost what he had acquired during his prolonged captivity). Like Victor and the wolf girls, he missed out on crucial social input at an important stage of his development and was never able to recover from it. In contrast, Springer had learned her pod's dialect by the time of her capture, sufficiently to make her identifiable to both researchers and, presumably, other whales as a member of the A4 pod. She had also learned about salmon, as she fed enthusiastically on live salmon, provided by fishers of the local First Nations groups and introduced to her sea pen while she was captive. The contrasting stories of these two killer whales represent an inadvertent (and not perfectly controlled) experiment on what happens when killer whales are deprived of access to their culture while growing up. The consequences appear to be as disastrous for killer whales as they do for humans, and to us this speaks volumes about the importance of culture to both.

We will end this section with the intriguing case of the "52-Hz ghost whale" of the North Pacific.[30] The U.S. Navy, with its arrays of hydrophones around the world, has been listening to whales for decades, as background to their real interest: submarines. Scientists have been given access to the recordings and have used them to make some of the remarkable discoveries that we summarized in chapter 4—the original Bermuda recording of the song of the humpback whales and the global decline in the frequency of blue whale songs being the two most prominent examples. These scientists have learned to classify the various whale sounds heard on the Navy hydrophones into species and populations. Nearly everything they hear fits one or other of these classes. But starting in 1989 they heard an unusual song at about fifty-two Hertz. The movements of the singer were tracked from 1992 to 2004, between August and February each year: the whale sang as it moved over the northeast Pacific between the waters of Mexico and Alaska. The song over all these years seems to have been made by just one whale and is unlike that of any known baleen whale, and its track was unrelated to the movements or presence of any of the other baleen whale species that the scientists were monitoring with the Navy hydrophones. They do not know and cannot guess its species. It could be a hybrid. It seems to be an individual who has not picked up either a culture of song or a culture of migration. Unlike Keiko, the ghost whale lived, at least until 2004. It had its own songs, its own movement patterns, but they are not culture, and without culture it was alone, the "52-Hz ghost whale."

Critiquing Cetacean Culture: "Perfectly Absurd"

Labelling ape behavior as "culture" simply means you have to find another word for what humans do.[31]

This quote from Jonathan Marks, an American anthropologist, exemplifies the second bone of contention. It is the complaint from the anthropological wing of the animal culture debate that what we call culture is not culture at all. This is not so much a critique of cetacean culture specifically but a general criticism of the whole idea of nonhuman culture, be it in chimpanzees (as the quote makes clear), dolphins, bats, birds, or rats. So at this point, "we" includes ourselves, who focus on whales and dolphins, and all the other scientists who share our broad perspective on culture and think that some things that nonhumans do are usefully termed "culture." These critics attack our usage of the word because of the differences between human culture and anything that animals do—because culture, in humans, is so much more than simply information and behavior: it encompasses abstract components like shared systems of meanings, moral codes to govern behavior, and symbolic markers of identity.[32] Preserving the word "culture" for humans alone emphasizes what, from this perspective, is an absolutely crucial gap. They believe our definition of culture is far too simplistic and hence far too broadly applicable. Definitions are of course a matter of perspective and choice. As we discussed in chapter 2, we think that our broad definition of culture is useful. Nonetheless, we must live with its consequences, and by this definition culture is fairly widespread among animals. We are comfortable with this as it allows us to get on with asking, and sometimes answering, the important questions about what is actually going on in cetaceans and how it compares with what is going on in other species.

To what extent can our broad concept of culture relate to one that includes abstract concepts such as "meaning," which these critics believe to be central to their concept of culture? If "meaning" is defined as "the attribution assigned to cultural knowledge by minds," then there initially appears little hope for overlap.[33] Our entire consideration of culture in cetaceans is rooted in observables. Meanings in cetacean culture, should they exist, are not accessible to us. We are not alone in this, as meanings are likewise unavailable to those who study any nonhuman cultures, and they are really only partly accessible to students of human culture—specifically, that part of an interview where the interviewee is telling the truth. To be blunt, the concept of meaning appears to hold little relevance to us in our studies. If culture must include systems of meaning like languages, then cetaceans (and other nonhumans) cannot have culture on current evidence, and it is not clear how we would ever obtain such evidence in the future. In our view, laid out in chapter 2, shared systems of meaning (i.e., language) are not necessary for culture, although they are a striking component of what anthropologist Jerome Barkow calls human "hyperculture."[34] Our understanding of cetacean culture has a long way to go before we can incorporate meanings on an empirical basis, and knowledge of meanings is at best peripheral to the evolutionary questions that motivate us. Nor is it clear to us why it is cultural meanings that are the essential ingredient of culture, over and above the fact that it propagates socially. We haven't seen any arguments to that effect beyond simply stating that they are part of human culture, so must be part of any concept of culture.

We should be quite clear that while we might call some of what nonhumans do culture, we do not intend to equate it with human culture. Psychologist Michael Tomasello, a consistently clear thinker about the real differences between human culture and what other animals do, puts it like this, referencing a paper published in support of chimpanzee culture by William McGrew: "McGrew (1998) claims that nonhuman primates engage in social activities that are best characterized as cultural in that they share all the essential features of human culture. I agree with this. Nevertheless, at the same time I insist that human culture has, in addition, some unique characteristics (as may the cultures of other primate species)."[35]

Although we might quibble with his rather exclusive focus on primates, there seems to be much sense in what he says. Human culture has unique features, such as language and cumulative technology, which set it apart. However, nonhuman cultures might also have unique features of their own, but what unites them all is being a shared collection of socially transmitted knowledge and behavior. Why should we not be as comfortable with this as we are, for example, with a broad concept of locomotion in animals, which can incorporate such diverse attributes as flight, four-legged sprints by cheetahs, the ambulation of limpets on rocky shores, and our own bipedal gait? Each has unique features. For example, we can't outrun a cheetah over short distances, but it seems bipedalism makes us rather good at exhausting prey over marathon chases.[36] To us as evolutionary biologists these different ways of moving around have evolved for the same reason: to allow animals to relocate to their advantage. Similarly, what we call culture in humans and cetaceans is united by the common feature of being knowledge that travels between individuals and, perhaps more important, across generations independently of genes—and therefore has the potential to influence, perhaps profoundly, the dynamics of the evolutionary process itself.

We could be accused of wanting to have our cake—use the word "culture" with all the human cultural cachet that comes with it—and eat it, by not having to defend the claim that whales and dolphins have culture like humans have. We try and avoid this by being open that our concept of culture is broad and by accepting the consequences of that breadth. A similar cake-based critique can be made of the position of some anthropologists, though. How long have humans had full-blown symbolic culture? Jonathan Marks has recently argued that human evolution has been so influenced by culture that it cannot be understood without reference to it and specifies that he means "culture" "in the anthropological sense of a symbolic, linguistic, historical environment, limited among extant species to *Homo sapiens*."[37] His view is that culture, used in this sense, has been present in the human lineage, and actively coevolving with our genes, for the last 2.5 million years. His evidence? The existence of lithic technology—stone tools—in the archaeological record. His argument is that the existence of those tools proves that culture "in the anthropological sense" was present. This is where the anthropological and psychological critics of nonhuman culture clash—to the latter, it proves nothing beyond the making and using of tools.[38] Some Galápagos finches modify twigs to make tools for probing crevices where insect prey hide, but social learning is not necessary for this behavior to develop—so why should we accept the existence of stone tools as evidence for symbolic culture?[39] Michael Tomasello, for example, has a different view of when the crucial changes happened that propelled human culture into the symbolic realm anthropologists would recognize, which admits uncertainty: "in the range of 2 million to 0.3 million years ago—with my own theoretical bias being toward the smaller of those figures."[40] Kim Hill goes further, arguing that the archaeological record suggests humans may have only had what he calls "complete 'culture,' including social norms and ritual and ethnic signalling" for the last 160,000 years—half the lower end of Tomasello's estimates.[41] It seems to us that if one wishes to make the claim—which we would agree with—that culture has been intertwined with human evolution for 2.5 million years, as Marks does, one has to accept a reasonably broad concept of culture, very similar to ours. It doesn't seem tenable, on the evidence currently available, simultaneously to have your cake, by claiming that you can only have culture if you have symbols, and eat it, by asserting that culture has been present in its exclusively human form since long before the origin of anatomically modern *Homo sapiens* some 200,000 years ago. One either has to accept a broad concept of culture that is therefore not exclusive to humans or find another word for what our ancestors did prior to about 300,000 years ago.

Something (and we're not really sure what or when) changed in human evolution to give us a new way of understanding each other—and so to create and exchange through shared attention and cooperation—that allowed us to start building cumulative human cultures. Thus each generation could benefit both from its own aggregated knowledge and that of the generations that went before. By this process we went from caves to the moon, and we are the only species that did. It is not anthropocentric to highlight these very real differences.[42] The mystery of how these changes came about is nonetheless one of the reasons we want to understand the kinds of culture that evolution has produced in whales and dolphins because it helps us to understand how we got our culture.

Cetacean Culture and the Meaning of Identity

My personal definition is that culture is the collective programming of the human mind that distinguishes the members of one human group from those of another.[43]

This concept of culture, from the influential Dutch social scientist Geert Hosfstede, encapsulates one area where we might think about bridging the gap between some anthropological notions of culture and the evidence we have for cetaceans. Kim Hill puts it more graphically: "If animals have culture, why do they not form imaginary social boundaries around arbitrary groups of individuals that fight thousand-year wars and recruit suicide bombers to kill infidels?"[44] Although it might seem a bit strange to hold up these rather awful features of human culture in order to argue that it is so special, these notions illustrate one of the principal fault lines that make some anthropologists so resistant to the idea of nonhuman culture. Susan Perry, an American anthropologist who has spent many years studying the behavior of capuchin monkeys in the wild, also highlights that broad definitions such as ours miss out on the "emotional salience that links particular symbols, artifacts, or behavioral traits to group identity, and the moral pressure to conform to a suite of traits."[45] In short, the claim is that only humans have ethnic markers. One of the reasons that these are important in human culture is that they serve as indicators both for cooperation and its reverse, hostility.[46] In Glasgow, Scotland, for example, people have been murdered, by complete strangers, for wearing the wrong soccer shirt in a city where sports allegiances parallel a sectarian divide.[47] Ethnic markers are a key way in which culture can affect social structure.[48]

In her work, Perry has described extraordinary ritual-like games that capuchins perform with each other. These include hand sniffing, sucking of body parts, taking a partner's finger and putting it into one's eye socket, and turn-taking games such as the "finger-in-mouth game":

> In this game, monkey A puts its finger in monkey B's mouth—this can start in the context of grooming the inside of the mouth. Monkey B clamps down on A's fingers, apparently not enough to hurt but hard enough that A cannot readily remove its fingers. Monkey A goes through various contortions to extract its finger—this may involve using both hands and feet to pry open B's mouth or putting the feet on B's face for leverage while pulling hard on the captured hand. Once A's finger has been successfully extracted, either A reinserts its hand in B's mouth for another bout of the game or B inserts its fingers in A's mouth so that the game continues with the roles reversed.[49]

Or the "hair-passing game":

> In this game, monkey A bites a tuft of hair out of the face or shoulder of monkey B. Then monkey B attempts to extract the hair from A's mouth, using the same techniques described for the finger-in-mouth game. Once monkey B succeeds in recovering the hair, A tries to get the hair back from B's mouth. The game continues, with A and B reluctantly passing the hair from mouth to mouth, until all the hair has been accidentally swallowed or dropped. Then A bites another tuft of hair out of B to start the game anew.[50]

These conventions were only found in a few groups, practiced by a few animals, and died out after some years. When she found that these games varied between groups, Perry initially "was intrigued by the thought that these might be ethnic markers of sorts—rituals performed by members of a group that would serve not only to strengthen their own bond but to advertise their solidarity and clique membership to nonclique members."[51] But with more observations and reflection, she came to the view that they were not. The hand-sniffing, hair-passing, and finger-in-mouth behaviors occur in private, quietly and unobtrusively—not good characteristics for signaling to others about anything, let alone your group identity. However despite the apparent lack of ethnic markers in her capuchins and other apes and monkeys, Perry did not rule out the possibility that such ethnic markers might exist in nonhumans. She suggested that we might look at societies with important cooperative relationships, social learning, and enough physical movement of individuals within them that it could be useful to identify potential cooperative partners from among the range of individuals you might bump into. All these features are pretty evocative of cetacean societies, something that Perry noticed, too, when she wrote that "I find it more plausible that ethnic markers might exist in cetaceans, birds, or bats that in nonhuman primates."[52]

Perhaps we can chip away at the divide by examining the overlap between meaning and identity in some of the best-known cultural behaviors of whales and dolphins—their vocal communication. For example, we presented in chapter 5 the evidence that bottlenose dolphins can recognize each other through signature whistles and engage in whistle matching in the wild. Vincent Janik suggests that these whistles may be used to refer to individuals and has also provided evidence that this is the case in the wild—so in that referential sense the whistles used may be said to have meaning.[53] However, the relationship between this and the cultural processes occurring in dolphin populations is not yet clear—do bottlenose dolphins learn the particular set of meanings linking particular whistles to particular individuals in the communities they are born into? We would answer probably yes, but we don't know for sure. The deeper question is whether these differences between whistles of different groups are barriers or otherwise to various forms of cooperation, like joint foraging. Away from the vocal domain, is the separation that arose between the trawler and nontrawler dolphins in Moreton Bay motivated by a disdain, for example, for those dolphins that would lower themselves to eating the fish entrails left behind the boats? These are questions we can't answer, but they are surely worth asking.

But do the vocal variations that define signature whistles in dolphins and dialect groups in sperm and killer whales hold meanings for the animals themselves, in the sense that anthropologists talk about symbols having meanings in human culture? If they do, that meaning likely concerns identity—for example, the identity of the caller's group—with respect to the listener's group. Whether this would qualify as a "system of meanings" for anthropologists we do not know. We suspect not on current evidence—the function of these dialects is currently still the stuff of speculation, so again we cannot answer now. However, we now have the technological tools at hand with advanced recording tags and in-water speakers to start doing playback experiments with wild groups that might start prizing the lid off this can of questions. If future research can pin down the function then we might also start to work out what meanings they hold for the animals. We can in the meantime think about the idea of cultural identity in whales and dolphins.

Populations of killer and sperm whales appear to be structured along cultural lines, and this structure comes with a whole package of habitat and feeding knowledge essential to survival, so that we can make the provocative suggestion that the importance of a cultural identity might not be something unique to humans. Despite contact between cultural groups and, in the case of sperm whales, transfer of individuals between them, the groups remain sharply defined by their vocal dialects. In humans, dialect boundaries are often important social boundaries and can serve as ethnic markers that may also be barriers to altruistic acts.[54] The description of interactions between fish-eating and mammal-eating killer whales in chapter 6 doesn't suggest they would be likely to hunt cooperatively together, and what appears to be strong cultural conservatism in these killer whale populations might be linked to what Lance Barrett-Lennard calls their xenophobia.[55] So killer whale call dialects look like symbolic markers, at least on the surface, with considerable relevance to things like cooperation and apparent hostility.

Sperm whales, meanwhile, have been observed endangering themselves to rescue group members from attacking killer whales.[56] When individual sperm whales leave the relative safety of the concentrated group to guide isolated and injured group mates back to the safer collective defense formation, "all who watch are shaken by these acts of apparent altruism," in the words of biologist eyewitnesses Robert Pitman and Susan Chivers.[57] We know that sperm whale social units associate only with others having the same vocal dialect, but is this kind of helping behavior limited only to members of the same social unit, or does it extend to any members of the same clan that might be associating that day? If, and it is still a big if, clan dialect boundaries are also barriers to altruism in sperm whales, then we might draw very close parallels with the evolution of human linguistic variation and perhaps move closer to being able to suggest that vocal dialects are functioning as ethnic markers. Again, these are questions that are surely worth asking.

Culture Justified?

Jonathan Marks, as we learned near the start of this chapter, would have very little truck with the idea of a whale having something called culture. Obviously we disagree. We nonetheless think he is correct when he asserts that biological evolution alone cannot explain human behavior without recourse to the centrality of culture in human affairs—what he terms "biocultural evolution" but what we would call gene-culture coevolution.[58] What we hope has come across in our book thus far is that a rather similar assertion could be made about whale and dolphin societies, whether you call it social learning (which of course applies to all human culture, too) or tradition or go for the more controversial label of culture.

Perhaps it is our evolutionary biology backgrounds that lead us to give particular importance to certain characteristics of culture over others. When a body of socially transmitted information has such an impact on the lives of individuals within a species that it structures their populations, defines their ecological relationships, and ultimately, as we shall go on to explore, feeds back to affect even the genetic evolution of populations, then to us it is informative to label it culture. To some this is a futile semantic exercise. What, Jonathan Marks asks, is the point of shifting words around? It doesn't get you away from the existence of the basic divide between nonhuman and human cultures but simply rearranges the deck chairs of a ship sinking inexorably after colliding with the icebergs of symbolic meanings and cumulative technology.[59] Kim Hill will at least entertain the question of whether we gain or lose by adopting a broad concept of culture, but he thinks we lose: "My concern is that the loose application of the term 'culture' for all socially learned behavior may obscure our ability to understand the evolution of what appear to be very unique characteristics of *Homo sapiens*."[60]

While this is a laudable concern, we think it is misplaced. In fact we think the effect of a broad concept of culture will be the precise opposite of what he is worried about. By having such a broad concept and qualifying it in the human case we move to a situation where our language more accurately focuses on that which makes human culture unique. It acknowledges that which is similar—social learning of shared behavior—and by doing so focuses on what differs. So instead of humans having culture and other animals having something different, we describe some nonhumans as having culture and humans as having symbolic, cumulative culture. And if we so choose we can ask how these cultures are similar or different—but that is not the only point of the exercise. In studying the evolution of various forms and features of culture, precise identification of what you wish to explain the evolution of is a critical first step, and we argue that a broad perspective on culture helps us do just that. This seems to us more satisfactory and arguably more accurate than having a cultural Rubicon that only we have crossed. Not everyone will agree. Whether you do or not depends largely on what you consider the most important characteristic of culture—that it is a flow of information across generations outside of genes or that it is a system of shared symbolic meanings.

You will have noticed that most of the examples of whale and dolphin behavior we have discussed fall into the "likely culture" category. There is still uncertainty, even if the scales of evidence are considerably tilted in the culture direction. If we were to wait until we could move everything under "definitely" and "definitely not" headings before writing this book, we wouldn't be the ones writing the book; it would instead be some lucky whale biologist of the future with research equipment currently only found in science fiction. When will we be able to successfully cross-foster sperm whale calves, even if we thought it ethical (which we don't)? We don't know, but it isn't anytime soon, and in the meantime we in our societies need to be making decisions right now about how we treat, conserve, and manage populations of whales and dolphins. If we don't get them right, with our limited understanding of the forms of culture present in these populations, our future whale scientist won't be a biologist—she will be a paleontologist. We need to evaluate the evidence available today.

So what is gained by studying culture in whales and dolphins? The promise of a cultural cetology—a study of whales and dolphins informed by an understanding of the role of culture in their lives—is manifold. It lies partly in a correct understanding of cetacean behavior. Simply getting our facts right about their nature will also be crucial if we are to make the right decisions about how to coexist with whales and dolphins. Over and above that, we get excited about the potential for comparing how culture has evolved in different ways in different species; cetaceans have a variety of social structures, inhabit a broad range of ecological niches, and have varying cultural systems. Thus we can hold out the hope of a greatly increased understanding of the impact of social and ecological factors on the evolution of culture and the capacity for culture. Such understanding will inform those seeking to understand humanity's own biological and cultural evolutionary history.

9 | HOW THE WHALES GOT CULTURE

We hope we have succeeded in convincing you that quite a lot of whale and dolphin behavior is acquired by social learning, is shared by groups, and therefore is culture—precisely how much is uncertain. In the previous chapter we outlined a minimum pool—definite culture—and a fuzzy area around it—likely culture. As science has so far only managed to get its hooks into a few areas of behavior for a few cetacean species, we are certainly missing much of the cultures of whales and dolphins. That huge slab of unknown whale culture is tantalizing, and it drives our own research. Where does it reach? How does it link: from cultural behavior to cultural behavior; from cultural animal to cultural animal; from cultural species to cultural species? And how does culture affect sociality, ecology, conservation and evolution? How can we find our way into all this?

There is also much uncertainty about the extent of culture elsewhere outside humans. We must be missing much of the culture of chimpanzees, crows, and elephants. There is fish culture and rat culture that we don't know about. However, the extent of this uncertainty is generally much less than with whales. Most terrestrial mammals and birds are visible and live in an environment similar to our human environment, one that we feel that we innately understand pretty well. Many of these animals can be studied in controlled, captive conditions. So we get a better feel for how their lives work. With the exception of the whales and dolphins, and perhaps elephants (see chap. 12), science as a whole has achieved what might be called a basic understanding of the extent of culture among animals.[1] Culture seems remarkably restricted. While there appears to be a cultural contribution toward some behavior in many social species, such as numerous songbirds, coral-reef fish and even some insects, there are fewer for whom culture seems important in a number of facets of life and for whom their offspring would simply not be viable without significant cultural inputs at crucial times during their development.

This becomes something of a puzzle. If culture is useful, then why don't more species use it more often? And if it is not useful, why do a few species use it a lot? In the words of Peter Richerson and Robert Boyd: "The exceptional nature of the human species deepens the puzzle—if culture is so great, why don't lots of other species have it? One of Charles Darwin's rare blunders was his conviction that the ability to imitate was a common animal adaptation."[2] From this perspective, it is the rarity of imitation that explains the rarity of culture, a consequence of imitation being a central plank of some definitions of culture. However, we have hopefully made it clear by now that we do not think imitation is the only way social learning can lead to persistent complexes of behavior that we might usefully identify as cultures. For us then, the rarity of culture is likely due to a combination of culture only being of value in rather particular environments and the fact that animals need prerequisites—some smarts and a proclivity to pay attention to the right things—to use culture effectively. Whales, then, live in these particular environments and have these prerequisites.

So, in this chapter, we ask three questions. When is culture useful? What properties does a species need to make good use of it? And how, given this framework, might the whales and dolphins have acquired the capacity for culture?

When Is Culture Useful?

Life depends on information moving from one organism to another. Without such inheritance, there is no biology, no life, and no evolution. On Earth, the primary method of transferring this information is through DNA or RNA, what we call genes. Most life forms use these biochemical formats almost exclusively to move the information about. The evolution of genes was utterly crucial to life on Earth. Like other parts of each organism's phenotype—a phenotype is the composite of all that we can observe of an individual organism—the behavior of early animals was controlled by the products of genes. So members of the same species had similar behavior because they had similar genes.

Like other elements of the phenotype—such as the body plan—behavior changed through genetic evolution, with the most important form of evolution being Darwinian natural selection. In a changing world, natural selection adapts the phenotype of a species to the environment. So if winters become colder, genes that promote burrowing behavior may be selected because those animals that have these genes survive better and have relatively more offspring, who also tend to have these genes, and in consequence burrowing behavior increases in the population.

There are two problems with phenotypes being determined only by genetic evolution in a changing world. First, adaptation through natural selection to environmental change is slow; it occurs over timescales of generations. Although this can work over the course of often surprisingly few generations, adaptation through natural selection can still be problematic if the environment is changing fast and the generations are long.[3] The pool of genes in a population may not be adapted to current conditions—approximating, perhaps, what made sense a generation or two ago—so the organisms that make up the population lag behind the times and suffer for it. This is a problem for the population. In some cases old-fashioned behavior may threaten a population's existence in a new and different world. The second problem is more important and is specific to an individual. If you have the wrong genes for the environment that you are living in and genes totally determine behavior, there is nothing you can do about it.

The obvious solution is to adapt to changes as they occur. If it gets colder, burrow; if warmer, forage. This is technically known as phenotypic plasticity because the phenotype of an animal can alter without any change in the underlying genes. Phenotypic plasticity makes a lot of sense for any organism living in a variable environment, and it is found all through the living world.[4] In an unpredictable world, plants and animals that can adapt their growth, shape, physiology, or behavior to different environments generally do better than those than cannot. Thus the second-order trait of phenotypic plasticity itself is chosen by natural selection.[5] But phenotypic plasticity is not as universal as one might at first expect; some organisms do not make seemingly sensible changes to their phenotypes as the environment changes. Humans do not grow more hair in winter and shed it in the summer—or even grow more hair if brought up in Scandinavia rather than in Italy. These would seem sensible ways to adapt our phenotypes to the environment, and other mammals do just these things. But human physiology is constrained. It is constrained because phenotypic plasticity is expensive. One needs extra, costly systems to have one's phenotype change with the environment.[6]

In this evolutionary world of trade-offs, natural selection searches out the cheapest and most effective ways of achieving an end. "Cheap" and "effective" often work against one another, as the best tools are often the most costly. A particularly effective form of phenotypic plasticity is learning: puzzling out cognitively, perhaps through trial and error, what is best in the current environment. But individual learning is costly. One needs a brain or some other quite sophisticated, and energy-consuming, decision-making system, and individual learning itself may also take valuable time or expose the animal to dangers, like poisoning or predators. With learning by trial and error, the errors are the problem: if an error results in you being eaten, this is a drawback. However, in many cases the trade-off was and is worth it. Animals living in variable environments developed brains and learned for themselves.

At this point a shortcut may become available. Instead of learning directly from the environment, individual learning, why not learn from other animals—what we call social learning? It will often be quicker, safer, easier, and, if the model is chosen well, comparably effective in picking up suitable behavior. The brain infrastructure for learning is already in place—being used for individual learning—and the animals just use it in a different way, concentrating on each other rather than the environment itself.[7] If lots of animals are learning from each other, their behavior becomes homogenized and they have culture.

This, then, is an outline of a generally accepted process by which the capacity for culture evolved: environmental variation driving phenotypic plasticity and individual learning and, then, social learning providing a cheap shortcut. However, as noted in chapter 2, social learning also has its risks, as there is less of a guarantee that the cheap information you get is good.[8] What kinds of environment promote the changes from genetic determination to individual learning and from individual learning to social learning?

The first transition requires an environment that is sufficiently unpredictable that an animal with an inflexible, or preprogrammed, behavioral repertoire is often doing things so poorly that its contribution of genes to the next generation is reduced, compared with an animal that can effectively track the changing environment. For instance, a rodent might spend a fixed time burrowing each day that is more than needed when temperatures are warm, so it loses foraging time, and that is less than sufficient when it is cold, so that it may die of exposure. To make the evolutionary transition, individual learning must sufficiently improve this subpar behavior so that the costs of the individual learning—a brain, time, and risk—are paid back with interest. Generally, for individual learning or some other form of phenotypic plasticity to pay, the environment must vary in such a way that individuals actively altering their phenotypes have their fitness—basically their expected number of grandchildren—increased on average compared with animals that do not actively vary their phenotypes.[9] Thus, if changes in the environment do not make much difference to fitness, there is no need for individual learning. In contrast, if the environment is oscillating a lot *and* this affects fitness, then we expect individual learning or some other type of phenotypic plasticity to evolve to deal with the problem.

The second transition from individual learning to social learning seems to be prompted by a particular type of environmental variability. The environment must vary, so phenotypic plasticity is profitable. But if this variation is unpredictable over the time span between learning and behaving, then copying others will not help. So our burrowing rodent could learn from its parents how much to burrow, but if it experiences a warmer, or cooler, climate than its parents' generation, then this information is of little or no value. If, conversely, temperature stays pretty constant over timescales of a few generations, then copying mom may make sense. Horizontal social learning—using short-term information from neighbors, whether kin or otherwise—is less affected by medium-scale environmental variation. But if things are changing very fast, no one can know what is going to be right, and no social learning helps.

We can formalize these kinds of differences in environmental variation using a concept borrowed from the physical environmental sciences, like meteorology: "1/*f* noise."[10] The *f* here represents frequency—how often, per unit time, the environment changes. The concept of 1/*f* noise is a method of mathematically describing a range of ways in which an environment can vary. We introduced it in chapter 3 when describing variability in the ocean, but at this juncture we are going to be a little more specific. At one extreme is white noise, under which the environment is totally unpredictable: what happens now has no bearing on what happens later, and all frequencies of variation are present in equal amounts. White light is white noise—hence the name. So is radio static (in the audio realm). But add a little predictability—so that what happens today is a bit more like what happened yesterday than what happened last week—and we have pink noise. Moving farther along the spectrum, and it really is "the spectrum" if we are talking about light, there is red noise. In red environments, the majority of the variation is over long timescales so the world stays pretty constant over short scales but can change a great deal over longer ones.[11] We can quantify this axis using a parameter (technically the slope of the inverse of the power frequency spectrum) usually called ω (Greek omega), so $\omega = 0$ is white noise, $\omega = 1$ is pink, and $\omega = 2$ is fire-engine red.

Social learning flourishes in a red-noise environment, where there is substantial change, but change that is sufficiently slow that learning from others helps—an environment where ω is greater than about one.[12] Hence, from the point of view of a species wondering, evolutionarily speaking, how to control its behavior, we can compare three types of environment: low noise, where normal variation in the environment doesn't make much difference to fitness; white-noise environments that change fitness but in ways that are almost completely unpredictable; and red-noise environments, where there is change but it happens relatively slowly. In low-noise environments animals can generally rely on their genes to give them sensible behavior; with white noise, individual learning or some other kind of phenotypic plasticity is adaptive; but in a red environment, social learning may become the most efficient way to get useful behavior. Therefore, these red environments are where we would most expect to find animal culture.

This theory is all very well, and rather nice since the effect of environmental variation on fitness (substantial or not) and its redness (is ω greater than one?) predict evolved learning strategies in a simple way. But the key predictor, environment, is not specified beyond being some one-dimensional factor that could affect fitness. What is this environment for a real species? To define "environment" we need to look at the question the other way around. What factors outside the animals themselves affect their fitness? These factors are all elements of the environment. They form the dimensions of its ecological niche. When looking for culture, we seek species where there is a factor whose natural variation has a large effect on fitness, and variation in this factor is red.

Temperature is a natural candidate, perhaps the premier candidate, for a key factor of the environment for most animals.[13] Animals of any particular species do well at a certain range of temperatures and poorly outside this range. Many also have genetically controlled mechanisms to deal with predictable changes in the environment, such as hibernation. It is the variation around these predictable norms that can be the problem. For very short-lived animals, such as some insects and zooplankton, temperature is likely to be pretty stable during their brief lives, and they don't need costly phenotypic plasticity to deal with it. But for many other species whose lives span weeks or more, temperature can be a killer, literally. Phenotypic plasticity or some kind of learning—such as burrowing or heading into the shade as the temperature changes—makes sense. What about social learning? Given humanity's preoccupation with the weather—temperature is vital for us too!—it is not surprising that there have been lots of studies of the structure of temperature variation over all kinds of timescales.[14] Over the shortest scales, temperature is pretty red with ω at about 1.5. However, over longer timescales, this changes: ω decreases to about 0.5, in the light pink range and well below the threshold for social learning. So if temperature is the key source of environmental variation, we might expect culture only among short-lived creatures. However, for these fly-by-night animals the unpredictable part of the temperature variation is often too small to select for phenotypic plasticity in the first place, and, as we will discuss in the next section, there may be other constraints on the evolution of social learning and the capacity for culture that militate against culture in the (usually small) short-lived species. Thus temperature variation does not seem to be a good candidate for the proximate driver of the evolution of cultural capacity.

All animals depend on other species for their sustenance. So the plants or animals that one eats are a crucial element of the environment, and variability in the abundance of those plants or animals is a potential problem. It is obvious that some resources vary enormously and unpredictably, for example, the abundance of flying insects. This variation is itself driven by weather, often temperature. So are we back where we started? Not quite, because at each step up the food chain variation tends to grow redder, and therefore ω generally increases as we move from variation in temperature, to the abundance of flying insects, to that of insectivorous birds, to hawks that prey on the birds.[15] In addition, larger species have generally redder population trajectories, as do mammals when compared with other vertebrates.[16] So high-level predators on big mammalian species may live in particularly red environments, in terms of the availability of their prey, with ω much greater than one and, consequently, in conditions that generally favor social learning.

What else makes for a red environment? We first became interested in the color of environmental noise on reading a paper by the oceanographer John Steele who, in 1985, compared the fundamentals of terrestrial and marine ecological systems.[17] He showed that the ocean is naturally a much redder environment, in terms of temperature and other physical properties, than the land. For instance, we noted that, for temperature, ω shifts from about 1.5 (pretty red) at short timescales to 0.5 (pale pink) for longer periods. This transition happens at scales of about one month on land but a year in the ocean.[18] So over scales of a few months the contrast between the two environments is particularly telling: over seasonal scales, terrestrial environments are reasonably predictable, marine environments much less so. Steele argues, then, that marine species should have generally redder population trajectories, with higher ω, than those on land.[19] Ecologists Pablo Inchausti and John Halley have since examined the redness of the population trajectories of 544 populations of 123 species.[20] They found that, in general, aquatic organisms had less reddened population spectra, contradicting Steele's hypothesis. However, in their database, the five aquatic species that are secondary carnivores—and thus on at least the third trophic level—had much the reddest populations, with ω about 1.5, well above those of any other group of species, and up in the region where social learning is helpful.

We have touched on the structure of the physical environment and the biological environment of species. But for some animals, variation in social environment may be at least as important. Think of all the unanticipated social situations you may have to deal with: unexpected visitors, intensely rebellious teenagers, sudden deaths of important family members, infidelity, and so on. The culture that you have amassed from your parents, friends, and the media is vital in getting through these difficult times. A female sperm whale living in her highly communal social unit may have to deal with some of the same problems. Infidelity is probably not a concern for whales, but visitors, teenagers, and sudden death in the family may all be challenges.[21] What she learns from her unit members, and their unit-specific or clan-specific behavioral norms, may be essential. So social learning can be vital when confronted by variation in the social environment, and this will particularly be the case for animals whose social life is intense and complex, animals like humans, elephants, bottlenose dolphins, and sperm and killer whales.

These ideas have underpinned some of the speculations about how cultural capacity evolved. For instance, Peter Richerson and Robert Boyd have suggested that humans got culture in a big way, as well as much larger brains, in response to the challenge of an extremely variable climate during the late Pleistocene, between about eight hundred thousand years and eleven thousand years ago.[22] We will return to the possible relationship between culture and large brains in the next chapter, but for now we want to draw a pretty obvious conclusion from all this. If social learning is favored in red environments, and the reddest environments are those made up of populations of animals that are large, feed at high trophic levels, live in the ocean, or are mammals, then social learning, and culture, should be especially favored among the predators of such animals, especially when social life is also important and unpredictable. So other marine mammals, and especially killer whales, become the prime candidates for culture.

Prerequisites for Culture

But, wait a moment . . . where is the supremely cultural great white shark? They are at the top of the marine food chain, sometimes eating large marine mammals. Their physical and biological environments are very similar to those of killer whales and will have similar patterns of variability. Sharks may have some cultural behavior, but it is not apparent, and almost certainly it does not match that of the cetaceans or some land mammals. To evolve cultural capacity, a species clearly needs more than a suitably red environment.

First, the animal must be able to learn. But if it is already individually learning—and sharks can learn—then making the jump to social learning would not appear to be particularly hard.[23] In many circumstances it would seem easier to copy someone else rather than figure it out for yourself. In the evolutionary scenario sketched in the previous section, this is one of the advantages of social learning.[24] However, to use lots of social learning effectively likely takes sophisticated cognition.[25] When should I learn socially? Who is the best model? Who should I learn which behavior from? How should I combine what I experience from the different models and what I learn individually to give me the most effective behavior? These are not easy tasks, even for us humans. This cognitive constraint gives us an explanation for the apparent commonness of a little bit of culture—the many species that use culture for one or a few bits of their lives—but the rarity of sophisticated cultures. Shaping one's song so that it closely resembles the songs of other birds in the audio neighborhood is not cognitively trivial, but it can be broken down into a series of fairly straightforward steps. In contrast, one needs pretty good high-order reasoning to use social learning effectively if there are many potential interacting behavioral types and models.[26] Think about the task of assimilating from other members of one's group all the different aspects and possibilities of building a yurt or the difficult and dangerous but profitable business of stranding on a beach to catch seals.

Second, the social learner needs a model and therefore must be around other animals. Some of these may not be suitable models, so a set of strong social relationships will give better learning opportunities—one will know who is a good, or bad, source of a particular kind of information—and the form of social structure will affect the nature of cultures, with more social structuring generally leading to more cultural structure.[27] If the behavior is complex, such as the beach stranding of killer whales, the exposure to the model or models may need to be long, so long-term relationships are also beneficial.

Frans de Waal argues that it is not just about being around other animals but also that social learning is deeply meshed with social structure.[28] Social learning is driven into high gear by both conformism and a motivation to act like others, either a particular "other" with whom the individual has, or wishes to have, a strong bond or a general "others"—the members of its community. De Waal calls this bonding- and identification-based observational learning (BIOL), "a form of learning born out of the desire to belong and fit in."[29] With this kind of social learning, the drive to copy the model exactly and to conform is very strong. When animals are in its thrall, they will learn socially in a wide range of areas, not just those where social learning is obviously adaptive. So they may develop common foraging methods, dialects, and movements, and they may play the same games. All of this is culture, and all driven by strong social relationships with each other and identification with the group itself. Importantly, and in contrast to other theories of social learning, this cultural homogenization of behavior through BIOL may not have any nonsocial benefits for the animals.[30] De Waal gives the example of Nose, the matriarch of a family group of rhesus monkeys, part of a captive colony in the United States.[31] Nose developed an unusual method of drinking water, in which the whole arm was soaked and then licked. Her family adopted it, but none of the members of the other families living in the same enclosure did. Nose's way of drinking had no particular benefit—there was plenty of easily drinkable water for everyone—but it was the way her family did things. We see strong parallels in some of the culture of the matrilineal whales discussed in chapter 6—pushing dead salmon around may have been mainly about bonding and identity.

Related to this is the feature that we introduced at the end of the previous section: variability in social behavior driving culture. A family group of wild elephants, or killer whales, meets all kinds of other family groups, some of which are close associates, some of which are very foreign. Dealing effectively with all this social intricacy is not easy. Karen McComb and her colleagues showed that elephant groups navigate this social uncertainty much more effectively when they have an older matriarch in their group, presumably because she contributes her substantial knowledge and leadership to making the right social decisions.[32] This elephant example is a clear signal that in complex societies strong cultures can be vital.

So there seems to be a strong feedback system between sociality and cultural complexity. Complex and important social systems—lots of unpredictable but important variability in the relationships between an animal and who it encounters—drive complex cultures that allow animals to navigate the social labyrinth. And complex cultures need strong social relationships between models and learners.

Now back to the great white shark. While sharks live in the same red environment as killer whales, and do have social lives, their societies are undoubtedly much simpler than those of killer whales.[33] We can also reasonably suppose that variation in the social environment is less complex and that this variation has less impact on fitness than social factors do for killer whales. Fundamentally this is because sharks do not have the prolonged mother-offspring relationship that forms the bedrock of most mammalian social systems, whether simple or complex. There may also be cognitive limitations on the sophistication of their social learning, perhaps partly, as discussed in chapter 3, because without a consistently high body temperature, the evolution of shark brains is constrained in ways that mammal brains are not.

The Evolution of Cultural Capacity in Marine Mammals

The marine mammals came into the oceans with warm bodies and high metabolisms that could maintain large and complex brains, as well as prolonged, suckling-based mother-infant bonds. There, as we have described in chapter 3, they found a habitat in which there were no refuges, where resources and territory could rarely be defended economically and are generally not worth squabbling over. These are ideal conditions for the development of complex and important, as well as largely cooperative, social systems. On entering the ocean, the first marine mammals also met the red environments in which social learning is so advantageous. So culture became important to them. This, in outline, is the way we see the evolution of cetacean capacity for culture. In this final section of this chapter, we will fill out the scenario, mostly, we admit, through speculation. But first let us consider the other marine mammals.

Ancestors of seals and sea lions, of manatees and dugongs, also entered the reddish-pinkish ocean with nearly all the mammalian benefits and baggage of cetaceans, yet there is little discussion of seal or manatee culture. Why?

First the sirenians, the manatees and dugongs. They are social, and this sociality seems, as with the cetaceans, to be based on the mother-calf bond, which lasts for one to two years.[34] However, Paul Anderson, perhaps the first scientist to really think about sirenian sociality, describes their social behavior as "apparently simple," and their physical and biological environments also seem simpler, as well as less red, than those of the cetaceans.[35] They eat sea grasses in shallow waters. The sea grasses are a relatively predictable and stable resource in comparison to the pelagic fish and squid that are the staples of most cetaceans—so their source of sustenance is not particularly red in its pattern of variation. There are parts of their environment that are unpredictable, such as dugong-eating sharks and manatee-debilitating cold snaps. They may learn from each other, and probably especially from their mothers, ways to combat these threats, for instance, by swimming to warm power plant outflows in cold weather.[36] So culture may have an important role in their lives. The calls of manatees resemble those of their mothers, but it is not clear whether this is because of shared genes or social learning.[37] Because the sirenians do not possess the tight matrilineal groups of the larger whales, their cultures, if they exist, are going to be hard for us to penetrate.

While the sea grass beds that sustain the sea cows do not have particularly red variation, many of the pinnipeds—seals, sea lions, and walruses—make their living in the deep or deepish ocean, where resources and the environment are inherently unpredictable over the long term. Often they feed alongside cetaceans. But in successfully making their way from the land to the ocean, the pinnipeds ditched one of the in-built mammalian traits that the cetaceans and sirenians kept, and it was a vital trait. Pinnipeds have their pups on land or ice, nurse them there, and then leave them there. With one exception, there seems to be no substantive postweaning mother-offspring relationship in the pinnipeds. This severely limits the range of their social systems. Like sharks, they provide excellent internal care for their pups and go a step further by providing nutrition for the first part of the pup's life—but only until the pups become free swimming. For seals and sea lions, as well as for sharks, there is no prolonged period of following mother about, seeing what she gets up to, and using these observations and experiences as a template for the young animal's behavior. Without the opportunity to learn about their physical, biological, and social environments from their mothers, young pinnipeds must rely largely on instinct and individual learning. Social learning is still a possibility, but it will be much more horizontal—among peers—and ad hoc than in structured matrilineally oriented social worlds, like those of some cetaceans. We mentioned an exception: it is the walrus. Unlike other pinnipeds, walrus pups follow their mothers in the ocean, staying with them almost continuously for the two or more years of nursing and then remaining associated for years afterward.[38] We know almost nothing about the nature or extent of walrus culture—there has been very little study of the behavior of wild walruses—but they do make complex, stereotyped songs.[39]

It is perhaps telling that, outside the cetaceans, the marine mammal for which we have the best evidence for social learning is the sea otter. We know a lot about sea otter foraging because they bring each item to the surface to process, where an observer with a telescope can make an exact tally of who ate what. In contrast to the sea cows munching away on sea grasses, sea otters eat a wide range of foods—abalones, sea urchins, crabs, lobsters, and snails, among other things—some of which require quite complex processing. Unlike the pinnipeds, sea otter mothers and pups have important postweaning relationships, in which the pup learns from the mother one or more of these foraging techniques.[40] Thus mothers and their pups share distinctive ways of making a living because of social learning. Whether this is culture is debatable; if the only important social learning going on is between mothers and pups, and unrelated otters with similar foraging strategies do not preferentially associate, then sea otter foraging does not fulfill the second part of our definition of culture, it being a communal activity. The sea otter is an important example. It demonstrates the significance of the mother-offspring relationship for social learning but also indicates that, to have an important culture, a population needs other, nonmaternal, social bonds.[41]

Whales and dolphins, though, are generally communal, and they learn socially, resulting in culture. This is a sketch of how cetacean culture might have happened. It is a sketchy sketch. We don't really know how humans "got" culture, and when studying humans we have several considerable advantages. We know our bodies well, our current cultures well, and most of the human cultural explosion happened quite recently, within a few hundred thousand years. Because an important part of human culture is material and some of the materials we use are durable, there are physical relics of past human cultures. Stone axes have told us a lot about the development of human culture.[42] From its appearance about one and a half to half a million years before present, one form of the ax, the Acheulean, stayed remarkably consistent in size and shape over about a million years of human prehistory and over large geographical areas. This can be interpreted as humans of that period not having important evolving cultures. Then, about 350,000 years ago, axes started to become more complex and diversify dramatically. It seems, as we discussed in the previous chapter, that human culture had now got its head.

Cetaceans don't make hand axes, or anything else that endures, so how can we chart the development of cetacean culture? Well, one of the pieces of evidence that is used in deliberations about the origins of human culture is the development of our brains. Large brains are a feature of the most culturally sophisticated species, and the human brain increased substantially in size just about when the axes grew complex and diverse.[43] Whether this correlation results from complex brains being a prerequisite for complex cultures or a consequence of complex cultures is, as we discuss in a few places in this book, unclear. The two are strongly correlated though, so it seems reasonable to use the development of the cetacean brain as a tracer for the development of cetacean culture. Brain size is one area where we do have paleontological evidence.

The neuroanatomist Lori Marino and her colleagues have used CT (computed tomography) scans to look into the skulls of fossil whales and measure their brains.[44] They found that cetacean brain size, relative to body size, increased substantially about thirty-eight million years ago when the odontocetes evolved from the ancient archaeocetes.[45] Overall, the animals became considerably smaller, while their brains grew larger.[46] They also changed in structure. Marino discovered from her CT scans that, with the emergence of the odontocetes, "cerebral hemispheres started to get more bulbous and take on the shape of the modern dolphin brain (though not nearly as expanded) and the olfactory bulbs started to recede." She suspects that there were also increases in the convolutions of the cerebral cortex, although these don't get preserved in fossils, and concludes: "So not only was the brain getting bigger at that time but it was starting to change shape in a way that suggests that the odontocetes were becoming more 'cerebral.'"[47]

What drove these changes? It does not seem to have been the transition to an aquatic existence itself as that occurred about fifty-five million years ago and brains stayed at roughly the same relatively small size relative to body weight as the archaeocetes made their gradual entry into the ocean. A better hypothesis is that the increased brain size of the odontocetes thirty-eight million years ago was driven by the evolution of echolocation.[48] The early odontocetes had inner ear bones that were good at picking up high frequency sound, which suggests that they had developed a form of sonar. Lori Marino thinks "that echolocation came on line and then got co-opted for social communicative purposes."[49] In this scenario, the odontocete brains increased in relative size to deal with the acoustic information itself, as well as, perhaps, a new perceptual system based on the data from the returning echoes.[50] But, as Marino and her colleagues—who included both of us—pointed out, the change may have been even more profound: "This may indicate that the large brains of early odontocetes were used, at least partly, for processing this entirely new sensory mode [echolocation] that evolved at the same time as these anatomical changes and perhaps for integrating this new mode into an increasingly complex behavioral ecological system."[51]

We suggest that this new "behavioral ecological system" was key to the evolution of complex cetacean cultures and that it might have happened something like this. As far as we can tell, the early archaeocetes made their living a little like crocodiles, sitting, waiting, sneaking up on prey that came close, and grabbing.[52] This mostly took place in the shallow waters close to shore. Later archaeocetes were more active, perhaps hunting more like pinnipeds, individually chasing prey, but still usually in shallowish waters.[53] With echolocation, a whole new world opened up. Potential prey could be sensed much farther away. They could be tracked precisely and imaged at range. Now, according to our scenario, it made sense, in terms of maximizing the difference between energy gained from prey and the energy spent getting it, to search actively for prey, to move into deeper waters where food comes in larger quantities but takes more finding. With lots of fast travel, perhaps it paid to be smaller. At this point the benefits of sociality really kick in. Social companions can help in scouring the deeper ocean for large balls of prey, as they do for some of today's dolphins.[54] Also, with relatively small size, avoiding predation becomes more of an issue. There were sharks. Initially those big, mean archaeocetes were still lurking. Later on, new predators arose from their own odontocete line.[55] And, as we noted in previous chapters, sociality can be an effective way to combat predation for a smallish cetacean.[56] These new ecologies, movement patterns, and societies were the new "increasingly complex behavioral ecology" posited by Marino and colleagues. The brains of these odontocetes were different from their ancestors not only because they had their sonar systems to manage but also because a whole new ecological and social world had opened up.

Thus, from about thirty-five million years ago, we have these early odontocetes zipping around the ocean with their echolocation, becoming more and more social. They are depending on each other to find and capture food, as well as to identify and respond to predators. In such a social environment, social learning makes sense. They exchange information with other members of their group. Perhaps this is when the cetacean facility for vocal learning, absent in most other mammals, evolves.[57]

Perhaps this also the start of what has been called Vygotskian intelligence.[58] The Soviet psychologist Lev Vygotsky argued that the human child develops a cognition that is shaped by its cooperation with other humans and that cooperation is embedded in, and includes, the culture of the child's community.[59] Children, as they develop, begin to share with each other, as well as with adults, a focus on the same things. A shared focus drives cooperation and collaboration. This cooperative perspective on the evolution of social intelligence contrasts with the more standard Machiavellian intelligence hypothesis, in which intelligence evolves to deal with the complexities of social competition.[60] Henrike Moll and Michael Tomasello use data on the development of sociocognitive skills in humans and other great apes to suggest that while the Machiavellian intelligence of competition drives the development of intelligence in other primates, the Vygotskian intelligence of cooperation is paramount for humans.[61] The nature of the ocean, with diffuse resources in three dimensions, makes competition less structured. Animals still compete but not often in the direct contests for this or that in which Machiavellian intelligence has its role; instead the competition tends to be a scramble for what is available, and during scrambles the manipulation of others is less effective. The structure of the ocean, without refuges but with predators and unpredictable resources, can put a big premium on cooperation. Hence we hypothesize that about thirty-five million years ago, as sonar and sociality developed in the early odontocetes, a Vygotskian intelligence also evolved to make best use of the knowledge and assistance of community members. If substantial culture *causes* the evolution of large brains, there is the potential for feedback between culture and evolution, as we discuss further in the next chapter. Complex culture places a premium on traits such as imitation and Vygotskian intelligence, which are selected for in evolution, and these in turn lead to more sophisticated cultures. So we speculate that the cetaceans have had substantial culture for more than thirty million years.

Odontocete brain sizes increased again roughly fifteen million years ago with the emergence of the Delphinoidae, the ancestors of modern dolphins, porpoises, killer whales, belugas, and so on.[62] This may signal a second step up in social and cultural complexity. However, one of our quintessentially cultural cetaceans, the sperm whale, was on the wrong side of this bifurcation. And the data on the evolution of brain and body sizes amassed by Lori Marino and her colleagues suggests that the evolution of the Delphinoidae did not represent a quantum change in brains and behavioral ecology. Instead the data suggest a gradual general increase in brain size since the emergence of the odontocetes thirty-eight million years ago, with the Delphinoidae containing most of the relatively largest brained species over the last fifteen million years (see fig. 3.1).[63] In this interpretation, the emergence of the Delphinoidae was not marked by a leap in behavioral, social, cognitive, and cultural complexity like the earlier appearance of the odontocetes. Rather, these dolphin-like creatures were best able to use most of the niches where large brains were useful—perhaps including the chasing-around lifestyle—and relegated the other odontocetes to the fringes, where brains were generally less important. These fringes are where we find the non-Delphinoidae odontocetes today: the river dolphins in freshwaters, the beaked whales in the deep subsea canyons, . . . and the sperm whale.

Something very special happened to the sperm whale. We think it was the evolution of the spermaceti organ.[64] This extraordinary nose—a spectacular sonar system—allows the sperm whale an unprecedented and unequalled perspective of its world, giving it dominance in the strange deep waters. In the struggle for the squid resources of the deep, sperm whales were mainly competing with other sperm whales. When competing against one's own kind, the general rule is to be careful, slow, and steady. Reproductive rates drop as parents put increasing amounts of effort into each offspring as well as into protecting themselves.[65] Animals survive better and live longer. Species on this slow, careful, life history track tend to become more social, building up important relationships to help them survive and reproduce. For sperm whales there is a particular benefit for sociality: babysitting. Calves have to be left at the surface while their mothers feed at depth, so having permanent companions around as babysitters, who are often kin, is a huge advantage. The food-finding efficiency of the spermaceti organ allowed sperm whales to get big. And size improves safety, both because the number of potential predators decreases and because large animals can better survive poor conditions by living from their body stores. So they live even longer, put even more care into their offspring, and become even more social. Within this network of relationships, they learn. This social learning is the essence of their cultures. The social relationships and culture promote the development of large brains, which the sperm whales can easily house within their large bodies, and the brains help longevity, sociality, ecological success and culture. This, then, is our proffered positive-feedback scenario for the evolution of the weird and wonderful sperm whale.

How do the mysticetes fit into this? We devoted a whole chapter, chapter 4, to the culture of the baleen whales. There has been no systematic analysis of the evolution of mysticete brain size, so we are deeper in the dark than with the odontocetes. While the mysticetes have large and complex brains, in relative terms their brains are smaller than those of most odontocetes.[66] Even in absolute terms, the biggest mysticete brains—belonging to the blue and fin whales—while larger than almost all other brains on Earth, are smaller than those of relatively petite sperm whales. It seems that mysticetes and odontocetes evolved separately from small-brained archaeocete ancestors, basilosaurs, so the evolution of the large brains of the baleen whales seems to have been independent from the sonar-driven odontocete trajectory.[67] We suppose that with the evolutionary development of baleen and its associated structures—such as the jaws to hold it in place—the early mysticetes were able to make a new and better living. They gradually moved into deeper water where the schools of prey are generally larger and denser, better for bulk feeders. They did well. As in the sperm whale scenario, efficient foraging allowed the baleen whales to grow large bodies that could hold large brains. Their lives slowed down as they became dominant, and sociality became more important. Sometimes they worked with each other to find and catch food and, maybe, to escape predators. Fred Sharpe believes that the cooperative bubble netting of the humpback whales that he watches in southeast Alaska "provides a beautiful example of Vygotskian intelligence."[68] In this cooperative society, they learned from each other. Culture took hold. But, in an important contrast to the sperms, the baleen whales did not need to dive very deep or for very long to find their prey, so babysitting was not a big issue, and the females did not experience the pressure to form large, permanent groups. This made for a quite different form of culture: the large, population-wide cultures of chapter 4's baleen whales, as compared to the tight, group-specific cultures of chapter 6's matrilineal whales.

Thus, we think that cetaceans have been cultural animals for a very long time: until the emergence of modern humans in the last few hundreds of thousands of years, an evolutionary blink, the most significant cultures on Earth were in the ocean. But this does not mean that cetacean cultures were immutable. We have seen how within a decade humpback whale foraging culture changed and how over a few weeks killer whales took up pushing dead salmon around, then gave it up. However, have the general types and levels of cetacean culture stayed constant for over twenty million years? We suspect new structures evolved, perhaps with the delphinoid origins about fifteen million years ago and with the emergence of tight, matrilineal groups of the larger toothed whales, whenever that was. Another driver of cetacean culture that we have alluded to in chapter 6 is the arrival of killer whales and their tremendous predatory power ten million years ago. Culture may have been the vehicle for defensive strategies in those species that survived the new threat.

That is our sketch of the evolution of cetacean culture. It is, we readily admit, rather intensely speculative. Some details will be wrong. The scenario may be wrong in more fundamental ways. Perhaps cetacean culture is much less ancient than we suppose. As science gallops on, we will learn.

10 | WHALE CULTURE AND WHALE GENES

Culture changes everything. When significant amounts of information begin flowing within species independently of their DNA, evolutionary processes can be affected in profound ways. As described in chapter 2, we can see this written in the history of our own genome. Now that we've explored cetacean culture in some detail, we are going to consider whether and how it might have similarly affected their evolutionary history. In this chapter we will address the controversial possibility that culture can drive genetic evolution in the whales and dolphins.

Evolutionary Transitions: Gene-Culture Coevolution

In 1995, John Maynard Smith and Eörs Szathmáry wrote *The Major Transitions in Evolution*, a book that was as groundbreaking as its title suggests.[1] The two evolutionary biologists describe eight transitions in which the complexity of life increased dramatically and how at each of these transitions a slew of new evolutionary processes came into play. These transitions introduced novel methods of storing and/or transferring information, for instance, from replicating molecules to something like RNA—transition number 1. In conjunction with many of the transitions, entities that were previously autonomous started working together, as in transition number 4, the shift from prokaryotes to eukaryotes: eukaryote cells contain organelles with their own genetic material, which is distinct from that of the organism as a whole. The transitions also introduce specialization, as when, following transition number 3, information was stored by DNA and bodies were built by proteins—two jobs previously done, less efficiently, just by RNA. Another example of specialization was the introduction of males and females after transition number 5, the evolution of sex. The final transition is from populations of social animals to populations of social animals with culture, "sociocultural evolution." Maynard Smith and Szathmáry characterize this eighth transition as from "primate societies" to "human societies, and the origin of language."[2]

After this transition, information is primarily transmitted culturally, rather than genetically. Culturally marked groups of individuals, which may be genetically unrelated, cooperate with one another and may compete against other groups. And there may be division of labor, as segments of societies hold, and use, particular sections of cultural information.

Following the eighth transition, evolution works on the two major forms of inheritance, genes and culture, as well as on their interactions, gene-culture coevolution. In the words of Peter Richerson and Robert Boyd: "The symbiosis between genes and culture in the human species has led to an analogous major transition in the history of life—the evolution of complex cooperative human societies that radically transformed almost all the world's habitats over the last ten thousand years."[3] So the results of this final major evolutionary transition, as seen in modern humans, are both spectacular and unpredictable.

What brought about this eighth major transition of life on Earth? This is one of the great puzzles, one that has challenged scholars from many disciplines. The archeological data that we referred to in previous chapters suggest that it happened between 350,000 years ago when we got out of the Acheulean hand-ax rut and 50,000 years ago when we blossomed technologically and in other ways.[4] During that period, we became "modern" humans, got our large brains, moved into new habitats, tamed fire, and developed symbolic culture. But why did all this happen? Peter Richerson once wrote to Hal: "As far as I can see, if you gave a smart, knowledgeable undergrad a beer for every coherent scenario consistent with the data as we know it, the kid would be quite pissed [as in drunk, rather than angry] before the stories stopped making sense."[5]

Maynard Smith and Szathmáry, along with a range of other scholars, believe that language was key.[6] Hence the title of their eighth transition includes "the origin of language." The development of a referential language by modern humans allowed culture to be transmitted much more efficiently, and in much more complex ways, than was possible by prehuman hominids or other species whose cultures were therefore necessarily constrained. Language also led to new ways of dealing with other humans and with the natural world.

If language is key to the eighth major evolutionary transition, then, as Maynard Smith and Szathmáry describe, no other species, including the whales and dolphins, have made it. However, as we alluded to in chapter 2, other schools of thought do not conclude that language is the central key to human culture. For instance, the psychologist Merlin Donald argues for three major cognitive transitions that led to modern humans: mimetic skill (i.e., advanced social learning), language, and external symbols.[7] In this scheme, language is an enormously significant part of the human cognitive process but not the trigger for it. Among the arguments against the central role of language in the explosion of human culture are that deaf-mutes manage pretty well in many nonlinguistic areas of human culture, that we ourselves learn many things better by observation than by being told, and that human language itself may have primarily evolved to assist in the complexity of social relationships—who did what to whom—rather than to convey specific cultural knowledge.[8] Then there are those linguists who, as we explained in chapter 2, suggest that you need a significant history of cultural evolution to generate proper language, which, if correct, flips the causal arrow right over.[9] However, when culture becomes really complicated, in stories and laws, for instance, then language must at least be a major asset in transmitting it.

Humans, with more or less assistance from language, made that eighth transition into a radically new way of life, a sociocultural life. Have some whales or dolphins also made this transition? If so they are in an entirely different evolutionary space, a space where there are two primary modes by which information is transmitted, the genetic and the cultural, that may interact in ways that are both complex and unpredictable. Looking at gene-culture coevolution is difficult—very difficult. Until quite recently, there was only a little evidence of it in humans. The connection between dairy farming and adult tolerance of lactose that we mention in chapter 2 was often dragged out, but there did not seem to be much else.[10] However, in the last few years huge genomic databases have become available for cross-cultural studies. While the genomic study of the relationships between genes and culture is just beginning, the extent of human gene-culture coevolution that is emerging is remarkable: many human genes seem to have evolved in response to cultural pressures, especially from the introduction of agriculture, changing agricultural methods, or the increased rates of disease transmission among the denser populations, and closer relationships with animals, that agriculture made possible.[11] Because changes in agriculture through time and space are better mapped than most other human cultural processes, gene-culture coevolution is most clearly recognized here. For instance, human populations with a high proportion of starch in their diet have more copies of the salivary amylase gene, a gene that likely improves the digestion of starchy foods.[12] Other nonagricultural influences such as changing culturally driven social environments are also likely to have driven human genetic evolution. Anthropologists Gregory Cochrane and Henry Harpending make the case for cultural practices driving human genetic evolution in their 2009 book *The 10,000 Year Explosion: How Civilization Accelerated Human Evolution*.[13] As an example, they suggest that there was selection for those genes that through their effects on the workings of the brain improved business skills among the Ashkenazi Jews of medieval Europe. As the Ashkenazi population was quite small and insular, the diversity of genes affecting brain development was reduced compared to that in the wider European population. This left the descendants of the Jewish bankers and traders more vulnerable to a range of neurological disorders.

And the whales? Uncovering the extent of gene-culture coevolution will be slow and harder than for humans. First, as we argued in the previous chapter, most whale culture is likely old, manyfold more ancient than the modern human, so the genes affected by emerging cultural practices will usually have been scrambled over the millennia, perhaps becoming universal. Second, we do not know whale culture anywhere nearly as well as we know human culture, and finally, there has been no comparable research seeking the statistical signatures of recent selection in whale genomics. As we write, there have been no large-scale examinations of functional genes in whales, those that actually do something useful. Almost all research on whale genetics has examined what are thought to be neutral parts of genomes, whose variation makes no practical, phenotypic difference to the animal itself. This is because those are the parts of the genome that accumulate mutations most rapidly, because they aren't selected out, and so give maximum resolution for detecting population structure, which is the aim of most whale genetic research. Nevertheless, there are two principal lines of argument that culture has directly influenced the genetic evolution of whales, and we believe there to be other traits of the whales and dolphins where culture has left its mark.

Killer Whale Radiations

We've described how killer whales in the eastern North Pacific and Antarctic, and probably elsewhere, exist in several different ecotypes that share their ranges but have completely different ways of making a living. For instance, in the waters of the eastern North Pacific, from at least California, through the waters off British Columbia, Alaska, and into the far east of Russia, resident killer whales eat salmon, while transients eat mammals. The ecotypes are now so different in a variety of ways that, some argue, they probably should be recognized at least as subspecies—and possibly as species.[14]

So how did this happen? Rüdiger Riesch and his colleagues—Lance Barrett-Lennard, Graeme Ellis, John Ford, and Volker Deecke, some of the most experienced killer whale biologists in the world—laid out a scenario in which cultural traditions have led to ecological speciation.[15] It is a scenario that makes sense to us. Ecological speciation occurs when members of a species start to use different environments or niches, then natural selection molds them for these environments, and eventually the attributes of the groups diverge sufficiently that individuals stop mating across groups, and we have two species. Riesch and his coauthors believe that the ancestors of the different ecotypes were groups of killer whales that specialized on different foods—for instance, seals or salmon—and passed these preferred diets on culturally from mother to daughter. The descendants of the groups became more distinct behaviorally and are now mostly isolated reproductively, at least in the female line.

They suggest four ways this might have happened. The first of these processes, which we suspect may have been most important, is the xenophobia that Lance Barrett-Lennard sees as one of the defining behavioral characteristics of the killer whale.[16] Socially learned characteristics of their own ecotype, such as the preferred diets themselves, greeting ceremonies, and especially dialects could form cultural badges of ecotype membership. Consequently, the killer whales can recognize who are members of their own ecotype and who are not. In this scenario, the significance of the badges becomes sufficiently strong that animals only mate with other animals possessing the same badge or set of badges. These mating boundaries could be reinforced by the distinctive ways the different ecotypes use a habitat, so that animals rarely meet potential mates of other ecotypes even though they are in the same general geographical area. For instance, the residents concentrate along the salmon migration routes, the transients around seal rookeries.

The second process inhibiting mating between ecotypes is the difficulty that immigrants would face in switching ecotypes. Having learned a particular way of making a living from their mother and maternal relatives, they would now need to assimilate into a community where things are done very differently. There is no record of any killer whale achieving this. In chapter 8 we met Keiko, the *Free Willy* killer whale, who failed to make social connections with other killer whales after he was released from captivity; he ultimately died alone.

The third factor restricting mating between ecotypes would take many generations to operate. Once the ecotypes had spent some time with little genetic connection through either mating or migration, they will, through genetic drift, have developed different genotypes—that is, sets of genes. These may be, at least partially, incompatible so that hybrids are either less viable or infertile. The small population sizes of the killer whale ecotypes—

just a few hundreds or thousands of animals in each—will increase the rate of genetic drift and, thus, the importance of this effect.

Riesch and colleagues' final isolating mechanism is gene-culture coevolution itself. An ecotype's culturally determined way of life may select for different genes, for instance, those that make stronger jaws for the eaters of large mammals—there is a little bit of evidence that the mammal-eating transient killer whales have more robust mouth parts than do the fish-eating residents—or, maybe, good vision out of water for the pack-ice killer whales who spy out their prey on top of ice floes.[17] Migrants from other ecotypes will tend to lack these genes and qualities, and hybrids would only partially possess them.

Through some mixture of these four mechanisms, killer whales in the wild do not usually, and perhaps do not ever, switch ecotypes or mate with members of other ecotypes. The ecotypes can, however, mate and produce hybrid offspring in captivity that are themselves fertile, so the genetic isolation is not complete.[18] Because this does not seem to happen in the wild, some ecotypes are genetically distinct, especially in the maternal line, and appear to be well on the way to becoming separate species.[19] In the Pacific, the mammal- and fish-eating killer whales are some of the most strongly divergent ecotypes, in terms of both behavior and genetics, and it appears that the genetic separation resulted partly from a historical period of geographic separation between the lineages. We do not know, however, whether the ecological separation into fish and mammal eaters began before the geographic separation or after it ended.[20] In the Atlantic, though, there also appear to be whales that concentrate on fish or mammals, and there may also be generalists.[21] Studies of the chemical traces of diet that get left behind in teeth and tissue samples suggest the separation between ecotypes may not be as strong as in the Pacific, so perhaps the process is in an earlier stage, although we do not have anything like the amount of behavioral observation data from the Atlantic. No matter the timing, Riesch and colleagues argue that it is social learning and culture that set the ecotypes off along separate roads and then kept them going.[22]

There is even a possibility that nonecological cultural traits could have started the process.[23] In this scenario the different groups, rather than being primarily defined by distinct diets, had other distinctive culturally determined behavior that precipitated the divergence. For instance, among the resident fish-eating killer whales off Vancouver Island, members of the northern resident *community* and the southern resident *community* both eat salmon, but they have community-specific greeting ceremonies and other behavior. The two communities have largely distinct ranges, so they do not often meet. But if differences in greeting ceremonies or other social behavior prevented mating when they did meet, then the communities may be en route for speciation: sociotypes rather than ecotypes.

Whether ecotypes or sociotypes, we think Riesch's scenario is probably right. Cultural evolution drove genetic divergence of killer whale ecotypes. Now, genetic and cultural evolution operate separately on each type, steering them on independent courses. Riesch and his colleagues consider the possibility that culturally separated ecotypes could get back together again.[24] There may be occasional interecotype matings, but they have been very rare. In general, it seems that the ecotypes stay separate.

This brings up a potential puzzle. If killer whales, with their characteristic pickiness, xenophobia, and culture, are destined to splinter into ecotypes that head off down the road of speciation, why aren't there lots, or at least a few, really obvious killer whale species out there? Killer whales have been around for about ten million years, but of all the living dolphin species—the thirty-eight or so species of the Delphinidae—*Orcinus orca*, the killer whale, is the only one without a living sister species joined by a common ancestor less than ten million years ago.[25] The eight currently identified ecotypes around at the moment seem to have diverged from one another over the past 150,000–700,000 years.[26] So again, why do killer whales with such an effective means of diversification—culturally driven ecological speciation—and ten million years to work with, not have lots of real and unambiguous species?

An important piece of the puzzle may be that the ecotypes are precarious. They are vulnerable to extirpation and local extinctions. With time they tend to become more and more specialized, making a good living in a distinctly particular way, and then an especially good living in an even more particular way. Techniques will have been honed through cultural evolution—the optimal way to tip a seal from an ice floe or to intercept the Chinook salmon migrations—as well as perhaps gene-culture coevolution, such as stronger jaws for mammal eaters or better above-water eyesight for the seal tippers. The ecotypes themselves will sometimes have split into even more specialized ecotypes. This increased specialization leaves the ecotypes vulnerable in two important ways. First, any ecotype that is dependent on a particular resource is vulnerable to the disappearance or diminution of that resource. This can happen through various processes, such as changes to the breeding habitat of Chinook salmon or the appearance of a new Chinook predator in the rivers, or through the efforts of the killer whales themselves. As we suggested in chapter 6, humans are probably not the first species to destroy their own resource base. With increased efficiency and specialization, this may have become more likely for killer whale ecotypes. A second factor is population size. Killer whales are top predators anyway, and the "trophic pyramid" forces top predators to have relatively small population sizes. Specialization diminishes population size even more. Small population size greatly increases the risk of extinction because of low genetic diversity and consequent vulnerability to disease, the increasing importance of random environmental events, and other effects.[27] The AT1 transient killer whales of Prince William Sound, Alaska, who were decimated by the Exxon Valdez oil spill, seem destined for extirpation.[28] So the ecotypes, as they became more specialized, also became increasingly vulnerable.

In this scenario, we see killer whale evolution as a slowly changing patchwork of ecotypes, splitting, evolving, specializing, and disappearing: a dynamic, vivid pattern of culturally driven diversification. The least specialized ecotypes, such as the eastern North Atlantic "generalists," may not usually be as successful as their highly honed neighbors, but they are more resilient—they have a wider range of diverse prey, larger populations, and generally less impact on a particular prey type.[29] They endure and, thus, likely provide the material for the next series of ecotype radiations. This unusual mode of evolution is driven by the importance of culture in the lives of killer whales. Killer whale evolution is bound up with killer whale culture.

Cultural Hitchhiking in Whales and Dolphins

Our second example of gene-culture coevolution in cetaceans starts with the observation that the matrilineal whales—the killer whale, sperm whale, and the blackfish species—have a genetic anomaly. The diversity of their mitochondrial genomes, which they inherit from their mothers, is remarkably low, about a tenth of that of other whale and dolphin species with similar populations.[30] So, if we take four hundred base pairs of the much-studied "control region" of the mitochondrial DNA, each of the matrilineal species has less than twenty haplotypes (i.e., different arrangements of this stretch of DNA), while other cetaceans such as the fin whale and common dolphin with similar geographic ranges and population sizes possess thirty to fifty haplotypes.[31]

When whale geneticists first discovered these unexpectedly low diversities in the late 1990s, they invoked the two usual explanations for low genetic diversity in any species. The first is the "genetic bottleneck."[32] When a population is very small (i.e., the bottleneck), there are only a few individuals carrying genes. Some of these will be related and so necessarily have many of the same genes. Thus, during the bottleneck, the total number of different genes in the population is few. Later, the population may increase, but the number of different genes, which all come from the same few individuals that lived through the bottleneck, stays low. Hence a bottleneck can reduce the genetic diversity of a population. However, there are two problems with invoking the genetic bottleneck as an explanation for the low diversity of the mitochondrial genes of sperm whales, killer whales, and their like. The first is that, for the bottleneck to produce the kind of effect seen in the matrilineal whales, the population must get incredibly small or the bottleneck must last an exceedingly long time—about a hundred animals for a hundred generations or a thousand animals for a thousand generations—and there is no nongenetic evidence that any of the matrilineal whale species was reduced to these levels by whaling, glaciations, or any other factor, whether natural or human-caused.[33] In fact, and this is the second problem with the bottleneck theory, the nonmatrilineal whale species, such as the fin and humpback, were brought much closer to bottleneck numbers by modern whaling than were the matrilineal species, and yet they still have higher mitochondrial diversities than the matrilineal whales.

The second of the "usual" explanations for low genetic diversity is selection. This argument posits that there is little variation in the genes being studied because those genes are important to the organism. If individuals have the "wrong" type of gene, they do not prosper, producing few offspring to carry on the maladaptive gene type, and thus that type gets eliminated from the population. Genetic variation in the control region of the mitochondrial DNA, which has such low diversity in the matrilineal whales, is generally concentrated in so-called hypervariable sites—mutational hot spots that are assumed to be functionally neutral, but maybe in this case they are not.[34] However, why is there this exception exclusively in the matrilineal whales? There is no good reason that we can think of.

A third potential explanation for the low mitochondrial genetic diversity of the matrilineal whales looks better because it invokes the matrilineal nature of their populations. This demographic explanation has been proposed in a number of forms, but the essence is that the expected diversity of neutral genes in a population goes up with population size: more individuals, more genetic variation.[35] In a matrilineal population, because all members of the same matriline both carry the same mitochondrial genes inherited from their mother and live their lives together, the population size from the genetic perspective is the number of matrilines, not the number of individuals. As there are many fewer matrilines than individuals in a matrilineal population, genetic diversity is lower than expected. This sounds plausible, but it turns out that for these demographic effects to reduce genetic diversity, individuals of the same matriline not only need to live together, they must also, to a large extent, give birth and/or die together. Although breeding events are a little correlated among killer whales of the same pod, and pilot whales of the same matriline can mass strand together, these occasional synchronicities in life processes are nowhere near strong enough to reduce genetic diversity to the extent we observe.[36]

This brings us to a fourth explanation for the low mitochondrial genetic diversity, which Hal proposed in 1998: cultural hitchhiking.[37] Hal called it cultural hitchhiking as it is analogous to genetic hitchhiking in which a gene that gives its holders a selective advantage can increase the incidence of neutral gene variants, if those neutral variants are linked on the chromosomes to the genes being selected for and, thus, transmitted together. As a result, genetic diversity in the neutral variants may be reduced.[38] As with the demographic explanation, it depends on matrilineal groups doing the same things, but rather than coordinated births or mass deaths, it just requires that these shared behaviors affect the life processes of individuals. In cultural hitchhiking, smart, as well as stupid, cultural behavior passing through the matrilineal societies drags neutral mitochondrial genes with it. To illustrate cultural hitchhiking, we'll use the case of sperm whale vocal clans in the Pacific that we described in chapter 6. Imagine a sperm whale clan that gets a good idea, good in the sense that it generally improves reproduction or survival. Perhaps it is a productive foraging technique—we saw in chapter 6 how the different sperm whale clans have characteristically different ways of foraging.[39] Perhaps it is a better method of communal defense when attacked by predators; sperm whales in the Pacific have been seen using two quite different defensive methods when assailed by killer whales, either facing outward toward the predators or bringing their heads together, radiating out like the spokes of a wheel, and defending themselves primarily with their flukes.[40] We do not know whether these different defensive formations have different effectiveness or are clan specific, but they might be. Such good ideas will lead to members of one clan averaging more surviving offspring than another, and again this is what our evidence shows: females of some clans produce consistently more offspring.[41] Then, if the good idea is transmitted through the generations of the female sperm whales—to young females from mother, grandmother, aunt, or more generally from the matrilineal social unit—and the consequential difference in reproductive success persists, then the clan with the good idea will increasingly outnumber the clan lacking it. As the mitochondrial genes in the "good-idea" clan are being passed along in parallel with the good idea, the genes, too, will spread. Eventually the sperm whale population will contain mostly genes that were originally in the clan with the good idea. And as clans have less diversity in mitochondrial genes than the population as a whole, the overall diversity of the population will have declined.[42] The mitochondrial genes have hitchhiked on the good idea because the genes and the ideas travel in the same vehicles—individuals—along the same road—through the maternal lineage, with the cultural variants driving. Similarly, bad ideas will reduce the frequency of the mitochondrial genes being passed along in parallel with them. This is how cultural hitchhiking may have reduced the genetic diversity of the matrilineal whales.

Hal's mathematical models have shown that cultural hitchhiking works—at least in computers. And to us it makes intuitive sense. Not to everyone though! This explanation for the low genetic diversity of the matrilineal whales has been criticized on a number of grounds. Much of the criticism has focused on whether the sociocultural systems of the whales are sufficiently stable.[43] Cultural hitchhiking takes many generations to reduce genetic diversity. What happens if females occasionally switch clans, moving genes across?[44] What happens if a neighboring clan copies the good idea, or gets a better one, or the original clan with the good idea—say, an efficient foraging technique—then takes up a bad idea, perhaps a suboptimal way of defending against killer whales? A second, and more extensive, set of computer models has shown that porous social barriers and multifaceted, evolving cultures do not halt cultural hitchhiking. Good—and bad—ideas may arise, animals move occasionally, and still, generally, the matrilineal linkage between culture and mitochondrial genes reduces genetic diversity.[45]

We think that cultural hitchhiking is the most plausible explanation for the low genetic diversity of the matrilineal whales—it fits with what we know of the lives and behavior of the animals better than any of the alternatives—but it is far from proven. Pinning down evolutionary histories from current patterns is not easy, and this is even harder with coevolutionary histories. The recent science of genomics, in which huge genetic samples give substantial power to genetic analyses, gives us ways to test the cultural hitchhiking hypothesis, and the first results are positive. Alana Alexander and her colleagues sequenced over ninety-three million base pairs of genetic data from the mitochondrial genomes of seventeen sperm whales. The sequences they found are not consistent with the hypothesis of selection on the control region of the sperm whales' mitochondrial genome, but they do support the bottleneck or cultural hitchhiking hypotheses.[46] Unfortunately, Hal's models suggest that genetic signatures indicating cultural hitchhiking, as opposed to bottlenecks, will be generally subtle, so going further may be hard, but genomics is extremely powerful—and becoming more so almost day by day.[47]

For now, we believe that the genetic diversities of the matrilineal whales show signs of the influence of culture. This is one of the first suggestions of gene-culture coevolution outside humans and is another indication of the importance of cultural processes in these animals.[48]

There has been a recent suggestion that genetic sequences have hitchhiked on vertically transmitted cultural traits in dolphins.[49] It involves our old friends, the spongers of Shark Bay. Geneticists studying the maternally transmitted mitochondrial DNA of the bottlenose dolphins in western Shark Bay found genetic structure over small scales of just a few kilometers. Two haplotypes (types E and F) predominated among the animals using waters greater than ten meters deep and those with a third haplotype (type H) were found almost always in shallow waters. The deep and shallow waters are right next to each other, and dolphins can easily travel between them. This genetic structuring is then quite remarkable. How did it happen? Anna Kopps and her colleagues suggest that it is because the dolphin foraging strategies, some of which work in deep water and others in the shallows, are being passed from mother to offspring in parallel with the mitochondrial DNA. For instance, the spongers, who use deep water, are all of haplotype type E in western Shark Bay. Once again the genetic structure of the population is being molded by—hitchhiking on—culture.

Coevolutionary Drive

Our thinking about killer whale speciation and cultural hitchhiking in the whales and dolphins is founded on actual genetic data, albeit for supposedly neutral genes. However, a lack of gene sequences has not inhibited us from speculating further. Generally, if culture is important for a species we would expect the evolution of characteristics that make best use of all that information. And as we look at the cetaceans, we see intriguing characteristics that may be acting as cultural enablers.

The size and structure of the social units of the matrilineal whales may assist the efficient transmission of ideas, on top of the more standard benefits of cooperative foraging and defense. The multiyear nursing periods of some cetacean youngsters may be about optimally transferring information from mother, as well as about milk. In both of these cases, there is a double hurdle to proving gene-culture coevolution: first, we would need to untangle the cultural benefits from the noncultural ones and, then, we would need to show that these extra benefits resulted from genetically caused changes in behavior, not just cultural ones. These are high hurdles.

A better case can be made for menopause. Menopause can be defined as females routinely living a substantial part of their lives after their last birthing. While many large female mammals, including elephants and chimpanzees, reduce their reproductive rates as they age, in none of them is there an age at which reproduction essentially ceases, with many females living on for decades following their last birth—the defining characteristic of menopause. It is often said that humans are unique in that females routinely live a large part of their lives postreproductively.[50] This is wrong. Females of two species of Cetacea—two matrilineal whales—have very similar reproductive profiles to human women. Short-finned pilot whales and killer whales first give birth at about age ten to twelve and cease reproduction in their early forties but can live into their eighties.[51] In the case of the short-finned pilot whales aged fortyish, as with menopausal women, there are physiological changes precluding conception.[52] Whether killer whale menopause is also physiological is not yet known. Some other whale species may also show menopause.[53] For instance, in the data set compiled by Peter Best and his colleagues from whaling off South Africa, the oldest pregnant sperm whale was forty-one years old while there were twenty-two other females aged between forty-two and sixty-one.[54] You may have noticed that all the known and candidate menopausal whales are matrilineal. Interesting! Baleen whale species mostly have comparable lifespans to the matrilineal odontocetes, but show no sign of menopause.

Menopause does not make obvious sense from an evolutionary perspective. Even though reproductive systems may not be as effective for an elderly female, surely she should persist with birthing even if her offspring may not be as viable or healthy? That way she at least has a chance to get some more genes into subsequent generations. Menopause seems like Darwinian abdication. Consequently, evolutionary biologists find menopause most intriguing and have come up with various potential explanations. One of them is that menopause is basically an artifact of the much improved longevity of modern humans: in the hunter-gatherer societies within which human lifestyles evolved, hardly any women lived past forty so it did not matter physiologically what happened after that age. This line of reasoning was suspect from the start in light of quite large numbers of postmenopausal women in technologically simple societies—and was consigned to the scrap heap when we learned of the routine presence in cetacean societies of menopausal females, who do not have the benefits of modern medicine or any other of the changes that have made our lives longer over the past few centuries. Much more convincing is the "grandmother hypothesis": an elderly female, by not spending effort trying to reproduce herself, has the time and energy to help raise her current children and her grandchildren.[55] By not subjecting their bodies to the rigors of pregnancy, these females will likely live longer, and so this assistance will persist. In Darwinian terms, as grandchildren possess some of the genes of their grandmothers, this assistance will be promoted through natural selection at the expense of dangerous and not-very-efficient reproduction at the end of life. The theoreticians Rufus Johnstone and Michael Cant have shown why menopause makes sense in the quite different—but both unusual—social systems of humans and matrilineal whales.[56] The key is that, in these societies, as females get older, they tend to become increasingly surrounded by close relatives. There is evidence among humans that the presence of grandmothers helps grandchildren.[57] Tellingly, there is also evidence that, in resident killer whales, a living but postreproductive mother significantly increases the survival rates of her offspring, especially in males and especially as they get older.[58] With her sons' survival comes more potential grandchildren for her, particularly as older males are more likely to become fathers.[59] This is mothering rather than grandmothering, but the logic is similar, and our current data sets aren't long enough to know whether the effects persist in the grandchildren's generation.

A key word in the grandmother hypothesis for menopause is "help." But how do the grannies help? In the standard version of the grandmother hypothesis, it is provisioning, helping the grandchildren get enough to eat.[60] However, other than lactation by some possibly menopausal females, there is not much evidence for provisioning by grandmothers in the matrilineal whales.[61] Human grandmothers, though, do more than provision, and in fact, they often do not do much provisioning at all. Jared Diamond tells of visiting Rennell Island in the Solomon Islands.[62] Around 1910, the island was devastated by a huge cyclone, the "hungi kengi." It destroyed almost all the gardens that normally provide food. The inhabitants were driven to the brink of starvation. The people survived by eating new wild sources of food, not part of their usual diet. To do this they needed to know which plants were poisonous, which were not, and how poisons could be removed. Where did this knowledge come from? Such devastating cyclones are very rare. Diamond writes:

> When I began pestering my middle-aged Rennellese informants with my questions about fruit edibility, I was brought into a hut. There, in the back of the hut, once my eyes had become accustomed to the dim light,

was the inevitable, frail, very old woman, unable to walk without support. She was the last living person with direct experience of the plants found safe and nutritious to eat after the hungi kengi, until people's gardens began producing again. The old woman explained to me that she had been a child not quite of marriageable age at the time of the hungi kengi. Since my visit to Rennell was in 1976, and since the cyclone had struck sixty-six years before, around 1910, the woman was probably in her early eighties. Her survival after the 1910 cyclone had depended on information remembered by aged survivors of the last big cyclone before the hungi kengi. Now, the ability of her people to survive another cyclone would depend on her own memories, which fortunately were very detailed.[63]

For the Rennellese, as well as for the matrilineal whales, who live with the possibility of major, unpredictable long-term changes to their environment, the information held by elders can be vital. Like the elderly women described by Diamond, the older female killer and sperm whales may hold knowledge of how to deal with conditions, rare conditions—perhaps like the "super El Niño" that devastated the eastern Pacific in 1982–83—which the younger members of their societies have not experienced.[64] So menopause, by prolonging female lifespan, increases the probability that a group holds that important information. The young Rennellese faced with hungi kengi, or young sperm whale confronting a super El Niño, might paraphrase Red Riding Hood and exclaim: "Grandma, what big ideas you have!"[65] It seems, then, that menopause may be wrapped up with culture and has evolved in both humans and the matrilineal whales because cultural information is so important in both.

And what about the large brains of the cetaceans—especially the brains of the most apparently cultural of cetaceans? The sperm whale has the largest brain on Earth and, as a percentage of body size, bottlenose dolphin brains are only surpassed by humans. The cetacean brain has a particularly large cerebral hemisphere and is highly convoluted (with many folds, an indicator of complexity), and at least some have spindle neurons, which seem to be important in navigating the intricacies of complex societies.[66] So far, spindle neurons have only been conclusively found in humans and the other great apes, elephants, and a number of species of cetaceans, including sperm and killer whales.[67] They seem to have evolved independently in the three groups of mammals with exceptionally large brains and very complex societies. Among groups of mammals, both cetaceans and primates have particularly varied relative brain sizes.[68] There are some primates and cetaceans with relatively huge brains—humans and bottlenose dolphins—and some with relatively tiny brains, like lemurs and fin whales. In contrast there is less variation between species among cats and bats and bears. We hypothesize that culture can quickly drive brain evolution, and this is what has happened in the most social, and cultural, cetaceans and primates.

Using cultural information adeptly can be hard, but in a cultural world it is a massive advantage. That other source of information, the genome, will have been alerted to this through evolution. Natural selection will tailor the parts of the phenotype that it controls, such as the reproductive system and the brain, so that the animal becomes as effective as possible in using and manipulating this culture. That is the theory, a theory we find very plausible. Thus we suspect, but do not know, that culture has shaped the genetic evolution of whales and dolphins.

11 | THE IMPLICATIONS OF CULTURE
Ecosystems, Individuals, Stupidity, and Conservation

In the previous two chapters we have looked at cetacean cultures from the long-term perspective of evolutionary biologists. In chapter 9 we considered the why: Why, over the millions of years of their evolutionary history, have the whales and dolphins ended up relying so much on social learning? What factors drove them to become cultural creatures? In chapter 10, still wearing our evolutionary biologist hats, we looked at cetacean evolution the other way round: Given that these animals are cultural, how did this affect the evolution of their genes? Now, in this chapter we are taking the shorter-term perspective of ecologists and ethologists, who are interested in the relationships between life forms in real time. Who eats who? Why do populations wax and wane? Who associates with whom? Why do some animals behave one way, and others in a different way? Ecology and ethology are theoretically based, empirically driven sciences. They are often difficult, imprecise sciences because systems of living things are complex and behave in complicated ways, but sciences they are.

Ecology and ethology are closely related to evolution because ecological and behavioral relationships are at the heart of many evolutionary forces, while behavior itself has consequences for natural selection. Traits that are successful in the ecological and ethological domains are selected by evolution, and those same evolved traits give the agents of the ecological and behavioral dramas their character. Hence, ecology and ethology use the ideas of evolutionary biology.

Ecologists and ethologists are also interested in humans: ethologists in human behavior and ecologists in humans' relationships with the rest of the ecosystem. So ecology has another connotation. It is nearly synonymous with the Green political movement, and many ecologists are also passionate conservationists. Some prefer to call themselves conservation biologists and collaborate with social scientists in trying to get humans to conserve the biosphere.

In this chapter we explore the ecological and ethological implications of whale culture. How are ecosystems affected by culture? Does culture change the role of the individual? Can culture cause behavior that is not in an individual's interest? How does culture change conservation biology, and how should it affect conservation policy?

Cultural Species in Ocean Ecosystems

In chapter 9 we argued that the evolution of the capacity for culture was driven by the advantages it provides in dealing with a variable environment. Thus, it is no surprise that culture changes the ways that a species interacts with its world. Culture can give the members of a species a potentially huge evolutionary advantage. In the words of Peter Richerson and Robert Boyd: "Other organisms must speciate in order to occupy novel environments, whereas humans rely mostly on culture."[1] Human culture has dramatically changed the biosphere, and the evolutionary forces acting on its other inhabitants, mainly because our culture has accumulated new, and sometimes devastating, ways to exploit or incidentally degrade the resources of the earth. Culture produced factory-freezer trawlers, strip mines, PCB's, cigarette butts, and climate change.

What about cetacean culture? The killer whale is the top predator in the ocean, and this is the species whose cultures we probably know best. There is compelling evidence that killer whales, through their predatory behavior, affect the populations and behavior of many other species, even whole ecosystems. We laid out some of this evidence in chapter 6. Most dramatic is the otherwise unexplained disappearance of about half the species of cetaceans, pinnipeds, and sirenians on planet Earth after killer whales appeared about ten million years ago. More recently, James Estes and his colleagues have suggested that the inclusion of sea otters in the diet of a few killer whales totally changed the structure of the nearshore ecosystem off the Aleutian Islands.[2] Also of enormous consequence are the ways the risk of killer whale predation might have affected the migratory behavior of baleen whales and the social structure of sperm whales.[3] The former moves hundreds of millions of metric tons of mammalian biomass seasonally into the tropics, while the latter seemed to spark another cultural blossoming in the ocean: sperm whale clans. A warming climate is today allowing Northern Hemisphere killer whales to extend their range northward into areas previously inaccessible to them, and the effect on the other marine mammals there has been noted. The increasing appearance of killer whales in the Arctic has the Inuit worried. According to the *Nunatsiaq News*: "What the Inuit described was a voracious killer that is sweeping through the northern seas, cutting some marine populations by as much as a third."[4] Just the appearance of this most fearsome predator instigates concern about ecosystem changes.

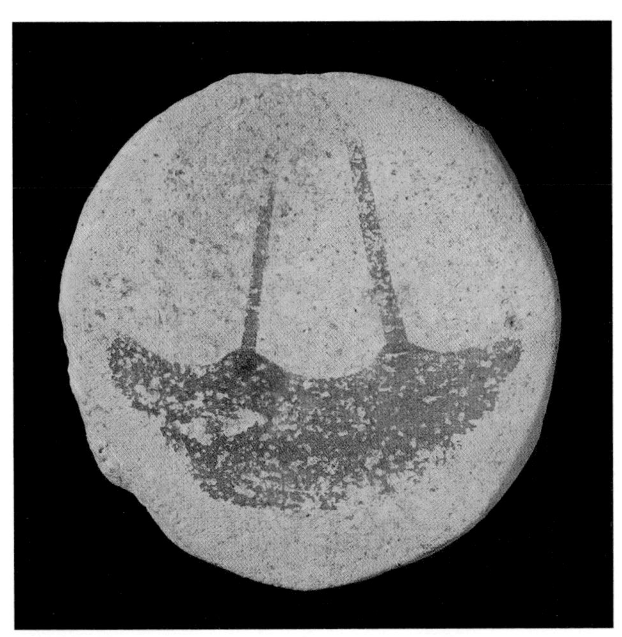

1. Image of a boat on a stone disk from about seven thousand years ago found in Kuwait. Photograph by Mr. Mohammed Ali, copyright held by British Archaeological Expedition to Kuwait.

2. The twelve-meter cutter that the authors use for their sperm whale research, here off Chile, illustrating the state of the cultural evolution of sailing boats in 1973 when the boat, a Valiant 40, was designed.

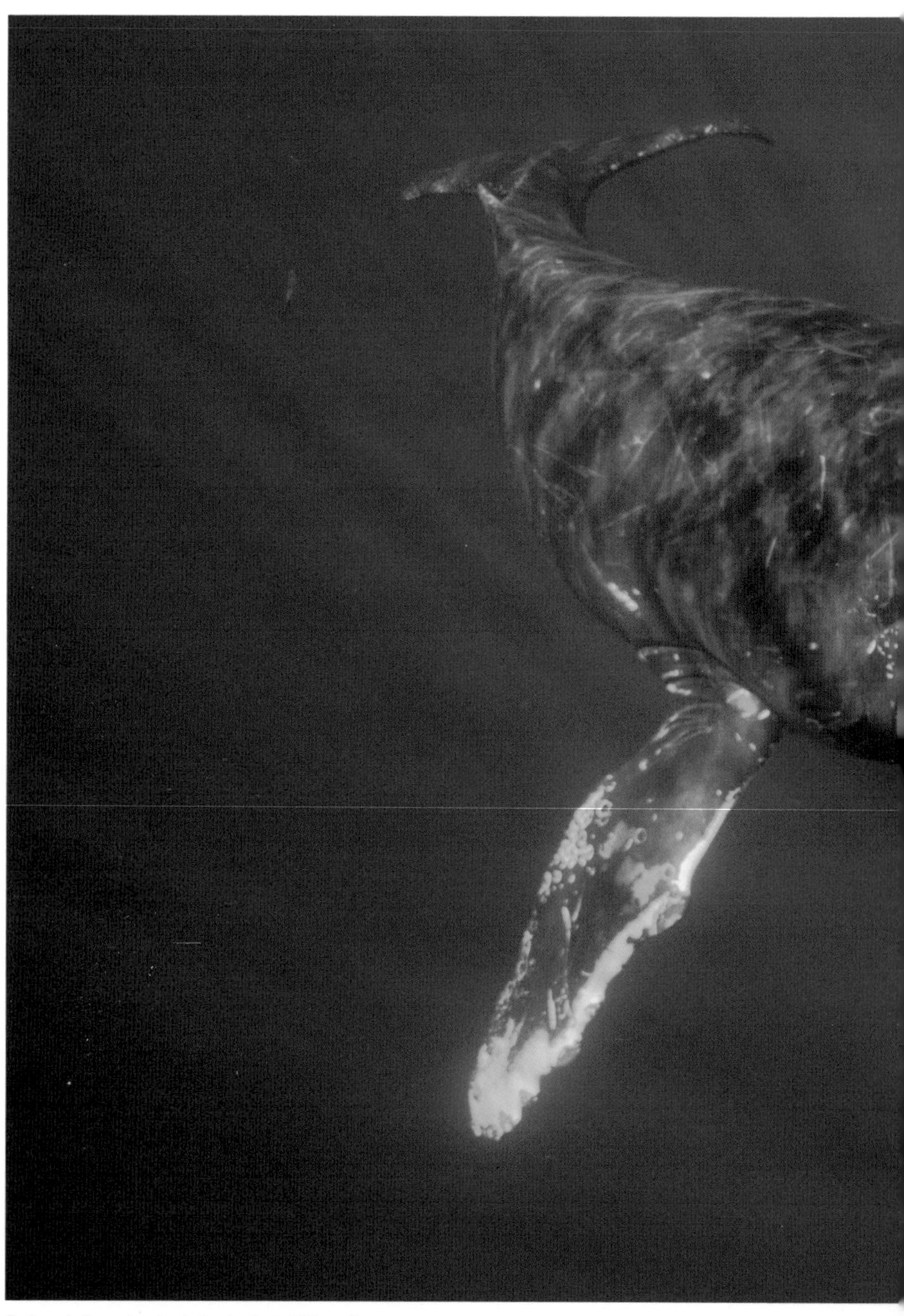

3. A male humpback whale singing off Hawaii.
Photograph courtesy of Flip Nicklin/Minden Pictures.

4. Synchronous lunges by southeast Alaskan humpback whales feeding on herring. Photograph courtesy of Flip Nicklin/Minden Pictures.

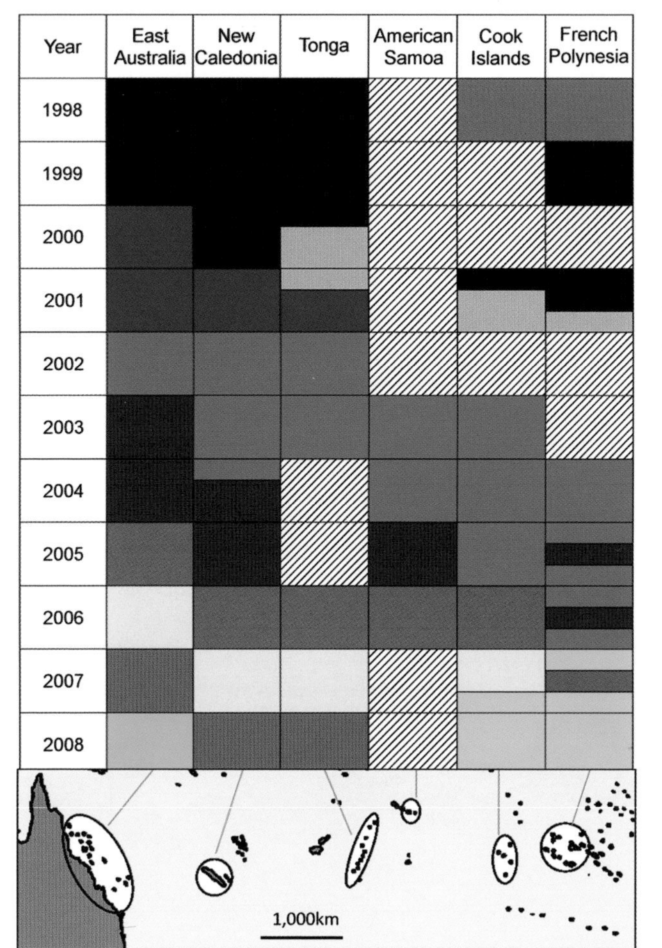

5. Progression of humpback whale song across the South Pacific. Each color represents a different song type; cross-hatching indicates that no songs were recorded. Modified from Garland et al. (2011).

6. Blue whale: aerial view of a twenty-four-meter blue, Sea of Cortez, Mexico.
Photograph courtesy of Flip Nicklin/Minden Pictures.

7. Mud ring feeding dolphins in Florida Bay. To the right of the picture, a bottlenose dolphin has made a ring of mud around a fish school. The fish jump out of the ring, into the mouths of waiting dolphins. Photograph courtesy of Leigh Torres, taken under NMFS permit 572-1639.

8. Sponger: bottlenose dolphin in Shark Bay, Australia, with a sponge on its rostrum. Photograph copyright Lars Bejder, Murdoch University.

9. Wave tailwalks off South Australia. Wave, a bottlenose dolphin, learned the technique from Billie, who learned it from performing dolphins during a brief period in captivity. Photograph courtesy of Mike Bossley.

10. Fisherman at Laguna, Brazil, waits while bottlenose dolphin drives mullet toward him. Photograph courtesy of Fábio Daura Jorge.

11. Common dolphins off southern California. To what degree have these dolphins integrated dynamic information about their environment and each other in a "sensory integration system"? See the section titled "The Oceanic Dolphins" in chapter 5 for more on this concept. Photograph courtesy of Sophie Webb.

12. Social learning. Adult and calf pack-ice Antarctic killer whales assess potential prey. Photograph courtesy of Robert Pitman.

13. What happens next: this seal is about to be knocked off the ice flow by the wave produced by the killer whales, whose white body markings are visible beneath the water. They will then eat it. Photograph courtesy of Robert Pitman.

14. A family group of sperm whales off the Azores. Photograph courtesy of Wayne Osborn.

15. Mass stranding of long-finned pilot whales, Chatham Island, New Zealand.
Photograph copyright Department of Conservation, New Zealand.

How much of the ecological impact of the killer whale is due to culture? We suspect much of it. Large predatory sharks, like killer whales, have powerful jaws and bodies. They also have sensitive sensory systems (olfactory rather than acoustic). While sharks may have important ecological effects, most seem to be generalists, "take-what-you-can" foragers.[5] Faced with such a threat, potential prey can evolve defenses, maybe a hard shell, perhaps high escape speed, or antishark diving behavior. If effective, the species will survive as the shark moves on to other species of food, and a kind of equilibrium can be established.

Superficially, the killer whale is also generalist. The whole killer whale species, or species complex if you like, eats a very wide variety of food, as we listed in chapter 6. However, a particular killer whale, or group/ecotype of killer whales, is typically highly specialist, going after just one species of salmon or seal, with its predatory techniques being honed culturally rather than genetically. Cultural evolution works considerably more quickly and efficiently than the genetic evolution that molds shark predation.[6] An innovative new technique quickly spreads through the group. Changes in the prey's gene frequencies have little ability to counteract this cultural overdrive quickly. Sharks do not leap onto beaches after seals or tip ice floes. Culture, together with fussy specialization, makes the killer whale potentially devastating. They go after a particular prey—and keep going after it, improving their skills. Both the fossil record and present-day observations suggest that many of the killer whales' preferred prey species have either succumbed to extirpation or had their population size and ecological role severely reduced.

This sounds like a creature from a horror movie, and for some of its prey the killer whale is exactly that. They even attract criticism from other Artic hunters—some Inuit interviewed in a study of traditional ecological knowledge complained that "orca appear to kill for sport on occasion, are picky eaters and waste food. . . . Some Inuit were happy to have the *maqtaq*, or skin and blubber, that the animals apparently eschew. Others were appalled at their wasteful ways."[7] But when looked at in a larger, ecological, context, the killer whale's huge role is probably positive. In general, top predators structure ecosystems, and their presence usually increases the diversity of life, what we call biodiversity.[8] This is because apex predators keep the elements of the next level down of the trophic pyramid in check, the so-called mesopredators, allowing lower levels to flourish. The killer whale is the "top, top predator," checking populations of marine mesopredators, like penguins, rays, seals, and porpoises.[9]

What about the other cetaceans? The sperm whale population with its huge food requirements and intense sonar must have had, and may still have, important effects on the squids of the deep ocean—a group of species we know very little about.[10] Some of this could well be due to culturally transmitted methods of catching the squid. However, the sperm whale is, both individually and as a species, a generalist among the mesopelagic squid eaters, and therefore perhaps less likely to have the severe effects on individual species that we postulate for killer whales.[11] But it is reasonable to suppose that they preferentially go for the large, fierce, mesopredator squid—jumbo and colossal squid, for example—rather than small, passive cephalopods, like the histioteuthids. Thus sperm whales, due to their numbers, size, and ubiquity could have a substantial positive impact on the lower-level biodiversity of the deep ocean. The bottlenose dolphin seems somewhat intermediate between the killer and the sperm, with the remarkable culturally driven feeding specializations that we described in chapter 5, but perhaps lacking the focused pickiness that can drive the killer whale into intense ecological embraces with its prey.

All these species have acquired huge ranges, larger than almost any other animal. The bottlenose dolphin uses waters shallower than it can swim in—when chasing prey onto beaches—as well as deep pelagic waters. We have seen them a few meters from the water's edge off the beach in Shark Bay, Australia, as well as off the continental shelf in waters several kilometers deep. We have seen them on the equator near the Galápagos Islands, as well as in the fjords of New Zealand and the firths of Scotland. The sperm whale rarely goes into waters shallower than a couple of hundred meters deep but can be seen close to the ice edge in both the Antarctic and Arctic.[12] In geographical spread, the sperm whale is just surpassed by the killer whale—the second most widely distributed animal on Earth—which leaps onto beaches and, unlike the sperm, goes into the pack ice at both poles. The most widely distributed animal on Earth is *Homo sapiens*. Our culture allowed us to live on the ocean, in the Arctic, and on the Antarctic continent—and to affect all three profoundly. That the second and third most widely distributed species are also significantly cultural is, we think, no coincidence. The dominance of other likely cultural species in their environments, such as starlings and elephants, reinforces this view.[13] Culture not only intensifies a species' footprint, it also spreads it, as cultural knowledge allows subsistence in what were untenable environments or niches when only genes controlled behavior.

Culture can also vastly intensify and accelerate the process of niche construction whereby animals, or plants, change the environment in ways that significantly change the pressures natural selection can bring to bear on them.[14] Human houses and beaver dams are two of the most obvious examples, but there are many others, including earthworms that, through their activity, improve the soil characteristics, with "improvement" defined from an earthworm perspective, as in helping them to thrive. Human houses are clearly cultural; earthworm activity is almost certainly not. But what about beaver dams? Unfortunately, we could find no good scientific information on the possibility of beaver culture. And what about the cetaceans? Perhaps the best-known examples are the gray whales that churn up the substrate of the Bering Sea as they filter feed on the bottom.[15] This "plowing" improves the habitat for the amphipods on which they feed. While a form of niche construction, gray whale plowing is not, as far as we know, primarily cultural. However, there are several known cases, and probably lots of unknown ones, where cetacean behavior is both niche construction and cultural. One example is the fishing cooperatives in which the animals change their environment—in this case the human part of it—to get a better meal.

Culture can allow a species to master a complex foraging method. If it had to be invented from scratch by each new animal, there would be many fewer animals using the complex technique, perhaps just the occasional very smart or lucky creature. The fish that are currently targeted by the sponging dolphins at the bottom of Shark Bay, Australia, or driven into fishers' nets off Laguna, Brazil, would be much safer without dolphin culture. While the Laguna fishers work cooperatively and profitably with the dolphins, many fisherfolk in other parts of the world would lead less stressful lives if it were not for cetacean culture. The species that are overwhelmingly implicated in stealing from trammel nets, gill nets, longlines, and trawl nets are the bottlenose dolphin, the killer whale, and the sperm whale—three species where the evidence for culture is strongest.

In the Gulf of Alaska and Bering Sea, Holly Fearnbach and her colleagues noticed that killer whales of the resident fish-eating ecotype that were observed depredating fishing gear tended to be social associates of one another.[16] They speculated that depredation behavior first spread within matrilines and then to associating groups that may have witnessed the fish stealing during these associations. In Prince William Sound, Alaska, members of AB pod of the fish-eating resident ecotype started removing sablefish, and occasionally halibut, from longlines in 1985.[17] They may have learned this from killer whales in the waters around the Aleutian Islands or Bering Sea, farther to the west, which had been depredating sablefish longlines for decades.[18] The AB was a large pod, about thirty-five whales, and they became very good at exploiting the fishery.[19] They would key in on the sounds of winches pulling up the line and then remove the fish from the hooks as they rose from the bottom. Craig Matkin estimated that the killers were taking about a quarter of the catch. Needless to say, the fishers were not pleased. They tried underwater explosives. These deterred the killer whales somewhat, but the explosives were dangerous and soon made illegal. Matkin, who was studying the whales, started noticing small round holes appearing in their dorsal fins. They were being shot. Six whales disappeared from the pod in one year, while only one death would be expected through natural mortality. Then in 1986, the sablefish season was drastically shortened and shooting was outlawed. The opportunities for depredation and the repercussions of that depredation were both much reduced. Things looked better for the AB pod for a couple of years. But then in 1989 came the *Exxon Valdez* oil spill, which killed, directly or indirectly, seventeen members of the pod, including a number of potentially influential older females. The AB pod split into two but they still have not recovered their total numbers from *Exxon Valdez*. In the late 1990s, the sablefish season was lengthened and depredation by the ABs in Prince William Sound began to become a problem again.

Among these fish-eating, resident killer whales in the waters off Alaska, we can even come close to documenting how these techniques might have spread. In the more enclosed waters of southeast Alaska, about a thousand kilometers from Prince William Sound, killer whale depredation was not seen until the mid-1990s, when members of one of the two resident southeast Alaskan pods, the AF pod, began to be implicated.[20] The AF pod had visited Prince William Sound quite often in the late 1980s and had associated with the fish-stealing AB pod based there. The AF pod was therefore more likely to be exposed to learning opportunities around longlines. In contrast, the other southeast Alaskan resident pod, AG, had not been seen visiting Prince William Sound before 1992, at which time there was little depredation, and it did not depredate longlines in southeast Alaska, at least by 2004.

Let us consider briefly the sperm whales feeding from longlines set for sablefish and other species off Alaska.[21] Sperm whales seem to catch on pretty quickly. Although fishers had been longlining off Alaska since the 1880s, things changed in 1995. Up until then, the management system had been a derby, in which everyone tries to catch as much as possible before a total quota is reached. As a consequence, the fishing season was about five days long per year. Now individual boats are given quotas, which spreads the effort over eight to nine months. This change made for a more orderly

11.1. An underwater camera catches a sperm whale removing a sablefish from a longline off southeast Alaska. Photograph courtesy of SEASWAP (see Mathias et al. 2009).

fishery but also a much better opportunity for the sperms—longlines became a much more permanent feature of their environment, which probably facilitated the initial learning episodes about this resource. Widespread depredation by sperm whales started to be reported in 1998. In other areas, there seem to be delays of about two to fifteen years between the start of longlining and consistent depredation by sperm whales.

Jan Straley and her colleagues have been looking for solutions to sperm whale depredation of longlines off Alaska. They attached a camera to a longline (see fig. 11.1) and watched:

> A whale approached the line, grabbed it with its mouth and created tension by moving up the line. This tension created by these actions popped a fish off the line and the whale extracted itself from the line and swam after the fish. . . . These amazing animals have many ways of removing fish from a line. When we saw the footage of this immense whale delicately and deftly maneuver around the longline, we realized that these animals are clever and crafty. The movements were deliberate.[22]

This quote comes from an article titled "Sperm Whales and Fisheries: An Alaskan Perspective of a Global Problem." In the global extent and rapid spread of complex behavior—the depredation of longlines—we think we can see the power of social learning. Once one or two individuals work out how to access the bonanza, the behavior spreads across an entire community, quickly reaching problematic proportions for the fishers. The prey of cultural whales face a more extreme version of the problem. Unlike the fishers that Straley is trying to help, most don't have cultural methods, such as technology or socially learned antipredatory behavior to try to counter the innovative and predatory whales. And, of course, their lives are on the line. From farther down the food chain, whale culture appears in a more positive light, keeping those nasty mesopredators in check so that more basal species can flourish. We suspect that the ecology of many parts of the ocean would look and work quite differently were it not for whale culture. Whether these differences are positive or negative depends on one's point of view.

The Roles of Individuals in Cultural Systems

Culture seems the antithesis of individuality. Culture homogenizes behavior. But dig a little deeper and the individual emerges, in two ways. The first is in complex culturally transmitted behavior with roles for individuals. Good examples are human musical ensembles and the barrier-feeding dolphins off Cedar Key, Florida. In rock bands, jazz ensembles, choirs, and orchestras, individuals play particular instruments or sing lead or backup, but together they produce music that pleases the human brain. In each of the Cedar Key dolphin groups, one individual, the "driver," consistently herds fish toward the others, who form a barrier.[23]

The other way in which particular individuals become important in cultural societies is as nodes in the network of information flow. Cultural information is transmitted through a community via the social relationships between its members. Thus, individuals who have important information or who are well connected socially may be especially significant in what cultural information the community possesses, in how it is structured among community members, and in how it changes with time. Think of the outsized influence of an inspiring teacher or trendsetter—Karl Marx, Mahatma Gandhi, Charles Darwin, John Lennon, or Yves Saint Laurent, say—in driving different aspects of human cultures. David Lusseau, an expert on dolphins and social networks, and his colleagues have made some interesting propositions about the roles of particular individuals in whale and dolphin networks. For instance, Lusseau found that, among the bottlenose dolphins living in Doubtful Sound, New Zealand, both the side flops that initiate travel and the upside-down lobtails that terminate it are predominantly performed by the most socially central dolphins in the community, males in the case of the side flop, females for upside-down lobtails.[24] Based on a network analysis of this population, Lusseau and Mark Newman suggest that particular individuals act as "brokers," linking what are otherwise fairly distinct subcommunities, and that they may be "crucial to the social cohesion of the population as a whole."[25] Rob Williams and Lusseau made a similar analysis of the resident killer whale societies of the eastern Pacific, again suggesting that particular individuals have a special role in maintaining network cohesion and, perhaps, in information exchange.[26]

We need to be careful here. The Department of Biology at Dalhousie University where we have both worked is a multilevel concrete structure with ecologists working on some floors, molecular biologists on others. If we noted conversations among the inhabitants of this building, we would find little communities of biologists with similar interests talking among each other. In a network diagram, in which the inhabitants of the building are represented by nodes joined by links showing who talked with whom, these foci of biological expertise would appear as interconnected clusters. In our building they would be linked by one individual with connections to all the groups. Is this the great integrative biologist, driving cross-disciplinary innovation? No, the apparent center of the social network is a garrulous janitor, passing the time of day with everyone as he sort-of-cleans their labs. Appearing essential to the network, he actually transmits little information of significance to the biological endeavor (although few pass up the titbits of gossip). Similarly, a killer whale or bottlenose dolphin who visits lots of different social units might be the bearer of important cultural knowledge, but she might also be an outcast from her native group or, possibly, an exploratory adolescent whose arrival in each group has no cultural relevance. In the killer whale analysis, juvenile females were most "central" and so were suggested to play important roles in maintaining network cohesion.[27] An alternative interpretation might be that they are socially promiscuous and as yet unable or unwilling to form strong bonds with any other individuals in particular. Thus, social centrality does not necessarily equate with cultural significance.

However, it could. Among elephants, whose matrilineal social system is remarkably congruent to that of the sperm whales, the older female matriarchs are vital information and decision-making hubs in their communities.

Elephant groups with older matriarchs show consistently more appropriate responses to droughts, unexpected social circumstances and predators than those led by younger females.[28] We have almost no such evidence for whales or dolphins. One possible example is from a stranding of short-finned pilot whales in Florida, in which the animals only left the beach after the two largest, and hence probably oldest, animals were dragged off.[29] This lack of evidence for leadership may be because the organization of their societies is more egalitarian than that of the elephants or because we have not been able to penetrate their social complexity to anything like the depth that the elephant scientists have achieved.

Culture and Stupidity

Acquiring information from others allows people to rapidly adapt to a wide range of environments, but it also opens a portal into people's brains, through which maladaptive ideas can enter. . . . In creating a simulation of a Darwinian system using imitation instead of genes, natural selection created conditions that allow selfish cultural variants to spread.[30]

Genes and culture can be transmitted quite differently. One of the consequences of this is that while genetic evolution almost invariably produces adaptation, in the sense of individuals doing generally sensible things given the environment in which they evolved, when cultural evolution is added to the mix, stupid behavior may emerge.

In what way stupid? Primarily, we mean doing things that are bad for the animal's genes. Usually though, the stupid behavior is "intended," by cultural evolution, to promote a cultural variant. So, while priestly celibacy will stop the priest's genes in their tracks, if practiced conscientiously, it may also allow him more time and resources to promote his religion or give him more prestige within his community and, thus, more power as a promoter of the cultural variant that is his doctrine. The genes are sacrificed for the culture. Similarly with suicide bombers or kamikaze pilots: no hope for future contributions from their genes but perhaps a boost to the cultural causes they give their lives for. At least that is the idea. In these cases, the maladaption—the stupidity—is seen as such only from the gene's perspective; from the point of view of the cultural variants, the behavior may be adaptive.

Rather more puzzling are cases where culture leads to behavior that is good neither for the genes nor for the culture. One of the most extreme examples took place in 1978, when Jim Jones lead over nine hundred members of the People's Temple into the Guyanese jungle, where they killed themselves using cyanide. That was the end of the People's Temple as well as the genes of its members. Or take the "Charge of the Light Brigade." In 1854 during the Crimean War, a brigade of British cavalry with about six hundred and seventy soldiers received an ambiguous order to charge into a valley that had enemy Russian soldiers and artillery positioned on both sides and at the far end.[31] Two hundred seventy-eight were killed or wounded. Again, this was a move that ended the genetic contributions of many members of the Light Brigade and set back the British offensive, the ostensible cultural objective. Most members of the brigade knew the charge was both suicidal and pointless, but they did it anyway and were immortalized in Alfred, Lord Tennyson's famous poem:

"Forward, the Light Brigade!"
Was there a man dismay'd?
Not tho' the soldier knew
    Someone had blunder'd:
Theirs not to make reply,
Theirs not to reason why,
Theirs but to do and die:
Into the valley of Death
    Rode the six hundred.[32]

Why? They did it because there were overriding cultural imperatives. In the case of the Light Brigade, following military orders was paramount. That imperative evolved, culturally, because armies in which soldiers followed orders generally did better than those in which they did not. In the aftermath of the charge of the Light Brigade, the British celebrated the cavalrymen who, from a dispassionate perspective, had behaved so stupidly:

When can their glory fade?
O the wild charge they made!
    All the world wondered.
Honor the charge they made,
Honor the Light Brigade,
    Noble six hundred.[33]

This glorification inhibited the evolution of the British Army into a more effective war machine, doing "much to strengthen those very forms of tradition which put such an incapacitating stranglehold on military endeavor for the next eighty or so years."[34] The maladaptive culture was conserved.

In all these cases, whether the behavior is maladaptive just for the genes or for both genes and culture, there is a strong identity with a culturally marked group, to a religion or nation or military unit. This group identification trumps promotion of genes or even, in some particular circumstances, behaving in ways sensible for the group's long-term prospects. So soldiers ride to their death because following military orders is more important than preserving their own lives or even winning the war.

Our examples have mostly been of war, death, and celibacy, but the same process can work in milder ways: people give money to their alma maters, money that could be used to promote their own genes or ideas, but instead is used to support a cultural institution that the individual has long finished using. Soccer fans battle rivals in a manner that risks their own arrest *and* brings the club that they are supporting into disrepute.

So we have run through a little of the stupidity that human culture can engender. What about the whales? There is one prime example of apparent stupidity in whales. And it appears to rank right up there with the charge of the Light Brigade and the mass suicide of the members of the People's Temple. Whales and dolphins sometimes strand themselves alive on shorelines—not just briefly to catch a seal or fish but with no apparent intention of leaving (see plate 15). The animals involved are in grave danger. They may be mammals and able to breathe air, but they face dehydration, their bodies, if they are large whales, will gradually crush themselves without the support of water, and they cannot feed. They will usually die.

These strandings come in three general types: single strandings when just one animal, or perhaps a mother and calf, is found on the beach; "typical" mass strandings when a number of animals strand at the same place (usually within a few hundred meters) and the same time (up to a few hours); and "atypical" mass strandings when the animals strand at roughly the same time but over a large area, maybe spanning tens or hundreds of kilometers.[35] Single strandings almost always involve sick or injured animals and presumably result from disorientation, an inability to swim, or a desire for the support of the beach. All species of cetaceans can strand singly. Atypical mass strandings took some time to work out, but it is now clear that they are usually caused, in some way that we do not yet understand, by loud sounds in the ocean that lead whales to end up on beaches over a wide area, as sound travels so well underwater and is so important to the whales. The main culprits are naval sonars and explosions, but seismic exploration using loud sounds to map undersea geology—usually in the search for oil and gas reserves—is occasionally implicated.[36] The usual victims of the atypical mass strandings are the beaked whales, which seem to be particularly sensitive to these loud sounds, but other species are sometimes involved. These atypical mass strandings were exceedingly rare before humans started making loud underwater sounds in the mid-twentieth century.[37]

In contrast, the typical mass strandings—several animals on the same beach at roughly the same time—have been around for thousands, and probably millions, of years. In 350 B.C., mass strandings of whales and dolphins puzzled Aristotle: "It is not known why they sometimes run aground on the seashore: for it is asserted that this happens rather frequently when the fancy takes them and without any apparent reason."[38] They puzzle us today.[39] Unlike the victims of single strandings or atypical strandings, the animals on the beach appear, for the most part, to be perfectly healthy. Observers speak of the whales typically not appearing panicked but, rather, swimming slowly and deliberately toward the shore.[40] There is a range of theories, with varying amounts of evidence behind them, as to why the whales got on the beach in the first place. They include storms, magnetic anomalies, oceanographic variation, and navigational failures in areas with complicated bathymetry.[41] Putting these together, we can infer that the whales were close to shore in a tricky situation in which the environment was not giving them easily interpretable information on where they were and where they should have gone.

However, as indicated by the adjective "mass," a key aspect of the typical mass strandings is sociality. Animals follow each other onto the beach and will not leave it by themselves.[42] Usually, little groups of animals, or individuals, arrive at intervals of a few minutes.[43] Then, if taken off the shore, they will usually return to the beach, unless a large proportion of the other whales have also been taken off. Furthermore, the species most likely to be involved in typical mass strandings are matrilineal whales: sperm whales, pilot whales, false killer whales.[44] Oceanic dolphins can also mass strand, though they do so somewhat less frequently. And although we do not properly understand their social structure, they are clearly highly social.

In these typical mass strandings, the imperative appears to be to stay with the group—the definition of "group" in this instance is murky because mass strandings usually involve far more animals than the typical family unit of, say, pilot or sperm whales. We simply don't understand enough about the dynamics of group cohesion, dominance, or leadership in cetacean societies to fully understand the phenomenon. Some reports talk of a marked change in a group's behavior once larger individuals are pulled off the beach, which is suggestive of some kind of leadership role.[45] However, the social cohesion imperative obviously overrides an individual's instinct to protect its own life and its own genes or even the survival of its "group." Why? There are well-known advantages to sticking with a group—protection from predators, for example—but we don't see anything like mass strandings in the multitude of species that form groups apparently for these reasons, and precisely because these advantages accrue to individual-level fitness, we would expect animals to abandon their group once the cost of sticking with it outweighed the cost of leaving. We would suggest that it is the need to remain connected to the knowledge collectively held in the group, probably by the more mature members, and perhaps a willingness to credulously follow those "elders" wherever they go that leads to this extraordinary, and often fatal, phenomenon. To persist, this cohesion imperative must, on average, be a better bet than trying to figure things out for oneself. Being largely shore bound, we are much better informed about the presumably relatively rare catastrophes associated with the shore than about how this cohesion typically pays off for individual whales in the deep.

So culture may have fatal consequences for large groups of matrilineal whales. But perhaps it is sometimes much worse. There are two species of pilot whale, long-finned and short-finned. The short-finned are the warm-water species, using tropical and subtropical waters right around the globe. The long-finned pilot whale favors cooler water, being found in the Southern Oceans and North Atlantic but not the North Pacific. This seems odd, as the North Pacific is larger than the North Atlantic, and archaeological evidence suggests that the long-finned pilot whale was indeed found in the northern North Pacific until about 1,000 A.D., but then it disappeared from there.[46] Pilot whales are widespread, numerous, and resilient creatures. They have sustained continuous, quite heavy, drive fisheries in the Faeroe Islands in the North Atlantic for centuries, for example. So why did they disappear from a very large ocean, where they were never the target of an industrial whaling operation? Well, if conformism takes hold of a cultural population, becoming an imperative, individuals will be paying much more attention to each other than their environment and therefore may be severely affected by changes in that environment, with their population imperiled. This is what seems to have happened to the Greenland Norse, dairy-farming Vikings of southern Greenland whose way of life became gradually unsustainable with the start of the Little Ice Age in the fifteenth century. Their fate contrasted with the Inuit (Eskimos) who lived in the same area but whose way of life was well suited for the cooler conditions.[47] The Norse were extirpated; the Inuit thrived. Part of the reason for the contrast seems to be that the Norse were culturally conservative and hung onto the practices of their homeland for too long in the face of the changing climate. Modeling by Hal and Pete Richerson shows that cultural conservatism coupled with an environment undergoing large long-term changes can, ominously, be disastrous for human populations.[48] Perhaps the fate of the Norse was an example of this. Did the North Pacific long-finned pilot whales also become victim to their own conservatism, sticking to outmoded behavior in a changing environment, and suffer the same fate? This is complete conjecture on our part, but it is a potential explanation for the mysterious disappearance of an otherwise cosmopolitan species. Since one goal of conservation is to prevent such absences in the future, it also leads nicely into our discussion of culture and conservation.

Conservation and Management of Cultural Species

It will be extremely difficult to develop an exploitation scheme for such highly social species of cetaceans without disrupting their cultural traits. Driving whole members of a group of short-finned pilot whales into a harbor for subsequent slaughter is quite likely to wipe out a cultural trait and diminish the species' ability to survive. The current exploitation of these species being practiced in Japan is only possible by ignoring the individuality of group members and functions of living in a group.[49]

This statement, variants of which might have come from a number of our like-minded colleagues, is exceptional because it was made by a Japanese scientist, Toshio Kasuya, in an exceptionally public forum.[50] Kasuya was receiving the Kenneth S. Norris Lifetime Achievement Award from the Society for Marine Mammalogy at their biennial meeting. In this speech, as well as in his writings, Kasuya was going against a significant part of his own culture. The Japanese government line is that whales can and should be exploited sustainably, and almost all older Japanese whale scientists toe it.[51] Not Kasuya, whose scientific excellence and courage made him an ideal recipient of the Norris award; he codiscovered the extirpated North Pacific long-finned pilot whales, menopause in the short-finned pilot whales, and much else. In this section we will explore, a little more broadly, the question that Kasuya brings up. How does their culture affect our attempts to manage and conserve the whales?

As we have outlined, culture can drive the population biology of a species—evolutionarily, ethologically, and ecologically. The consequences of this culture have bearing on how the whales and dolphins interact with us humans. For instance, their wide ranges and ecological success generally increase the rates and severity of their interactions with us. Wide ranges mean more areas of conflict, and ecological success translates into increasing competition for resources between humans and cetaceans. When individual whales have important roles in society, their removal, whether by whaling, by-catch in fisheries, or live capture for display, affects the livelihood of those who remain. Among elephants, the poaching of older matriarchs reduces the effectiveness of the surviving members of their social unit in responding to threats and, consequently, also reduces their reproductive success.[52] Thus these removals lower the resilience of the elephant population at a much greater rate than would be predicted from considering just the numbers killed.

Because of the importance of knowledge, cultural species will tend to have prolonged care, and perhaps instruction, of the young. Hence reproductive rates tend to be low. At the other end of life, culture may help animals survive into old age, and because older animals are important repositories of knowledge, attributes such as menopause may evolve to increase longevity. In sum, cultural species tend to reproduce slowly and live long. Their populations cannot grow fast, but the animals can usually deal better with much of whatever comes along.

This works well, even over long timescales (the red noise that we have discussed in earlier chapters), in a world that may be highly variable but is varying between limits—limits within which the animals' biology and culture evolved. If a sudden new threat arises outside the animals' evolutionary experience, be it whaling, pollution, or a consequence of rapid human-caused climate change, then a species with such a low reproductive rate generally has little resilience. A sperm whale population, for instance, cannot absorb any new threat that removes more than about 1 percent of the population per year.[53]

However, in dealing directly with a new threat, culture itself may be an advantage, if effective means of combating the threat can be learned socially. There is some evidence that some sperm whales learned methods of evading the open-boat Yankee whalers of the *Moby Dick* era, for instance, by swimming fast upwind.[54] As the whaleboats were rowed or sailed, this was pretty effective. We do not know whether the tactic was learned individually or socially, but, given the nature of sperm whale society, it seems highly likely that there was a social component to the behavior. Culture may well allow some cetacean species to adapt appropriately to the changes in their habitats that will come with climate change.[55] They may learn new migration routes or the whereabouts of new foraging areas from each other, routes and areas that are more appropriate in the face of manmade changes in the ocean.

Alternatively, culture may be a disadvantage in combating threats if animals are culturally conservative. For example, southern resident killer whales may continue to use noisy, polluted urban waters in Puget Sound, which now contain little food and are bad for their health, while better alternatives exist. They may do this primarily because "those are the areas we use" and, secondarily, because using those areas helps delineate "we."[56] The resident killer whales also so strongly prefer Chinook salmon that the health of their population mirrors the strength of the Chinook stock, even though there are nutritious salmon of other species around.[57]

The speed at which cultural species take advantage of new food sources can generate opportunities but also threats.[58] We saw in chapter 5 how Bec Donaldson and her colleagues related the spread of begging for handouts through a dolphin population in southwest Australia with the social network in the population, strongly suggesting a social component to the spread.[59] As they interact with fishing gear, perhaps gaining an easy snack, whales and dolphins can become caught in nets or longlines or be shot at and sometimes killed by angry fishers.[60] There are benefits and costs associated with any innovation. These benefits and costs may generate evolutionary selection pressures: animals with some characteristics may do better, others worse. If the gains from the fish stealing outweigh the risks, then boldness and aptitude for social learning may be favored. But if exploiting humans is dangerous, then timidity and conservatism may be favored. This is what happened when some of the most aggressive male baboons of "Forest Troop" started feeding from the garbage dump of a Kenyan tourist lodge in 1982 and, consequently, died.[61] We will describe this remarkable story in more detail in chapter 12, but the take-away in this context is that the Forest Troop was left under the control of more timid surviving males who had not fed at the garbage dump. A culture of passivity set in and persisted in the Forest Troop for at least a generation after the original timid males had been replaced by newcomers. This cultural revolution—the second documented nonhuman cultural revolution, following the radical shift in eastern Australian humpback whale songs described in chapter 4—was primarily an example of cultural selection brought about by the dangers of exploiting human waste. But there may also have been some genetic selection, if the timidity of the surviving males was genetically heritable so that offspring were like their fathers in this respect. This genetic evolutionary effect would not have directly maintained the passivity of Forest Troop, as males transfer between troops at maturity, but it might have lowered the general level of aggressiveness in the population. These same kinds of cultural and genetic evolutionary forces are undoubtedly at work among the cetaceans as they interact with our fisheries.

The case of the Moreton Bay dolphins and the prawn trawlers that we discussed in chapter 5 shows how human fisheries can affect cetacean social structures quite profoundly. It also shows the resilience of a bottlenose dolphin population to direct impact of humans on their habitat and, secondarily, on their social structure. We humans created an opportunity, dolphin culture responded, and dolphin society was reshaped. In Moreton Bay the sequence was reversible, but we doubt that this will always be the case, especially with more culturally conservative cetaceans, such as killer whales.

And then there is the stupid stuff that culture produces. Its effects on conservation are less obviously direct, but stupid behavior confuses us. How do we distinguish between whales harming themselves because that is part of their cultural nature and us harming the whales as part of our cultural nature? As we have outlined, whales mass strand on beaches and die for reasons that have nothing to do with us. But they also mass strand on beaches for reasons that have everything to do with us.

Sometimes human intervention is wrongly seen as the cause of a "natural" mass stranding. Following the 2002 mass stranding of fifty-five pilot whales on Cape Cod, a *Toronto Star* headline read: "Is New Sonar Driving Whales Ashore?" No, not in this case—this was a "typical" mass stranding, the "natural" kind that pilot whales have been performing for millennia. However, there are confusing cases, as in January 2005 when thirty-three short-finned pilot whales—matrilineal and frequent typical standers—came ashore near Oregon Inlet in North Carolina following offshore military exercises. Two dwarf sperm whales and one minke whale came ashore at the same time but at different places. So the incident had attributes of both "typical" and "atypical" mass strandings, and the official investigation by the U.S. National Marine Fisheries Service concluded that, "overall, the cause of [the mass stranding] in North Carolina is not and likely will not be definitively known."[62]

One of the hardest tasks for wildlife managers is deciding how to divide the populations they manage into the "stocks" that are the subjects of management actions, such as hunting quotas. A stock has its own hunting regulations, a neighboring stock a different set. Conservation biologists may suffer even more angst as they try to delineate what are often called evolutionarily significant units (ESUs), which may then receive different levels of protection appropriate to their perceived conservation status. The consensus among conservation biologists is that ESUs should be defined geographically, genetically, or by differences in the animals' phenotypes, caused by selection for the local environment.[63] Generally these units are mapped genetically, following the work of conservation geneticists. But deciding when genetic differences are sufficient to separate populations is often far from easy. The decision may be loaded if it has substantial impacts on human activities, for instance, when a proposed ESU lives in an area where there is a proposal for a habitat-altering development.

Culture complicates all of this. Most fundamentally, if culture is an important determinant of phenotype, then logically it should be an important determinant of the ESU, or so we have argued.[64] But, secondarily, as cultural variation and genetic variation are not necessarily correlated, the standard, genetic methods for delineating ESUs may well miss what are truly important distinctions for the species and how its members interact with our threats. For instance, of the cultural clans of Pacific sperm whales, the Plus-One clan appears as though it will respond better to ocean warming than the Regular clan. The clans have consistently different behavior—different phenotypes. If global warming is seen as an important threat to the sperm whales, then it follows that the two clans should be protected separately, as important components of the totality of life on Earth. But they are not separated in the all-important nuclear genes and are only a little separated in the matrilineally transmitted mitochondrial genes, so they probably would not emerge as distinct ESUs from the standard procedures of the conservation geneticists.[65]

So what to do? We and our colleagues have suggested a tweak of the definition of ESU to include any form of heritable phenotypic variation, whether through genes, culture, or, potentially, some other process of inheritance.[66] Alternatively, ecologist Sadie Ryan has proposed "culturally significant units" (CSUs) that should be considered, and perhaps designated and then protected, when considering species where there is substantial culturally determined behavior.[67]

This may all seem rather esoteric and distant from the practice of protecting biodiversity. But it all came to a head in the case of the southern resident killer whales, the fish-eating *community* that straddles the maritime border between the waters off southern British Columbia in Canada and Washington State in the United States. The southern residents, who number less than one hundred, are faced with a number of human-caused threats, including pollution, fisheries for the salmon that they eat, and overenthusiastic whale watching. The question facing both the U.S. and Canadian authorities was whether the southern residents should be considered the legal equivalent of an ESU.[68] If so they would be designated as "endangered" and receive substantial protection under the Endangered Species Act in the United States or the Species at Risk Act in Canada. Salmon fishing or whale watching might be significantly restricted. If not their own ESU, they would be considered part of a larger, and presumably safer, killer whale population, which would also contain the more numerous and less threatened northern residents, and possibly other ecotypes and *communities* of killer whale that use the North Pacific—Alaskan residents, transients, and offshores. There would likely be much less legal protection. As we summarized in chapter 5, the southern residents and northern residents have quite distinct vocalizations, ranging behavior, and social behavior. These differences are assumed to be maintained culturally—the members of the two communities differ genetically in only one base pair of the mitochondrial genome, the smallest possible consistent genetic distinction between two populations. Would the government agencies consider the cultural distinctions sufficiently important to give the southern residents their own ESU?

The Committee on the Status of Endangered Wildlife in Canada listed the southern residents separately from the northern residents in 2001, assessing them as endangered.[69] The Canadian government describes them as "acoustically, genetically and culturally distinct."[70] After this, the southern residents gained considerable protection under the Species at Risk Act on the Canadian side of the border. However, the U.S. government saw things differently. In 2002, the U.S. National Marine Fisheries Service ruled that the southern residents did not form a "distinct population segment"—in other words they were not a legally designated ESU.[71] The fisheries service dismissed cultural arguments on the basis that "there was insufficient evidence to indicate whether these 'cultural' traits were inherited or learned, and thus whether they truly signify an evolutionarily important trait."[72] There were protests and lawsuits, often invoking the "culturally distinct" argument. In 2005 the U.S. government reversed its decision "based on evaluation of ecological setting, range, genetic differentiation, behavioral and cultural diversity," and the southern residents became classified as endangered on both sides of the border.[73] In the pages of the U.S. *Federal Register* announcing the decision resides the prescient statement: "There are differences in cultural traditions, and the Southern Residents may have unique knowledge of the timing and location of salmon runs in the southern part of the range of North Pacific residents."[74]

Hopefully these measures will protect the southern residents and their culture. Unfortunately, during the most destructive part of humanity's relationship with the whales, there were no endangered species acts or anything remotely resembling them, in either law or sentiment. Several species were reduced to tiny proportions of their sizes before the whalers ended their work.[75] In cases of extreme population reduction, conservation biologists worry about the low genetic diversity caused by such bottlenecks.[76] Effectively, a tiny population possesses very few different genes—those in the few individuals fortunate enough to survive. The population becomes too homogenous, potentially leading to problems caused by inbreeding depression, such as elevated rates of congenital defects and a lack of the genes that might help it deal with new or old problems. This may especially be the case if the genes that helped an individual survive the bottleneck, for instance, those that promote shyness or small size, are not generally advantageous.

This bottleneck problem can also operate on cultural traits. A cultural population reduced to very small numbers may lose useful information, just by attrition. The native Tasmanians, living for many generations in a small population totally cut off from the aboriginals on mainland Australia, ended up with the simplest culture of any known human population, a culture lacking many of the standard attributes of the Aborigines of mainland Australia including bone tools, nets, barbed spears, and boomerangs.[77] What about the North Atlantic right whale? Down to less than a hundred animals in the early 1900s as a result of whaling, the species is having a hard time rebounding despite protection. While southern right whale populations are rebuilding at about 7 percent per year, the North Atlantic species, which has an unfortunate tendency to be hit by ships and get caught in fishing gear, increases by just 3 percent at the best of times.[78] And there are bad times when the numbers drop for a few years in a row. These drops can be traced to summer conditions in the lower Bay of Fundy in eastern Canada. When plankton is poor in the bay, then the rights have poor feeding success and the females raise few calves.[79] The population declines. This strong dependence on the Bay of Fundy appears recent. In the eighteenth and nineteenth centuries, right whales were killed during summer months in several areas of the North Atlantic where they are no longer seen or are very rarely found, including the waters off Long Island, New York, the Gulf of Saint Lawrence, east of the Newfoundland Grand Banks, and south of Greenland.[80] We can imagine them moving between these summer grounds depending on feeding conditions. With lots of dense plankton, stay; with little, move. We know that right whales do this over smaller scales of a few meters and seasonally move between feeding areas in the Gulf of Maine as plankton concentrations change.[81] Additionally, there could have been different population segments specializing on the different grounds, whose good and poor years compensated for one another. Either way the population would have been more resilient than it is today with its overreliance on the Bay of Fundy. Bruce Mate and his colleagues speculated that a right whale mother tracked with a satellite tag who took her calf on a tour through a number of the western North Atlantic feeding grounds was perhaps teaching, although she could equally have simply been searching for food herself.[82] But either way, her calf would nonetheless have learned things about its habitat that it otherwise wouldn't. The loss of cultural knowledge of those other feeding grounds, like the loss of genetic variation, makes the population more vulnerable.[83] Cultural diversity is important.[84]

Philip Clapham and his colleagues have generalized this argument in their discussion of the management of baleen whale populations in the aftermath of whaling.[85] They examined eleven subpopulations of five species of baleen whales, all of which were extirpated by commercial whaling, and none of which have recovered at all in the forty to four hundred years since the end of the whaling. They suggest communal loss of cultural memory of the habitat as a fundamental or contributory cause of these failures of recovery.[86] Management needs to include these cultural effects.

Hence, as Toshio Kasuya declared so bravely, culture complicates conservation and it complicates it in all kinds of complicated ways. Horizontal cultures can propel populations into problems with humans, but they can also help them deal with the problems. In contrast, vertical cultures can be disrupted by human activities and may inhibit adaptation to them. There are many reasons why we may wish to conserve a species or a population. Their culture should be part of the equation. Our culture frames the debate.

12 | THE CULTURAL WHALES
How We See Them and How We Treat Them

The Culturally Driven Cetacean

Therefore, it is my view that cetaceans must depend upon knowledge accumulated through past experience. Such knowledge is likely to be transmitted, by learning, to other members of the group. This is the culture of the community.[1]

Yet again we return to Toshio Kasuya's perspective from a lifetime of trying to understand cetaceans, typically both gently spoken and right to the point, highlighting the cetaceans' dependence on knowledge, the social learning of that knowledge, and the significance of the social community with its characteristic culture. Some cetaceans barely seem cetaceans outside their society and culture.

But among the different species of whale and dolphin, and sometimes within a species, societies and cultures can take very different shapes. At one extreme are the tight matrilineal family units of the large toothed whales, sharing food, caring for one another, and having specialized xenophobic cultures. At the other are the loose societies and ocean-wide song cultures of the baleen whales. In yet another corner are species for which we currently have little or no evidence for culture or complex social structure, such as harbor porpoise. But that is just the span of our knowledge. What about narwhals under the ice, the river dolphins, or, most intriguing of all, the oceanic dolphins in their huge schools?

We humans face important barriers when trying to understand the nature and scope of cetacean culture. The habitat of whales and dolphins is inhospitable for us. They are large and many cannot be held captive; some would argue that none of them should be.[2] They sense their environment differently from us, and it is a different environment: three-dimensional, buoyant, opaque to light but transparent to sound, barrier free, and uniform over small scales but varying hugely over large scales. Our human culture is absolutely vital to our lives, and we are naturally fascinated by it. There are several long-standing academic disciplines involving hundreds of thousands of academics devoted to studying human culture. But we still do not understand, or agree on, how human culture operates. How much further is the handful of scientists studying culture beneath the oceans from understanding it?

The evidence for some putative cetacean culture is vulnerable to criticism. This is especially the case for behavior associated with foraging. Can behavioral patterns be explained by genetic differences, or by individuals learning behavior independently in different environments? Often, with foraging behavior, both mechanisms are operating alongside culture so that isolating the social learning component is hard. This does not mean that, for whales and dolphins, culture is less important in determining foraging behavior than it is for vocalizations or play, just that we find it harder to delineate.

Despite these difficulties there is considerable evidence for culture in whales and dolphins. Some cetacean culture, like the humpback, bowhead, and blue whale songs or the call traditions of killer whales, is clear. Much, including the variation in behavior among the units, pods, and clans of the matrilineal whales is very compelling. Then there are those features that we expect in a cultured species, either because of theoretical predictions or because they are found in other animals with important cultures, especially humans. The whales and dolphins have large brains, prolonged mother-infant dependency, menopause, maladaptive (stupid) behavior, ecological success, wide habitat ranges, large-scale cooperation, and indications of the coevolution of genes and culture.

The evidence for cetacean culture is multifaceted, spanning foraging behavior, vocalizations, movements, social behavior, and play. In killer whales, at least, all of these seem to have a cultural component, and behavioral variations map onto different levels of their hierarchically organized social system. In bottlenose dolphins and the matrilineal large toothed whales, groups of animals with different cultures use the same waters, meet, and sometimes interact. So these animals have multifaceted, multicultural societies. Cetacean cultures are transmitted both vertically, from parent to offspring, as well as horizontally between peers. There are highly stable cultural features, probably lasting generations, as well as transitory fads. So there is much to the culture of whales and dolphins. And maybe they are nothing without culture.

That is a brief summary of the preceding chapters. In this final chapter we step out and look more broadly at whale culture. With our limited insight into their lives, what may we be missing? How does the culture of whales compare with that of other species, whether nonhuman or human? And finally, does our limited understanding of the culture of whales and dolphins tell us anything about how we should view and treat them?

What Are We Missing?

The animal kingdom is symphonic with mental activity, and of its millions of wavelengths, we're born able to understand the minutest sliver. The least we can do is have a proper respect for our ignorance.[3]

Even for the species we know relatively well, like killer whales and bottlenose dolphins, we understand just a tiny part of their behavior, cultural or not. For all the eighty-four or so other whale and dolphin species we know nearly nothing. We are surely profoundly ignorant of large parts of whale and dolphin culture. The culture that we have described all comes in the form of general behavioral patterns—foraging, singing, and the like—that are relatively easy for us to observe and measure. We are missing two whole realms where there may well be cetacean culture: in the detailed intricacies of behavior and in overarching themes that are expressed across a wide range of behavior in the way these animals live their lives.

Imagine closely following a spotted dolphin, the way she searches for food, responding to so many little cues in the water around her and the sand beneath her. Watch her interactions with other parts of the environment and, especially, with the other dolphins. Second by second, her behavior shifts as she perceives and then reacts to her surroundings. How many of those behavior changes are shaped by what she has learned from other dolphins? And how much of this information is shared with other dolphins in her community? So, how prevalent is culture? We would not be surprised if there is a cultural element to nearly everything she does, but we don't know. Despite Denise Herzing's years of study of spotted dolphins underwater in some of the clearest waters on the planet, off the Bahamas, and some fascinating indications of social learning, we know of no definite spotted dolphin culture.[4]

The detailed, second-to-second behavior of humans has a major cultural component. If we watch someone just walking in silhouette at a distance we could make a pretty good guess whether they were raised in Norway rather than Italy, and if we watched them meet and greet another human from their own cultural milieu, our guess would become a near certainty. It has been shown, for example, that Americans and Australians are distinguishable by their gait and, even more so, by how they wave.[5] So we wonder whether there are similar subtle but telling differences between the deportments of members of the Regular and Plus-One sperm whale clans as they socialize off the Galápagos Islands, or fish-eating killer whales from the different communities near Vancouver Island. Our problem is that we are not sperm whales, or even cetaceans. Humans have been attuned through evolution to pick up on the differences in the behavior that discriminate Norwegians from Italians. The appearance of a Viking raider or Genoese trader in the British Isles would have had very different implications for our ancestors. But the differences between gaits or breathing patterns of a Regular and a Plus-One sperm whale will be in dimensions that were irrelevant to human evolution and, perhaps, not within our perceptual horizon. New technology will help. Ultrasensitive DTAGs (digital acoustic recording tags) are attached to cetaceans with suction cups for periods of a few hours, recording their every sound and movement.[6] The data retrieved from the tags can be analyzed using the latest statistical data-mining routines. But, although impressive, such technologies are still generally inferior to the products of natural selection—the sensory and cognitive capabilities of the species—so it may be some time before we can pick up the subtleties of cross-cultural expression in cetaceans.

Moving to the broader scale of the major forms of behavior, our record is brighter, as are our prospects. For a few species in a few places we have been able to document cultural, or probably cultural, forms of vocalizations and foraging behavior. This body of knowledge is expanding almost continuously as scientists record new killer whale, humpback whale, bottlenose dolphin, or sperm whale behavioral variants. The pace will keep up, and probably accelerate, as cetologists hew deeper and longer data sets from the seas, improve their methods, and develop studies in new areas. Other species will join the mix. For instance, there are several research projects on long-finned and short-finned pilot whales in different parts of the world that will soon shed real insight into the role culture plays in the lives of two of the most social and vocal of all mammalian species.[7] The species in which cetacean culture was first discovered, the humpback, killer, and sperm whales and bottlenose dolphins, have very distinct cultures. The other species may follow these models but may be distinct in completely different ways from one another. With more than eighty unstudied species, the possibilities are vast and exciting.

We need to exercise a little caution here, though. Perhaps the most-studied species are most studied in part because they are most cultural. Their cultures have helped them into particularly important, unusual, or prominent ecological roles, roles in which they interact with us humans as top predators, collaborators in fishing cooperatives, or depredators of our fishing gear. They proliferate into more niches, so we are more likely to encounter them in circumstances where we can do our studies. Their cultural behavior—songs and tipping ice floes to procure seals—fascinates us. So we focus on these animals, while the less cultural species, with smaller, less successful, more concentrated populations as well as less dramatic behavior, swim by unstudied. We have already started climbing the major peaks represented by the humpback whale, sperm whale, bottlenose dolphin, and, probably above all, the killer whale. In this scenario the cultural behavior of humpbacked dolphins, sei whales, or Stegner's beaked whales might turn out to be underwhelming compared to what we already know. Even so, our general understanding about the evolution of culture will still be profoundly affected because in comparative evolutionary analyses the absence of a trait can be as informative as its presence.

So, at this scale of behavioral modes, our knowledge of cetacean culture is still deficient with respect to the principal areas of behavior—mainly foraging, movement, and singing—in the best-studied species. It is deficient in the lack of any knowledge about how the majority of cetacean species come to behave as they do. And it is also deficient in the highly restricted scope of our knowledge. We have smatterings of information on the culture of play in killer whales and bottlenose dolphins, on how sperm whales learn to care for calves, and on social conventions in killer whales. But there must be more to learn about these types of behavior, as well as other important spheres of life, such as birthing, weaning, courtship, resting, and communication.

While we know almost nothing about how culture might affect the second-to-second detail of cetacean behavior, we have come by some fascinating slices of the cultural components of the general day-to-day activities of the whales and dolphins. But there is another, higher-level form of culture that is perhaps most intriguing of all: principles that channel behavior, principles learned from others, often parents and other older relatives but also a whole society. Some human societies have cultures of conformism, others have cultures of creativity or compassion. There are warrior cultures, hedonistic cultures, and capitalist cultures.

Can we talk about groups of nonhumans in similar terms? Perhaps we can. The most extraordinary evidence we know of concerns the origin and persistence of a tolerant and pacific culture in a group not of cetaceans but of olive baboons, a story that we briefly mentioned in chapter 11. Biologists Robert Sapolsky and Lisa Share had been studying these animals in the Masai Mara Reserve in Kenya. In 1978 they began to study a group they called the Forest Troop that was based near a tourist lodge. The Forest Troop usually slept in trees about one kilometer from the lodge. In the early 1980s, the garbage pit associated with the lodge was significantly expanded as the tourist trade grew, and another baboon group, called the Garbage Dump Troop, essentially made it their home, sleeping next to it and eating almost exclusively from the food remains discarded there. This food also attracted members of the Forest Troop, but because the Garbage Dump Troop defended their patch, an equilibrium was established in which only the more aggressive males from the Forest Troop—those best able to stand their ground—were able to forage at the dump consistently. It lasted until 1983. Then, in a devastating example of the risks inherent in these kinds of human/wildlife interactions, the baboons were hit by an outbreak of bovine tuberculosis caused by infected meat on the garbage pile. Most of the Garbage Dump Troop died, and it basically ceased to exist. Moreover, all of the Forest Troop that foraged on the garbage also died in the following three years, wiping out 46 percent of the males in that troop and doubling the ratio of females to males. In 1986, the researchers noted that the behavior of the group had changed because the males that remained were the least aggressive of the original troop but they also ceased collecting detailed behavioral data, understandably choosing to leave that apparently artificially damaged group alone, apart from censusing it every year. They started again seven years later, in 1993. By this point, none of the Forest Troop males from 1986 remained. All the adult males present had joined the group in the intervening years but, remarkably, the behavioral changes remained. Using both the Forest Troop before the tuberculosis outbreak and another neighboring group they had studied in the intervening years as controls, Sapolsky and Share showed in a 2004 paper that there was much more grooming going on from 1993 onward and that dominance-related conflicts between the males were different.[8] Almost all the conflicts were between males quite close in rank in comparison to the other troops, where high-ranking males regularly attacked extremely low-ranking males that were no direct threat to them in the hierarchy. In humans we would call this bullying, and it causes great stress to the victims. It apparently does the same in baboons, as low-ranking males in the control groups had much higher stress hormone levels in their blood than did the high-ranking males. But in the post-1993 Forest Troop, there was no such difference—low-ranking males had the same hormone levels as the highest ranking. The scientists noted that, overall, new males joining the group received more grooming from resident females than in the control groups. The disease outbreak had dramatically altered the social dynamics of the Forest Troop males by removing the more aggressive of them, and that difference had, through social transmission, outlived all of the original males present when it was established. Sapolksy and Share had witnessed the emergence and persistence of what they called "a pacific culture among wild baboons."[9]

Currently, we just don't know whales and dolphins well enough to make the same kinds of group level comparisons as Sapolsky and Share made, built as they are on years of highly detailed observations of second-by-second interactions. We have hinted at some of these types of overarching characters of societies in our descriptions of the general personalities of the cetacean species: conservative killer whales, innovative bottlenose dolphins, nervous sperm whales. But we know neither whether these characteristics are cultural nor the extent to which they vary between groups or populations. Perhaps killer whales have a conservative gene, sperm whales a nervous gene, and so on. We would love to be able to compare the degrees of conservatism among the different communities of transient, fish-eating killer whales off the west coast of North America or degrees of nervousness among clans of sperm whales off the Galápagos. Such differences would likely be cultural, but if they exist, we are currently ignorant of them.

We will end this discussion of the extent of our ignorance by briefly dangling our toes outside the scientific comfort zone, into an area few students of animal behavior dare tread. Consider this description of chimpanzees staring and displaying at a spectacular waterfall, written by Jane Goodall, the world's most famous primatologist: "As they approach their hair may bristle, a sign of excitement. And then they may start to display, charging with a slow, rhythmic motion, often in an upright position, splashing in the shallow water at the foot of the falls. They pick up and throw great rocks. They leap to seize the hanging vines and swing out over the stream in the spray-drenched wind. For ten minutes or more, they may perform this magnificent 'dance.'"[10]

It is hard to interpret this as misdirection of behavior that is intended for another chimpanzee or as play. So is it evidence of some kind of shared spiritual sentiment, one of the most hallowed forms of human culture? And is there anything like this in cetaceans? The only possibility we know of at the moment is that cetaceans, like chimpanzees, elephants, and some other animals, will interact with their dead. They can carry dead calves around for days, but is this just a misdirection of maternal care, or something more?[11] There is so much to learn just by continuing and developing our studies of the foraging behavior, movements, and songs of the species we know best. But we also believe there are extraordinary riches out in the ocean if we can start to focus on other forms of behavior and on other species. We need to bring the detailed second-by-second behavior of the whales and dolphins into our consideration of culture. We also need to look for the overarching principles that guide their behavior. Do communities of cetaceans share things we might describe as beliefs or values? This will take courage if it leads to stepping out of the scientific comfort zone, but in the words of Jane Goodall: "It is important that science dares to ask questions outside the prison of the biased mind, dares to explore new areas of animal being."[12]

Nonhuman Cultures

How do whale and dolphin cultures stack up beside those of other animals? Why would we even want to ask this question? Well, a fundamental reason that we prefer a broad concept of culture is so we can make precisely these kinds of comparisons. If we find links between cultural features and other factors—high mobility, or certain kinds of ecological roles, or social structures, for example—it potentially tells us much more about the evolutionary forces that favor culture as a solution to life's challenges than what we can piece together just from the fossils and archaeological remains of our own ancestors. Often these kinds of comparisons can turn into contests—which animals are "more" cultural or have "more important" cultures. We have fallen into that trap ourselves at times, but it is self-defeating because it just ends up in arguments we can't resolve because we don't know enough, rather than stimulating insight into bigger questions.[13] We must also admit that any such comparisons we make here will be incomplete—for many of the groups we will discuss here, entire books of their own could be written and, in some cases, have been.[14] So our picture will by necessity be broad but shallow and miss a lot of detail.

To address the questions we pose, we need first a list of which nonhuman groups we should consider cultural. In 2003, Kevin Laland and William Hoppitt, having thought about this, wrote that "the answer, which will surprise many, is humans plus a handful of species of birds, one or two whales, and two species of fish."[15] This statement was particularly provocative, perhaps by design, as their paper was primarily aimed at primatologists and anthropologists, for whom discussion of nonhuman culture had long revolved almost exclusively around the nonhuman primates. Any list of "cultural species" will depend both on how culture is defined and on the standards of evidence needed. At one end of the spectrum, there are definitions or levels of required evidence so restrictive that only humans can possibly make the list. At the other end is John Tyler Bonner's "transfer of information by behavioral means" under which slime molds have culture.[16] However, Laland and Hoppitt used a definition almost identical to ours when they arrived at their short list of cultural species. Their standards of evidence, though, were tough. They wanted "reliable scientific evidence of natural communities that share group-typical behavior patterns that are dependent on socially learned and transmitted information."[17] Hang on a minute though—what is so tough about that? What have we been talking about this whole book if not reliable scientific evidence? What Laland and Hoppitt meant by this was experimental evidence that the same behaviors proposed in natural communities to be cultural did in fact rely on social learning, by which they meant translocation or cross-fostering experiments. They did let "one or two whales" through, accepting the overwhelming evidence for killer and humpback whale vocal traditions, but no nonhuman primates, which they admitted would probably have primatologists "up in arms." The problem for many species, including apes, monkeys, and bottlenose dolphins, is that while there might be evidence for group-typical behavior patterns in the wild and for the social learning of behavior in captive situations, the social learning evidence does not refer to the same behavior as that observed in the wild.

Laland and Hoppitt's short list was actually just their first answer to the question of which species we should consider cultural. They also gave a second answer, for which they relaxed the standards of evidence. Using the same definition of culture and their "knowledge of animal social learning, observations of natural behavior of animals, intuition, and the laws of probability, we would say that many hundreds of species of vertebrate have culture."[18] We are going to take an intermediate position between the "handful of species of birds . . . and two species of fish" and "many hundreds of species of vertebrate"; we'll see how whale cultures compare with those of fish, birds, rats, primates, and elephants. Each of these comparisons has a particular interest.

In chapter 9 we suggested how the aquatic habitat of the cetaceans may have had a strong role in driving the evolution of culture. In this scenario, culture is particularly useful in red-noise environments, where the greatest unpredictability occurs over the longest time intervals. The ocean is a particularly red environment. Fish live there too, so what about fish culture? Fish, both marine and freshwater species, can clearly learn socially in experimental conditions.[19] They can learn from each other how to deal with predators, where to go, where to forage, and with whom to mate.[20] The best evidence for fish culture in the wild comes from coral-reef species, which are particularly easy (and pleasant!) to study. While snorkeling above a coral reef, scientists can watch whole soap operas at a range of a meter or two in the clearest conditions and can then conduct powerful experiments to test their hypotheses. For example, Robert Warner found that members of populations of blue-headed wrasse use specific mating sites on coral reefs off Panama.[21] Over twelve years, or four wrasse generations, there were no sites that were abandoned, and no new ones added. Then he transplanted whole populations between different coral reefs. The new fish used, and then continued to use, a different set of sites from the previous residents, areas that were suitable in every way but ignored by the previous population. These sites were part of wrasse culture. Extrapolating from this and general knowledge of the natural history of fish, we see the clearest parallels between fish and cetaceans in the transmission of cultural information about movements and foraging techniques. However, it seems unlikely that fish have anything like the vocal or play cultures of the whales or the sociocultural structures of the killer and sperm whales. Fundamentally we think this contrast is because of the mammalian characteristics that the cetaceans brought into the ocean, perhaps especially the mother-calf bond based on lactation that sets up strong social relationships.

Rats are much harder to study in the wild than coral reef fish, but they are a staple in the laboratory of the experimental psychologist. Experimental psychologists have many times shown that laboratory rats can learn certain things from one another.[22] Rat social learning includes what to eat, as well as how to use space.[23] The experiments have concentrated on the social learning of food preference or avoidance by rats. In one important case, field researchers overcame the difficulties of studying secretive, nocturnal animals in the wild, linked their observations to experiments, and gave us a very clear picture of animal culture. This is the study of the pinecone-stripping rats of Israel that we described in chapter 2.[24] Not only do we know the pattern of wild behavior and its ecological significance, we have a pretty good idea of the social learning mechanisms by which it is transmitted. This is a standout study in research on animal culture. Using the loose version of Laland and Hoppitt's method of positioning animal culture—the "knowledge of animal social learning, observations of natural behavior of animals, intuition, and the laws of probability"—we posit that the cultures of rats and some other rodents, like those of fish, largely concern foraging and movement. Unlike cetaceans (and bats), rodents, and most other terrestrial mammals, do not seem to learn their vocalizations socially.[25]

Among birds, the pattern is reversed. In some respects, such as the neurobiological, we may know more about vocal learning in birds than in humans. Some birds are adept at matching their own vocalizations to what they hear. So parrots imitate humans, dogs, or telephones, as well as other parrots. Lyrebirds have become television stars thanks to their ability to imitate the sounds of predigital cameras, and males of both this and another Australian bird family, the bowerbirds, build their vocal repertoires by copying the sounds of other species, as well as each other, very accurately.[26]

But above all, birds sing. Birdsong is one of the most prominent parts of the natural acoustic environment on land. Although the songs of some birds—such as the nonoscine passerines—are entirely innate, as are elements of the songs of the others, much of the repertoire of many songbirds is socially learned. And the intricate variable sections of the most spectacular songs are all, as far as we know, socially learned.[27]

The culturally transmitted songs of the oscine passerines and those of the baleen whales have much in common. If sped up, maybe sixteen times or about four octaves, humpback whale song sounds quite a lot like birdsong.[28] Birdsongs are also structured, used in courtship and mating, vary geographically, and evolve over time.[29] The songs of baleen whales and songbirds can be complex—humpback whales and nightingales—or they can be simple—fin whales and bullfinches—and we humans find some of each beautiful. Bowhead whales, minke whales and many species of songbird can produce two quite different sounds simultaneously during their songs. Birdsong as a whole corpus is perhaps more variable in structure, use, and function than whale song—for instance, some female birds also sing, and there are some simply astonishing duetting and chorusing performances among the wrens.[30] Of course, no bird sings as loudly as a whale, and the geographical scales of birdsong variation are tiny by whale standards, but these contrasts are driven by attributes we might not be so interested in from a cultural perspective—whales are huge, so sing loud, and their songs travel far in water. Thus while we have marveled at the steady reduction in frequency of the blue whales' songs, all around the world, the closest parallel in birds would seem to be the steady drop in frequency of the song of one population of three-wattled bellbirds over twenty-five years in Costa Rica.[31] While many birds accumulate gradual changes in song over time, like the evolutionary mode of humpback song change, we found it hard to locate an example of an entire population completely throwing out their whole song repertoire in favor of a new one, which is what happens in the revolutionary mode of humpback song change documented by Michael Noad, Ellen Garland, and their colleagues.[32] While the scales may be different, the learning processes driving them might nonetheless be quite similar. Both bird and whale song are illustrations of how culture can drive behavior into intricate, and sometimes beautiful, configurations and produce remarkably similar effects in very different types of animal. As we noted in chapter 2, birds, as well as rats and sea otters—who use anvils to extract the nutritious parts of shellfish—have been described as "one-trick ponies," having only one overt expression of cultural behavior.[33] But birdsong is quite a trick!

The "one-trick pony" label is unfair to birds in another way. (It is also perhaps unfair to ponies!) While bird social learning studies have concentrated on vocalizations, evidence is growing that they also learn other behavior from one other.[34] For example, some birds learn aspects of their migrations socially—hence the successful idea of using microlight aircraft to lead captive-raised whooping cranes on their first migration to the wintering grounds.[35] Birds also learn feeding and breeding habitats from each other. The choice of some of the decorations adorning the highly elaborate courtship bowers of male bowerbirds, such as shells or plastics of particular colors, seem learned from neighboring bowerbirds.[36] Birds also appear to use social learning to recognize predators or other dangers, to develop and understand alarm calls, and in a range of feeding-related activities.[37] A particularly well-known case of avian social learning is of tits opening the foil or cardboard caps of milk bottles (which used to be delivered to doorsteps and thus left unattended) and then drinking the milk, a behavior that spread across Britain between the 1920s and the 1950s.[38] Another famous case is the New Caledonian crows that make leaves into elaborate tools used to extract insects out of cavities, the finer aspects of which appear to be socially learned.[39] As birds are reasonably easy to study experimentally, we know a fair amount about the transmission of some of this bird culture. For instance, it seems likely that the tits did not imitate each other in opening the milk bottles; rather, they were socially learning by local or stimulus enhancement—that is, being attracted to other tits feeding and, thus, to milk bottles that had been opened already, then working out by themselves how to open them.[40] But local enhancement is still social learning, and so the opening of the milk bottles is still culture, by our definition. We suspect there is more to learn about bird culture, especially given the recent explosion of research on big-brained birds like crows and parrots.[41]

Birds have simply been studied a lot more, and in much more controlled ways, than cetaceans, so we have a much better appreciation of the roles culture can play in their societies. We also know of examples of complex behavior where social learning is not essential—the tool-using Galápagos woodpecker finches, for example, unlike the New Caledonian crows, can learn to use tools by themselves.[42] But do whales and birds have cultures "of a feather"? Certainly there are similarities—foraging and migration figure large in both, for example. Is this related to the relative mobility of the fliers and swimmers? Perhaps culture becomes especially adaptive if you are able to move long distances relatively easily—after all, this means you can easily end up somewhere you haven't been before, so the ability to quickly pick up on what the locals are doing could be a real asset. In both whales and birds, vocal culture seems tied to social bonds, but the bonds are different. The fundamental unit of social structure in many bird societies is the mated pair, a bond that can last for a lifetime.[43] There is no evidence we know of for such pair bonds in cetaceans, which appear to have a diversity of fundamental units, from mother-calf pairs to the social units, pods, and clans of sperm and killer whales. Yet in both birds and cetaceans, vocal cultures seem focused around generating and maintaining those bonds. Why has evolution favored vocal culture in these cases, and in ours, but not in many of the other mammal species that also have long-term social bonds?[44] We do not know the answers, but these are the kinds of questions the broad approach to culture can generate.

Another group of species with complex and flexible social systems are the nonhuman primates. The cultures of apes and monkeys have generated scientific and popular interest, awe, and angst. They are more "like us" than any other species and are usually the first places psychologists and anthropologists look for the evolutionary roots of things considered to make humans unique. Also, they have fascinating cultures. This combination produces the controversy. The position that "culture makes us human" has been comprehensively revised by recent discoveries about our closest nonhuman relatives, and we have had to get much more specific about what it is that might set human culture apart.[45] In the most famous study of ape culture, Andrew Whiten and his colleagues documented thirty-nine chimpanzee cultural behaviors, common at some study sites while absent at others.[46] How dependent are chimpanzees on this culture? Well, chimpanzee and other primate females raised without exposure to their own mothers, or other maternal behavior, are substantially poorer first-time mothers than those with the appropriate social experience.[47]

As we described in chapter 2, the concept of culture in nonhumans, as well as its serious study, began in Japan with studies of Japanese macaques. The spread of sweet potato washing by monkeys on the island of Koshima was the first and most famous example.[48] But these animals have additional cultural traditions that incorporate other foraging techniques (such as fish eating and sluicing wheat to separate chaff), as well as traditions that have nothing to do with foraging, including bathing in hot springs and a behavior that seems bizarrely arbitrary: stone handling.[49] Primatologist Michael Huffman has studied this intriguing stone-handling behavior for decades.[50] It is known to occur in at least ten different macaque troops, four of them captive and the rest provisioned, so all having had considerable contact with humans.[51] The behavior is simply as its name suggests—the handling of stones—small enough to be picked up in the macaques' hands—in various ways. These include biting and licking stones, putting them in the mouth, carrying them in various ways, cuddling them, stroking them, pounding them into the ground, clacking them together, and wrapping them up in leaves.[52] Huffman himself made the first observations of the behavior in one group—the Arashiyama troop of Kyoto—which was performed by a juvenile female in 1979, but he didn't see it again for the next nine months of observation. When he returned to observe the group again in 1983, he was struck by the observation that stone handling had become a daily part of the troop's routine.[53] And the behavior has persisted and expanded in complexity since then. Huffman has gone on to record variations in the stone handling repertoires across the ten troops known to do it, as well as evidence that it is socially learned behavior.[54] Why they do it is something of a mystery, as it has no obvious tangible benefit, beyond speculative suggestions that it may relax or stimulate the animals in some way, like our worry beads.[55] There is clearly a lot of culture we don't understand in macaque societies.

Capuchins, the busy New World "organ-grinder" monkeys, also have a range of likely cultural behavior, including foraging technologies. But most famous, and bizarre, are the social conventions and games we described in chapter 8. Capuchin social conventions can be both weird and brief. But one thing they do not seem to do is shape capuchin society. As we described, Susan Perry, whose Costa Rica study site contains some of the oddest capuchin conventions, initially thought they might function to delineate groups in some way.[56] But the more she looked the less this seemed likely. While the pairs of animals passing hair or sniffing hands were fascinated by the game, the ritual was unobtrusive and no one else seemed the slightest bit interested. Not quite what one would expect from an ethnic symbol. It does not look as though these cultural behaviors drive social structure, rather the reverse: capuchin culture is most captivating as a reflection of their intense social lives, illuminating the fascinating relationship between capuchin culture (based on social learning that occurs within social relationships) and social structure constructed by the patterning of those social relationships.

Another primate species for which we know a fair amount about culture, the orangutan, is almost the opposite, in that, being a generally solitary creature, it has limited social structures. However, bucking the theme we have developed so far about the link between culture and social structure, the big red ape of the Southeast Asian jungles nonetheless has quite a range of behavior that seems to be cultural.[57] Likely orangutan culture includes using tools to poke into holes in logs for insects, using other tools for sexually stimulating themselves, making sunshades above their nests, using leaves as napkins, and "snag riding": jumping on a rotten tree, riding it as it falls, and then, at the last moment before the tree smashes to the ground, grabbing onto nearby vegetation, such as vines.[58] Bungee jumping for orangutans! Largely missing from the list of likely or very likely cultural variants are social behaviors. This is not very surprising for a species that is not overtly very social. But it does bring up the question: How do orangutans learn their cultural behavior? Infant orangutans stay with their mothers, and sometimes older siblings, for two or more years, and that is a potentially great opportunity for social learning. Much orangutan culture, though, seems to be horizontally transmitted. The populations in which individuals spent less than 20 percent of their time with nondependents had considerably fewer cultural variants than those in which association with others was more the norm.[59] However, the fact that quite solitary animals like orangutans can accumulate substantial cultures suggests that we cannot presuppose that less gregarious cetaceans, such as minke whales, are not cultural.

The cultures of macaques, capuchins, and orangutans are fascinating. However, in all the research and discussion about culture in nonhuman primates, they, as well as the birds, rats, and whales, are eclipsed by the chimpanzee. Chimpanzee culture has grabbed attention for three reasons. First, they are our closest evolutionary relatives and thus a natural focus when we ponder the emergence of human culture. Second, while not trivial to study, primatologists have worked out effective ways to observe wild chimpanzees consistently: basically the researchers habituate a group so that they ignore human presence—a lengthy task—and then follow the chimpanzees around watching what they do as individuals. Partly because their behavior appears similar to ours and so is easy to catalog, these primatologists have accumulated large and comprehensive data sets of what chimpanzees actually do in the wild. Unlike other "cultural species," chimpanzees are not nocturnal or aquatic, and some populations live in quite accessible habitat. In contrast the bonobo, or pygmy chimpanzee, similarly related to humans as the common chimpanzee but living in much less accessible habitat, is a cultural unknown. The third reason why chimpanzees are so prominent in discussions of nonhuman culture is that they have lots of it.

In 2003, a prominent proponent of the concept of chimpanzee culture, William McGrew, wrote an article titled "Ten Dispatches from the Chimpanzee Culture Wars."[60] The titles of the first two sections of that article headline the polarized argument that he had stimulated over his illustrious career: "Chimpanzee Culture? Absurd!"; "Chimpanzee Culture? Of Course!"

The cultural patterns of chimpanzees—identified as those kinds of behavior common at some sites and absent at others with no clear ecological explanation for the difference—come in several spheres.[61] They use a range of tools as clubs, pestles, hammers, anvils, and fly whisks, as well as for extracting insects from holes and for tickling themselves. Their culture includes different ways of drawing attention to themselves (by knocking knuckles or slapping branches) and particular ways of holding each others' hands above their heads when pairs groom one another. Chimpanzee culture, it has been suggested, includes rain dances and methods of self-medication.[62] Although experimental psychologists initially were skeptical of the social learning abilities of chimpanzees, they have now shown that chimpanzees can learn from one another in controlled laboratory conditions.[63] Compared to humans, chimpanzees, when learning socially, seem more focused on emulation (achieving the same goal as the demonstrator) than imitation (copying every action).[64] However they manage it, chimpanzees have built a substantial culture—or rather, different cultures in their different communities. The evidence is not without controversy though, as much of it relies on the risky method of exclusion approach we discussed in chapter 2, and recent research suggests a need for caution because the similarity of behaviors between chimpanzees of different communities correlates with their genetic similarity, and chimpanzee communities that are genetically very similar have few behavioral differences.[65] This does not mean that the behaviors are not culture, but it does mean that we have to be more careful in interpreting them. Perhaps some putative chimpanzee culture is not socially learned. At the same time, we have the very good experimental evidence that chimpanzees are amply capable of social learning, so on balance it is likely that much putative culture in chimpanzees is the real thing.[66]

The comparison between the cultures of nonhuman primates, especially chimpanzees, and human cultures has almost reached industrial levels among academics. Unsurprisingly, those with different perspectives reach pretty different conclusions. A recent review by Andrew Whiten, who has studied social learning in both humans and chimpanzees, strikes a balance within this diversity.[67] He notes a range of features shared by human and chimpanzee culture: numerous culturally determined behaviors; local cultures including characteristic suites of these behaviors, perhaps clustered into higher-level cultural types (such as using different kinds of tools, twigs versus leaves, as potential solutions to new problems); a range of social learning mechanisms including imitation and emulation; recognition of copying; tool use; foraging techniques; social conventions; and dialects. But in all of these shared features, human culture goes way beyond chimpanzee culture. There are other features characteristic of human culture where there are glimmerings of a possible presence among chimpanzees: cumulative cultural evolution and ratcheting; conformity; and teaching. And some features of human societies, such as political systems, agriculture, and symbolic marking, seem completely absent in chimpanzees, as well as in other nonhuman primates.

Lining up the chimpanzees and other apes and monkeys against the whales and dolphins is much harder than the human-chimpanzee comparison. We know so little of whales and dolphins, they live in a different habitat where they naturally do things differently, they use different senses, and they don't look at all like us or like chimpanzees. But there are areas where the evidence points to a clear difference. The chimpanzees and other primates use more tools, more often, in more complex ways and for a wider range of purposes than the cetaceans. The cetaceans, however, appear to be superior social learners, especially in the vocal domain, where it is an enduring puzzle why chimpanzees show nothing anywhere close to the vocal learning capacities of both humans and cetaceans. There is arguably better evidence in the cetaceans for communal behavior as well as some suggestions of symbolic marking that we will discuss in the next section. Chimpanzee societies are not multicultural anywhere near to the same extent as those of killer and sperm whales, where groups of animals with quite different cultures use the same waters. In other respects, such as the complexity of foraging techniques or the overall diversity of cultural behavior, it is hard to pick a winner. This is fine because picking winners is not our goal; we are trying to understand why culture ended up where it did.

The rarity of tool use by cetaceans and its prevalence among the primates fits with the lack of material for making tools in the ocean and a lack of thumbs, or even fingers, among the whales and dolphins. Conversely, sounds travel particularly well in the ocean and the whales and dolphins have evolved the plumbing to make complex sounds. Chimpanzee culture exists largely within well-defined groups that possess, and live within, defended territories. Thus meetings between groups possessing different cultures are rare and generally hostile; and while females regularly disperse between groups, the males do not. A consequence may be that cultural group selection will be relatively unimportant, as differences in cultural behavior between groups rarely influence success in contests for resources. This in turn lowers the potential for some kinds of gene-culture coevolution. It also minimizes the likelihood of symbolic markers of group membership as there is little need for them when you know your territory as well as all the individuals in your group, and that anybody else is likely to be trouble.

So far in this section we have compared cetacean cultures to those of the nonhumans whose behavior has been studied from the "is it culture?" perspective: the fish, rats, birds, and primates. There are lots of other species. Meerkats teach.[68] Bats—with their complex societies, mobility, and mother-pup relationships—are intriguing, and while we suspect there is much to learn, there is little current scientific data on bat culture, beyond some fascinating examples of groups of greater spear-nosed bats converging on shared calls that allow the identification of roosting group-mates.[69] In chapter 9, we discussed the evidence for culture in the other marine mammals, the pinnipeds and sirenians, and why they have so little of it compared to the whales and dolphins. We will end this section with a similar but larger puzzle, the elephant.

With the biggest brains of any terrestrial mammal, substantial cognitive abilities, and complex social systems much like that of the sperm whale, we would expect cultural complexity among elephants. Elephants can be studied much more easily in the wild than sperm whales. In some open habitats they can be followed around, and watched continuously, from a sufficiently rugged vehicle. Elephants are also held in captivity, where some work well with humans and so can be used in experiments. Yet, there is little scientific report of elephant culture.[70] A couple of years ago an eminent Harvard psychologist told Hal that the reason there are no reports of elephant culture is because elephants don't have culture. In contrast, another psychologist Gay Bradshaw, in her book *Elephants on the Edge: What Animals Teach Us about Humanity*, explicitly assumes the presence of an elephant culture "as complex as it is vast" but provides no empirical evidence for this.[71] So who is right? We suspect that it is Bradshaw. Elephants maintain large stores of knowledge about their social and physical world, and they can learn from each other their vocalizations, what to eat, and probably much more.[72] Something of a window has recently been opened by two scientists who, ironically, are primarily known for their work on primates, Lucy Bates and Dick Byrne, working with long-term data sets collected in the Amboseli National Park in Kenya by Cynthia Moss and Joyce Poole. Bates and Byrne have shown how elephants keep a kind of mental map of the whereabouts of family group members even when they are out of sight, by surprising the elephants with scents of individuals that they know but are in "unexpected" locations.[73] They have also shown that elephants are well able to identify, and react appropriately to, both the typical scent and typical garment color of young Maasai men, who have a cultural practice of demonstrating virility by spearing elephants, in comparison to similar cues from the much less dangerous Kamba tribe, who have a more agricultural culture.[74] The elephants can not only distinguish between the tribes but also between dangerous and benign members of the tribes—young males and females, respectively—solely by listening to recordings of their voices.[75] Perhaps most fascinating from our cultural perspective is the phenomenon of simulated estrus. Bates, Byrne, and their colleagues have shown that female elephants will sometimes perform the behavioral signatures of being in estrus (i.e., fertile)—distinctive postures and gaits, as well as a higher frequency of touching—even when they cannot possibly actually be in estrus because they are pregnant themselves, lactating (which suppresses fertility), or simply too old.[76] Why do they do this? Perhaps a hint is given by the fact that they do it more when in the presence of young female kin who are themselves coming into estrus for the first time. The most likely explanation, according to Bates and Byrne, is that older females fake estrus in order to demonstrate to blushing debutante females just how to handle themselves in adult society. Picking the right male is important. You want a big mature male in musth—an annual testosterone-driven state of sexual activity and aggressiveness—because these males not only will likely be the best quality but will also be able to defend the young female from the unwanted attentions of multiple younger males who haven't yet demonstrated their quality. This knowledge is apparently acquired through experience, and the simulated estrus behavior of older females appears aimed at helping them learn.[77] As a form of social learning, this is not on face value too far from teaching. The value of the information held by older individuals is indicated by the responses of elephant groups to a prolonged drought in Tsavo National Park, Tanzania, in 1993. Calves were much more likely to survive if their groups left the park during the drought, and they were much more likely to leave the park if their clan contained females old enough to have experienced the preceding severe drought in 1958–61.[78] So the elements of culture are there and we await, and expect, the descriptions of the behavioral patterns that elephants learn from each other, descriptions that will complement, and perhaps sometimes exceed, those for chimpanzees, orangutans, birds, and our own studies of whales. But perhaps not.

The quandary of the cultural elephant is an extreme version of our uncertainty about the extent and significance of culture in whales and dolphins. For birds, rats, fish, and apes, we have an idea both of some of the areas where culture plays a role and of those where it does not, with a bog of uncertainty between. The scientific literature on elephant culture—basically, the lack of it—leaves almost complete uncertainty as to how much of a role culture has in their lives. For some whales and dolphins we know that culture is important in some areas—primarily for songs, foraging methods, movements, and play—but we do not know the extent of its significance or in what other areas it may be important. There are indications of higher-level cultural processes, such as conformism and symbolic marking, and maybe much more. It is the "maybe much more" that particularly excites us and contrasts the cetacean picture with how we understand the culture of birds, chimpanzees, and fish. For these groups of species we have an idea, admittedly fuzzy, of where their culture does not reach. For the whales and dolphins, as well as elephants, we know that there are no complex material technologies or sophisticated syntactical languages, but in most other respects we do not know where the limits of their cultures lie. The indications that culture is affecting other areas of these animals' biology, such as gene-culture coevolution, menopause, and maladaptive behavior, reinforce our view that the best is yet to come.

Cetaceans and the Key Attributes of Human Cultures

For instance, on the planet Earth, man had always assumed that he was more intelligent than dolphins because he had achieved so much—the wheel, New York, wars and so on—whilst all the dolphins had ever done was muck about in the water having a good time. But conversely, the dolphins had always believed that they were far more intelligent than man—for precisely the same reasons.[79]

This is where it gets really tough. The very act of comparing human and nonhuman cultures is an anathema to some. But even for those of us willing to take the plunge, it is a big plunge when it comes to the whales and dolphins. There is the fundamental problem, highlighted by Douglas Adams in *The Hitchhiker's Guide to the Galaxy*, that what is important to their culture may not be important to us, and vice versa. So here we are, as humans, using our culturally derived scientific methodology to compare the culture that we all use and that shapes us with that of beings who lead completely different lives and for whom wheels, cities, and wars, as well as the scientific method, have absolutely no relevance. Furthermore, that culturally derived scientific methodology that we use, including reductionism, Occam's razor, and the like, leads us to regard the cultural explanation for behavior as the null hypothesis in humans—we have to prove genetic causation—and the alternative hypothesis for nonhumans—we have to prove cultural causation. The whole comparative process appears triply stacked against whale and dolphin culture. Neither, though, should we let the pendulum swing too far the other way and shy away from discussing obvious objective contrasts between us and them.

In chapter 2 we concluded that the key elements of human culture seem to be technology, the accumulation of cultures beyond any idea that one individual could innovate from scratch, morality, culturally transmitted and symbolic ethnic markers, and its effect on biological fitness. After some equivocating, we added language as a maybe key attribute. When we look at cetaceans, how many of these stand up as uniquely human?

We will start with the maybe key attribute, language. Language is a word with many meanings. Definitions of language that explicitly restrict it to humans are of no more use to us than definitions of culture that include the word "human." At the other end of the spectrum of definitions of "language," we do not mean communication. Whales and dolphins, as well as virtually all other animals and many plants, communicate. Whales and dolphins communicate a lot, mostly acoustically. Some of this communication is subtle and complex, but it is not, as far as we can tell, language. The key attribute of human language for those considering the evolution of modern humans is syntax. Syntax allows a finite number of words, together with a few rules, to express an infinite number of concepts in a wonderful, and extremely useful, system of communication. Dolphins and other animals, including bonobos and African gray parrots, can learn and use some syntax when taught by humans.[80] They can distinguish between "bring the ball to the hoop" and "bring the hoop to the ball." Dolphins appear to use signature whistles as quite specific labels to refer to individuals. But to our knowledge, wild whales and dolphins do not use syntax, or use it in only an exceedingly simple way. We might easily be missing some basic cetacean syntax. But there has been enough study of dolphin whistles, killer whale calls, and humpback whale songs that we can be fairly sure that their syntactical organization is very simple at best and does not usually convey important information. When we first heard the patterned codas of sperm whales, an immediate hypothesis was syntax. Maybe the "4+2" click coda meant "go to [4] my calf [+2]." But the more we studied codas and the way the whales used them, the more it became apparent that these strange patterns of clicks were important in bonding and other relationships between individual whales and groups of whales, and the less likely it was that they were conveying complex information syntactically.[81] Thus the whales and dolphins, like apes, monkeys, and birds, do not have the long narrative stories, lectures about the structure of their worlds, or sets of instructions for doing or making complicated things that our language gives us.

Why don't whales and dolphins have language? It does not seem to be a cognitive deficit if they can use syntax when trained by humans. In their open marine societies, language might seem more of a benefit than in the stifling dominance hierarchies so common on land. Michael Tomasello and his colleagues have suggested that pointing is important in the development of a referential language in humans.[82] But dolphins can point to things and use sounds to refer to one another.[83] They communicate frequently and enthusiastically. So we are left with a puzzle in the lack of a syntactical language, or much syntactical language, among the whales and dolphins. The evolution of human language has been called "the hardest problem in science."[84] But its converse, the lack of language evolution in other species, is also taxing. According to Szabolcs Számadó and Eörs Szathmáry, two Hungarian biologists, no current theory for the origin of human language adequately explains its uniqueness.[85]

Less of a puzzle is the cetaceans' lack of material technology. Technology is the product of the tools, techniques, and crafts that we learn. Human technology has a large material component. We make spears, electric guitars, and space shuttles. By contrast, in terms of the cultural manipulation of material, the cetaceans—with so far only the putting of sponges on the beak recognized as such—are not in any way comparable to even "primitive" human societies—or even to chimpanzees. Only if technology includes manipulation of the acoustic environment can we talk about it as a feature of cetacean societies. While the magnificent sonar systems of the toothed whales and dolphins are the product of genetic evolution, there may be a large cultural element in how they are used. We do not know.

One of the things pretty much everyone agrees makes human culture special is that it is cumulative. We rely on a cultural inheritance constructed over millennia, which it would take us untold generations to re-create were we to lose it. In the words of Pete Richerson and Rob Boyd, "No other species seems to depend on culture to anywhere near the degree that humans do, and none seem adept at piling innovation atop innovation to create culturally evolved adaptations of extreme perfection."[86] Accumulation is key to democracy, iPhones, and opera, all of which are spectacular adaptations—although perhaps only opera could be described (and even then, only some humans would do so) as possessing "extreme perfection." While cetaceans have nothing like any of these, neither did human societies of ten thousand years ago. Also, there is little doubt that cetacean cultures accumulate changes in some ways. The song of the humpback whale is always changing, and it seems unlikely that its complexity could have arisen solely through the independent innovation of one animal. The humpbacks' lobtail feeding seems to be the cumulative combination of two other types of behavior, lobtailing and bubble feeding.[87] It is also plausible that elements of the matrilineal whales' vocal and nonvocal cultures incorporate accumulated knowledge about their habitat and the periodic changes it might undergo, and it is hard to see how the complex behavioral sequence of cuttlefish processing by dolphins would have been invented all in one go. It is extremely challenging, though, even to consider how we might collect definitive evidence for this. Nonetheless, culture in cetaceans has clearly not accumulated in any way comparable to that of human technological culture, and there seems to be no evidence of cetaceans being able to take advantage of material cultural inheritance—passing on of money, homes, or grandfather's favorite pipe—or otherwise to embody externally the content of their culture—say, in writing, diagrams, or electronically. It is confined to their brains, as was our ancestors' for millennia.

Morality—norms of acceptable behavior—is another key feature of human culture. The evolution of morality is an area of immense controversy, and there is disagreement about how we would recognize such a thing in nonhumans if identifying morality means having access to the motivations, conscious or otherwise, underlying observed behavior. The debate takes roughly the following shape. Only humans can be moral, one side of the argument goes, because only they can introspectively and rationally monitor their own actions to decide whether they are right or wrong. This ability is rooted in our theory of mind but develops in a cultural context, rather than being something we are born with.[88] For younger children it is outcomes that trump motivations—Jane breaking three glasses accidentally is more heinous than Judy breaking one glass on purpose—but as they develop, the perception reverses, so that Judy becomes (correctly, from the adult perspective) identified as the "naughtier." So only humans can be moral because only they can rationally consider their own actions and only they develop in a cultural context that imbues a moral sense. Those in the opposite corner, for instance, Frans de Waal, Mark Bekoff and Jessica Pierce, as well as philosopher Mark Rowlands, reject the notion that morality is such an intellectualized concept.[89] They argue, instead, that human morality is deeply rooted in our evolutionary heritage, that in fact, in Rowland's words, "our natural sentiments—the empathy and sympathy we have for those around us—are basic components of our biological nature."[90] Their position is somewhat reinforced by recent studies showing that human infants begin making judgments about the behavior of others before they can even talk.[91] From this perspective, we should not be surprised to find things in nonhumans that both resemble and are directly ancestral to human morality.

They are supported by many examples of humans behaving in ways most of us would consider demonstrative of the highest moral standing, without any of the introspection some consider fundamental to human moral agency. Consider the following excerpt, from a study of selfless acts, of an interview with a Dutch woman who had worked to save Jews from the Nazis during World War II:

INTERVIEWER: It was just totally nonconscious?

MARGOT: Yes. You don't think about these things. You can't think about these things. It happened so quickly.

INTERVIEWER: But it isn't really totally quickly, is it? There's a tremendous amount of strategic planning that has to be done [to do what you did].

MARGOT: Well, I was young. I could do it. Today I don't know. I'd have to try it. But I was 32 years old. That was pretty young.

INTERVIEWER: You didn't sit down and weigh the alternatives?

MARGOT: God, no. There was not time for these things. It's impossible.

INTERVIEWER: So it's totally spontaneous? [*Margot nodded.*] It comes from your emotions?

MARGOT: Yes. It's pretty near impossible not to help.[92]

In February 2013, while we were in the process of writing this book, a sixty-three-year-old British man, George Reeder, jumped into the sea to save a drowning baby and stated afterward, "I didn't have time to think. I just jumped in and pulled the buggy back to the edge of the quay."[93] For those denying a kind of moral agency to animals, examples such as these don't matter—even if we often act morally without thinking about it, it is simply the fact that we are able to think about it that makes humans uniquely moral.[94] We find this position so dubious that it naturally leans us toward the opposing arguments.

Donald Broom, an emeritus professor of animal welfare at Cambridge University, has a straightforward perspective. For him, morality is having the sense that some actions are right, others wrong and then acting accordingly.[95] Philosophers equate "right" with meeting individuals' basic needs (food, shelter, health) and maintaining their rights (life, freedom, privacy), while "wrong" is equated with actions that impede these needs and rights.[96] Broom believes that there is considerable evidence for this sense of morality in nonhumans. We will discuss later whether whales and dolphins have, or should have, rights, but they certainly have needs. When whales and dolphins go out of their way to help other creatures with their needs that looks pretty moral, at least on the surface. Here is what happened after some Antarctic pack-ice killer whales knocked a seal off an ice floe in January 2009, as told by Robert Pitman and John Durban:

> The predators succeeded in washing the seal off the floe. Exposed to lethal attack in the open water, the seal swam frantically toward the humpbacks, seeming to seek shelter, perhaps not even aware that they were living animals. (We have known fur seals in the North Pacific to use our vessel as a refuge against attacking killer whales.)
>
> Just as the seal got to the closest humpback, the huge animal rolled over on its back—and the 400-pound seal was swept up onto the humpback's chest between its massive flippers. Then, as the killer whales moved in closer, the humpback arched its chest, lifting the seal out of the water. The water rushing off that safe platform started to wash the seal back into the sea, but then the humpback gave the seal a gentle nudge with its flipper, back to the middle of its chest. Moments later the seal scrambled off and swam to the safety of a nearby ice floe.[97]

To give another example, on March 12, 2008, Moko, a bottlenose dolphin, guided a mother-calf pair of pygmy sperm whales out of an intricate set of sandbars off the coast of New Zealand where they seemed hopelessly disoriented and trapped—rescue workers were considering euthanasia after the pair stranded themselves four times.[98]

And then there are the cases when dolphins, or occasionally whales, save humans. One that particularly appeals to us as sailors occurred during a sailing race in the Caribbean.[99] A sailor fell overboard in rough seas and was quickly out of the sight of the other crew on his boat. The racing boats stopped the race and crisscrossed the area but could not spot the swimmer, who kept treading water for a couple of hours. It must have been a horrible time, but perhaps there was some relief when dolphins came by, apparently to keep him company. At about the same time one of the search boats noticed dolphins approaching and then moving off in a particular direction. They did this several times. The sailors on the search boat wondered whether it was some kind of signal and followed. They soon found the swimmer and his dolphin companions. The preceding examples are just a very limited sample of the many purported cases in which dolphins or whales have acted in such a way as to help humans or other animals in mortal peril.[100]

Although it may have been crucial for the seal faced with the most dangerous of predators, the flipper nudge from the humpback whale seems relatively minor in the scheme of things. However, if a human had similarly used an arm to pull a dog to safety from heavy traffic, most would consider it to be a moral act. The stranding team (i.e., rescuers of stranded marine animals) who unsuccessfully tried to save the lost pygmy sperm whales subsequently piloted to safety by Moko the dolphin and the searching Caribbean sailors who could not find the man overboard without dolphin help were all "doing the right thing" by the moral standards of human culture, so it seems churlish to deny the same of the cetaceans who actually did the saving. But, the counterargument runs, how do we know that the dolphins did what they did because they thought it was the right thing? The answer is that we cannot. Then again, how can we ever know it for sure of our fellow humans? Again, we cannot. Perhaps everyone but you, dear reader, is a cold-hearted sociopath who only appears to act morally because they fear the consequences of not doing so, while professing to feel the same natural empathy that you yourself do. Day to day, we humans use a simple heuristic that if other humans talk, and more important, act in ways that we can explain with reference to our own internal moral sense, then we give them the benefit of the doubt.[101] If it waddles and quacks, most likely it is indeed a duck. If a human did what the cetaceans did in the episodes we just described, there would be no debate as to whether they were moral.

Morality includes both right and wrong. Among the most compelling anecdotes suggesting that dolphins have concepts of "wrong" behavior is Thomas White's description of how a human snorkeler observing Atlantic spotted dolphins off the Bahamas went outside the bounds of the norms of behavior expected by the dolphins of human observers at that site.[102] The swimmer approached a calf engaged in learning to fish with its mother, a no-no in the rules of engagement between swimmers and these dolphins built up over years. When this happened, the mother then swam not to the hapless trespasser but to the leader of the group of swimmers, whom she could identify, and tail-slapped, her displeasure apparently directed at the leader who had not controlled the behavior of those being led.

One of the reasons for focusing on interspecific acts is that when helpful they are the most obviously altruistic. The recipients are clearly not related to the helpers, nor are they likely to repay them. Such acts are often dismissed as misdirected maternal behavior. Pitman and Durban consider this explanation for the humpback whale who saved the seal but ended their article by saying: "When a human protects an imperiled individual of another species, we call it compassion. If a humpback whale does so, we call it instinct. But sometimes the distinction isn't all that clear."[103] Between-species morality is most easily recognized, but it is how animals treat their own that matters most. Watching, and listening to, captive spinner dolphins, Ken Norris realized there were "echolocation manners": "Kehaulani and her schoolmates never sprayed each other with loud sounds, but when they were in an echolocation bout they generally swam below the others, head angled down and away from their tankmates. . . . In one hundred passes, our echolocating dolphins never once sprayed the other animals directly with their click trains. This jibed with our earlier observation that wild dolphin schools almost always are arranged in staggered three-dimensional echelons, like fighter planes in formation."[104]

Being zapped by dolphin echolocation is unpleasant, as well as marginally invasive—because it is using ultrasound, the echolocating dolphin can check out one's internal organs—but sperm whale sonar is something else. It is lower frequency and about twenty decibels louder, or a hundred times more powerful, than dolphin echolocation.[105] We sometimes wonder about behavioral norms among the sperm whales that we study. Sperm whales spend most of their lives foraging together at depth, using the most powerful sonar in the animal world. Echolocation clicks directed at each others' ears could potentially incapacitate the ability to forage for some time and might cause permanent damage if the emitter was at close range.[106] Thus sperm whales foraging together are potentially dangerous to one another. And they do forage together. In the Pacific, where groups are large and fairly compact, there might be thirty or more sperm whales echolocating within a couple of kilometers of one another. Imagine thirty or so humans hunting together in a forest using guns. Without norms of use, shared agreements about the right thing to do, there would be many dead hunters and deaf sperm whales.

If we put all this evidence together, it is at least sufficient to give us pause when asking whether whales and dolphins do have an idea of what is right and what is wrong, both for them and, sometimes, for us. At least sometimes they seem to act according to these precepts. In consequence, if we accept Broom's concept of morality, it seems to exist in the whales and dolphins, although we do not know its extent. In the context of this book, there is an additional question: Are these moral actions cultural? The assistance provided to other species in peril, the norm about how human snorkelers should behave, and "sonar safety" procedures might all be instinctual, without any contribution from social learning. We suspect not, but there is no good evidence that any cetacean morality is socially learned and, thus, culture.

Related to morality is ethnicity. Ethnic boundaries may set, and be set by, moral codes. Symbolic marking is necessary for ethnicity, and as we saw in chapter 8, the apparent lack of this in nonhumans is a major factor in leading many to draw the cultural line between ourselves and the rest of the animal kingdom. When cultural elements come to define different social groups, then this can intensify social connections within the social groups and distinctions between them.[107] In our multicultural societies, we preferentially, or perhaps only, associate with those who speak "our" language or wear "our clothes." We fight behind flags symbolically proclaiming ethnicity. Symbolic marking is seen as a particularly human cultural attribute, allowing us to build up societies over scales far greater than achievable by the standard mechanisms—kin selection and reciprocity—that produce cooperation in other species. The lack of evidence for symbolic marking in the apes and monkeys emphasizes that symbolic marking can be a key element of human culture as well seeming to form a clear divide between *Homo sapiens* and the rest of the animal kingdom.[108]

However, it does look as though the matrilineal whales use their vocalizations as markers of some kind of social boundaries. Killer whale matrilineal units seem to shape their pulsed calls to maintain distinctions between neighboring units.[109] And the differences between the coda dialects of overlapping sperm whale clans in the Pacific are greater than those between the dialects of sperm whales in different parts of the North Atlantic Ocean where there do not seem to be overlapping clans.[110] These are both indications of "ethnic" markers, but the whole tenor of the matrilineal whales' social lives is such that we would not be surprised to find symbolic delineation among elements of the different levels of social structure by animals that share ranges and encounter one another frequently. The coda clans of the female sperm whales in the Pacific, overlapping and with thousands or tens of thousands of animals in each, are unlike any other social structures so far identified in nonhumans and, in some respects, look very much like human ethnicities. Perhaps the young Pacific sperm whale learns that there are things "we do," things "they do," and "we" do not associate with "them." In the coda and the pulsed call of killer whales we have good candidates for culturally transmitted symbolic markers of "us" and "them."

The final key attribute of human culture is that it affects biological fitness—basically, that a human's culture has bearing on how many grandchildren he or she is likely to have. This attribute allows culture to affect genetic evolution, and it clearly has in humans. What about cetaceans? The clans of sperm whales, while living in the same areas, seem to have different reproductive rates. The numbers of groups with calves, as well as the proportion of immature whales in each group, vary considerably between the clans.[111] There are several ways in which the cultural differences in behavior may result in contrasts in the number of calves raised by members of different sperm whale clans.[112] The clans have consistent differences in feeding success, so that females in one clan may be in generally better physiological health than those in others; the differences in their babysitting methods may reflect in the rates at which the calves survive; there might also be clan-specific ways of protecting against predators, again changing the average survival rates of the young; and there could be mating norms ("*we* only mate with males who . . ." or "*we* only mate when . . ."), resulting in different pregnancy rates.

In Shark Bay, Australia, the reproductive success of a female bottlenose dolphin depends partially on her cultural inheritance. A study by Janet Mann and her colleagues in 2000 found that calves of provisioned dolphins that came to the Monkey Mia resort beach to receive handouts from humans had over twice the probability of dying in their first year compared to those of nonprovisioned females.[113] In recent years, as provisioning behavior has become much more controlled, this difference has diminished. The original shortfall may have been due to diseases that the young dolphins acquired in the resort area, for instance, from sewage; less contact with their mothers who were intent on collecting the goodies; or changes in the availability of food or risk of predators in nearshore areas where the provisioning occurred. In any case, the cultural trait of using human handouts was not good for the reproductive success of the Monkey Mia females that possessed it. Mann and her colleagues have also looked at the possibility of a connection between reproductive success and the most famous cultural trait of the Shark Bay dolphins, sponging. Females who put sponges on their noses, despite having to spend more time foraging and less time socializing, raised calves at a somewhat higher rate than the average dolphin.[114]

The quantity that has been found to differ between the cultural clans of sperm whales and between provisioned and nonprovisioned dolphins is the reproductive rate—essentially the rate of producing offspring. Reproductive rate is not the same as fitness—essentially the number of surviving grandchildren—but the two are usually closely correlated. Therefore, unless there is a major compensating factor, such as much higher mortality for the animals in the group who have more surviving offspring, we can reasonably infer fitness differences between the clans of sperm whales or foraging types of dolphins. If these fitness differences are maintained consistently, generation after generation, then cultural group selection becomes possible—some clans will prosper while others wither, and their characteristic cultures will follow suit. Culture will be driving evolution because of its link with social structure.

Some key elements of human culture are rare or absent in the cetaceans, then. Only in the sponging of the bottlenose dolphins is there even potential evidence of material culture, whereas we are physically surrounded by our material cultural, and it accumulates spectacularly. There is a similar contrast with syntactical language. But there are at least indications that cetacean culture does include some other key attributes of human culture: perhaps the roots of accumulation, morality (although we do not know whether it is culturally transmitted), symbolic ethnic markers, and that it affects biological fitness. Particularly in the symbolic ethnic markers and in affecting fitness, as well as in the consequences of these two attributes, it seems as though there are striking parallels between cetacean cultures and that of *Homo sapiens*.

Our knowledge of our own human culture has helped us frame our enquiries into whale and dolphin culture. We look for tools, conformism, and symbolic marking. But the aid can be mutual. We are, not unreasonably, fascinated by our own culture. Why language, why flags of nations, why music? One of the key techniques that evolutionary biologists use to investigate "why" questions is what is called the comparative method. By comparing how a trait is expressed (or not expressed) in a range of species, or sometimes populations of a particular species, we can make inferences about when it evolved and how it evolved.[115] Thus, for instance, if we find, as we do, that within several groups of animals, the more social species generally have the largest brains, then this strongly suggests that sociality and brain size are linked in some way.[116] The comparative method is particularly important when we cannot perform manipulative experiments because of logistics (the time line may be too long, as with many evolutionary questions), ethics, or some other reason. But the comparative method depends on multiple comparisons. As input, it needs a range of species or populations with or without the trait or with different values of it. The correlation between social complexity and brain size is so compelling because it is found in birds, and primates, and cetaceans. In contrast, many aspects of human culture are perceived as unique to humans and universal among human populations. In such cases, the comparative method is moribund, stifling scientific inference. This is a principal reason why studying the evolution of human language is such a hard question. The cetaceans add a major new perspective to our musings on the evolution of human culture. Almost no tools, but conformism and symbolic marking (perhaps!). So what have humans, chimpanzees, and New Caledonian crows got that cetaceans have not got that led to tools? Conversely, what is common to humans and the large toothed whales, but not chimpanzees and baleen whales, that leads to symbolic marking of ethnicities? The cetaceans add a new window as we ponder the mysterious origins of human culture.

Peter Richerson and Robert Boyd "suspect that a sophisticated capacity for culture has only been adaptive for a short, recent bit of the earth's history and we are merely the first lineage to discover its advantages."[117] We suspect that a sophisticated capacity for culture has been adaptive for many millions of years in the ocean, and the cetaceans discovered its advantages about thirty-four million years ago, with the evolution of the odontocetes and a sudden increase in brain size but, for reasons that are still unknown, never translated it into an engine for generating the awesome body of accumulated skills, knowledge, and materials that characterize human culture.[118] Furthermore, during this time, evolution will have tailored whale psychology to deal with the implications of living in their cultural world, which despite frequent changes in specific content (song types and particular foraging strategies) may have been pretty stable in its general extent since well before the evolutionary split of the chimpanzee and human lines. Human culture has been and is evolving extremely rapidly. So perhaps there is a much better fit between whale psychology and whale culture than we currently experience between human psychology and human culture. If so, we might really push the boat out and suggest that cetacean psychology could be a distant landmark in the direction that human psychology, driven by our hyperculture, is heading. Perhaps.

Should Whale and Dolphin Culture Influence How Humans Treat Them?

The evidence for culture in whales and dolphins is flimsy in places, stronger in others. The flimsiness reflects our known unknowns. It may largely be a result of our limitations in getting at their culture, but it may also be that not all cetacean species are cultural to the same extent. The strength results from culture having such a major role in the lives of these animals that it shines through our crude methodology. We have discussed what this means for their evolutionary trajectories and their ecological relationships with other parts of the living world. We have also considered how this transition affects our attempts to exploit, conserve, and manage them. But does it mean more? We can ask two questions. First, what is the moral worth of culture? Second, should the culture that we observe in cetaceans affect our thinking about how we treat them?

Does culture have, in and of itself, a moral worth? It seems pretty clear that humans, by their actions, demonstrate that they think it does. Large amounts of time, effort, and money are devoted to preservation in museums and private collections of cultural elements that have long outlived their usefulness in the modern world. People today form groups to learn and practice the production of stone tools by flint knapping, a technology that was outdated possibly before we even had full-blown language. We get concerned about the loss of cultural diversity. We spend time and effort trying to preserve threatened languages. In 2001 the United Nations Educational, Scientific and Cultural Organization (UNESCO) produced the Universal Declaration on Cultural Diversity, article 1 of which asserted that "cultural diversity is as necessary for humankind as biodiversity is for nature." Similarly, seventy-eight nations have ratified the 2007 Convention for the Safeguarding of the Intangible Cultural Heritage, which states that: "This intangible cultural heritage, transmitted from generation to generation, is constantly recreated by communities and groups in response to their environment, their interaction with nature and their history, and provides them with a sense of identity and continuity, thus promoting respect for cultural diversity and human creativity."[119] Aside from the reference to "human creativity," it is not hard to see how this rubric could be equally applied to the culture of at least some cetacean species. Nonetheless, the authors undoubtedly had in mind human culture when they wrote it—cumulative, symbolic, and linguistic. It would be dishonest of us to lay out our broad notion of culture and then argue that instances where the word itself is used in a much narrower sense automatically apply.

For this reason, transferring these arguments for the conservation of culture directly from humans to other species makes some uneasy. Therefore, using some of the ideas in chapter 11, let us rephrase in the terminology of conservation biologists. Conservation biologists are interested in preserving the diversity of life. Diversity can be considered important for its own sake and because it gives the ecological systems that we depend on stability, resilience, and productivity, as well as aesthetic value.[120] Organisms are different from one another because of genetics, because of environmental variation, and for other reasons, including culture. Thus the conservation of nonhuman culture is an element of the conservation of biodiversity, as are the conservation of genetic and environmental diversity.

Of course, arguments for the conservation of whale culture run fairly directly up against those for conserving certain parts of human culture. Take, for instance, the question: "Since most Western nations are opposed to whaling, why doesn't Japan just abandon its tradition?" The Japan Whaling Association's reply to that is: "Asking Japan to abandon this part of its culture would compare to Australians being asked to stop eating meat pies, Americans being asked to stop eating hamburgers and the English being asked to go without fish and chips."[121]

The conservation of whale culture and the conservation of whaling culture are not easily reconciled. We do not know of pro-whaling forces directly attacking the idea of whale culture, but they do their best to portray the cetaceans as behaviorally unsophisticated, quoting, for instance, the British scientist Margaret Klinowska saying that "whales betray little evidence of behavioral complexity beyond that of a herd of cows or deer."[122] Presumably the whalers are trying to head off arguments like that of Toshio Kasuya, the Japanese scientist whose quote began this chapter. So why are whalers worried that their prey will be thought of as cognitively advanced or behaviorally complex? The whalers' concern seems to be that whales and dolphins might, because of these features, be given special rights that might preclude their being hunted.

The cetacean rights movement has been spearheaded by philosophers whose argument goes something like this.[123] In practice, humans separate life forms into "persons," who are given special rights such as life and liberty, and "nonpersons," who have a restricted range of rights that might include freedom from unnecessary suffering and from biological extinction. Persons have a different "moral standing" than nonpersons because they have greater needs and vulnerabilities.[124] There are various ways persons and nonpersons can be distinguished. The simplest is just that living humans and only living humans are persons. This is implicitly the argument of the Catholic Church against abortion and euthanasia, but it is unsatisfactory to the many who do not give the status of person to a newly formed fetus or to a brain-dead but physiologically living individual. So "person" is usually defined in a way such that not all living humans are persons. Unless this definition includes restriction to a subset of *Homo sapiens*, a logically unsatisfactory approach, the door is open to nonhuman persons. Philosophers considering "personhood" in humans use definitions that include factors such as being alive; being aware; having positive and negative sensations, emotions, and a sense of self; controlling one's own behavior; recognizing other "persons" individually and treating them appropriately; and possessing a variety of sophisticated cognitive abilities.[125] Philosopher Thomas White argues that the bottlenose dolphin, at least, meets these requirements and thus has "moral standing."[126] This conclusion depends on what is contained within the variety of sophisticated cognitive abilities, but if one accepts this argument, then dolphins are persons and should be given rights somewhat similar to those of humans. This is, perhaps unsurprisingly, a highly controversial conclusion, perhaps mainly because it challenges the nature of our "dominion" over the natural world.[127] The possibility of nonhuman persons is not the only reason, or even the principal reason, that this traditional perspective is being questioned. We are, after all, living in a time when the relationship between human economic culture and the ecosystem on which is depends is in question as never before, and alternative perspectives about the relationship between humans and the natural world stemming from other religions and philosophies are being increasingly valued. The personhood of the dolphin—or elephant or chimpanzee—is one more question to throw into the mix.

What does this have to do with culture? Culture is not included directly in the usual requirements for personhood, but it is seen by some as the milieu within which some of these factors operate.[128] For instance psychologist Howard Gardner argues that intelligence is "activated in a cultural setting to solve problems or create products that are of value in a culture."[129] Culture therefore could be regarded as a necessary, though not sufficient, ingredient of personhood, and studying it in cetaceans could directly inform some highly contentious moral debates.

So in answer to our questions, cetacean culture should affect the way we treat cetaceans in practical ways right now, when it comes to effectively conserving them, and as our knowledge grows it may also affect the way we view our responsibilities toward them and our relationships with them. In the last few paragraphs we may have moved further than what you are comfortable with, but if you are now finishing this book, we hope that we have alerted you to an extraordinary flourishing of culture in the oceans. We hope that we have convinced you that this has important implications for what the whales are, how they use their habitat, and how they interact with their environment, as well as, perhaps, how we should treat them. If you start books from the end, and have just picked this up, please read at least chapters 4–6. These will take you to the whales and dolphins in the wild and the extraordinary lives that they lead. Then try and get on a boat to see them at sea. You probably won't see much of the animals, but you will have a small link into their world. Perhaps you will become entranced with what else lies out there. We have been. We are off to sea again, where there is still so much to learn.

THIS BOOK CAME FROM AND IS BUILT ON . . .

The development, writing, and production of this book have depended on the direct and indirect assistance of many. By tracing the book's evolution first within our brains and then on our laptops, we will acknowledge much of this. But we have undoubtedly omitted some who contributed ideas, some who provided assistance. We apologize.

The foundation of this book is us sailing with whales, mainly sperm whales. While sometimes a little skittish, the sperms have been extremely tolerant of our sometimes bungling attempts to follow them in their milieu. Thanks to the whales.

As we sailed with the whales, we fleetingly saw the huge creatures—"beings" or "hunks of blubber" depending on one's perspective—on the surface beside our boat, we listened to the sounds that they made far beneath, and we wondered: What is it to be a whale? We never expected to know with any depth at all but thought just a little window into their world would be wonderful. However, behavioral ecology, the major academic foundation of our research, did not always seem sufficient to explain the whale behavior that we saw and the whale sounds that we heard. The whales appeared to be doing things that were not obviously designed to advance their genetic contribution to subsequent generations. Inspired by the ideas, discoveries, and musings of scientists like Ken Norris, Katherine Payne, and Roger Payne, who first implicitly and explicitly considered cetacean culture, we realized that culture might provide a useful way for us to see parts of the world of the whale, especially if culture was an important part of the whales' lives. Thanks to Ken, Katherine, Roger, and the other pioneers.

From this background, we started trying to make the link between the whales and culture more formally, so that culture could become a scientifically testable explanation for whale behavior. We wrote a review, "Culture in Whales and Dolphins," that was published in 2001, the same year that we helped Harald Yurk and Lance Barrett-Lennard arrange a workshop on marine mammal culture at the Biennial Marine Mammal Conference in Vancouver. Harald did most of the work—thank you Harald!

Both the review paper and the workshop generated lots of interest. Those who participated in the workshop as well as those who wrote commentaries for the review left us with much to think about. There were new data, new ideas, fierce and gentle criticism—much warranted. Our ideas and our work sharpened. Thank you, colleagues and critics.

Some whale and dolphin scientists shared our perspective on culture and have continued to apply it to their research programs. In the twelve years since the workshop and review, the advances have come in many forms. There has been much progress in the areas where we already knew something: humpback whale songs, dolphin sponging, fishing cooperatives, killer whale calls, and sperm whale codas. Some of the questions we asked are now answered. We now know, for instance, what happened to the trawler dolphins of Hervey Bay when the trawling ended, and the structure and evolution of the songs of bowhead whales has been described, as have the ecotypes of Antarctic killer whales. There have also been big surprises. We heard of the worldwide deepening of the song of the blue whale, the progressive eastward movement of humpback whale songs across the South Pacific, and the Brazilian "holidays" of the Antarctic pack-ice killer whales. Extraordinary! These developments, both the incremental and astounding, result from the dedicated persistence of teams of scientists, over months and often years of work, some on small boats, some in rough seas, some in the lab, overcoming the consistently negative interactions between electronics and seawater, and so many other frustrations. This book is inspired by those results and built on all that labor. Thank you to the scientists, students, and volunteers who achieved all this, as well as those who funded the research. And a particular thank you to the colleagues, students, and volunteers who worked with us on the sperm whales, as well as the Natural Sciences and Engineering Research Council of Canada (NSERC), the Natural Environment Research Council of the United Kingdom, and the National Geographic Society, which together mostly funded our sperm whale research.

We very much enjoy our research at sea, being with whales, and just plain sailing. But there are other sides: storms, engine problems, and bureaucracy. It is hard to foresee the tedium that can be involved in navigating a boat through the bureaucracies of the countries that it visits. However, on August 20, 2008, when Hal visited the customs house in Puntarenas, Costa Rica, to try and get his boat legally into the country, he went prepared. He brought his laptop and, during the interminable proceedings, wrote the outline of this book. So thanks are due to the unhurried pace of some Costa Rican customs procedures.

The following year we had a large slice of good fortune. Returning to port in between research trips, Hal discovered that NSERC had unexpectedly awarded him more research money than he had asked for and that, because of the delay in receiving the message, he had twelve hours in which to decide how to use the extra Discovery Accelerator Supplement. He decided that his research would receive most acceleration from enhanced collaboration with Luke. So, NSERC bought some of Luke's time—he was working with Kevin Laland's research group at the University of St Andrews, Scotland—and we increased our collaboration, particularly on this book. So thanks to NSERC—and to Kevin as well for endorsing the arrangement. During the writing, Luke moved to a position funded by the MASTS pooling initiative (the Marine Alliance for Science and Technology for Scotland) and their support is gratefully acknowledged (MASTS is funded by the Scottish Funding Council, grant reference HR09011, and contributing institutions).

As we wrote, we kept asking friends, colleagues and strangers for specific facts, to check pieces of text, or to discuss ideas. Some sent unpublished, or hard to access, manuscripts. All offered their help unconditionally. So thanks to Shelley Adamo, Ina Ansmann, Robin Baird, John Barresi, Lars Bejder, Neeltje Boogert, Dick Byrne, Mauricio Cantor, Nicola Davies, John Durban, Holly Fearnbach, Chris Gabriele, Howard Garrett, Juan-Carlos Gomez, Andrew Horn, Kelly Jaakkola, Meghan Jankowski, Anna Kopps, Lori Marino, Ron O'Dor, Peggy Oki, Noa Pinter-Wollman, Peter Slater, Jan Straley, Chris Stroud, Chris Templeton, Jeff Warren, Thomas White, Andrew Whiten, and Matt Wold. More specifically, Mark McDonald reviewed the parts of chapter 4 on blue whale song, and Fred Sharpe those on humpback feeding in Alaska; Denise Herzing checked parts of chapter 5 on spotted dolphins, Celine Frère the section about her work, and Mike Bossley the paragraphs on Billie the tailwalking dolphin; Craig Matkin reviewed parts of chapter 6 on killer whales off Alaska; José Abramson, Kelly Jaakkola, and Mark Xitco reviewed sections of chapter 7 describing their studies; Malene Simon reviewed our description of Keiko's reintroduction in chapter 8; and Lori Marino advised on brain evolution for chapter 9. John Barresi found a number of errors that we have corrected. We thank all of them for their efforts. We are particularly grateful for whole-chapter reviews from Amy Deacon (chaps. 1–6), Kevin Laland (chap. 2), Ian McLaren (chap. 3), and Dick Byrne (chap. 8). Of course, any errors that remain are entirely our responsibility.

We are neither artists nor photographers, and so we have depended on others for the illustrations of whale culture. Emese Kazár produced accurate, clear diagrams illustrating the evolution of the whales and dolphins. We assembled the color and black-and-white images with help from Lars Bejder, Mike Bossley, Robert Carter, Fábio Daura Jorge, Shane Gero, Daren Grover, Toshio Kasuya, Jake Levenson, Marina Milligan, Larry Minden, Michael Moore, Flip Nicklin, Wayne Osborn, Robert Pitman, Bonny Schumaker, Jan Straley, Leigh Torres, Sophie Webb, and Tonya Wimmer. The images were edited by Jennifer Modigliani.

The anonymous reviewers of our book proposal, as well as those of our draft manuscript, were most kind and encouraging—their reports buttressed our perseverance. Our editor, Christie Henry, has been consistently positive and forgiving throughout the book's extended gestation. As it neared completion, our manuscript was edited thoroughly and very carefully by Yvonne Zipter. Thank you.

Finally massive thanks to Claudia Ortiz and Jennifer Modigliani who supported us, and put up with us, when the writing took hold, as well as to our children, Ben, Steff, Sonja, Dylan, and Nico, for tolerating the whale obsession of their dads.

NOTES

Chapter 1

1 In this book we will use notes like this to indicate where our information comes from and, sometimes, to provide more background to a particular point. In this case, we refer to archaeological studies of ancient ships by Carter (2006).

2 We are not sure where the saying "culture changes everything" originated. It is quite widespread on the Internet but also seems to precede the Internet.

3 Maynard Smith 1989, 12.

4 Catchpole and Slater 2008.

5 To understand why biologists have forsaken the nature/nurture debate, we recommend Bateson and Martin (2000).

6 Pinker 1994.

7 Richerson and Boyd 2005, 12–13.

8 As Richerson and Boyd put it: "Culture has led to fundamental changes in the way that our species responds to natural selection" (Richerson and Boyd 2005, 12–13). The human genome project has now begun to reveal the startling truth behind that statement, as human geneticists catalog site after site in our DNA that shows statistical evidence of having undergone strong selection in the last twenty thousand years. The most plausible source of these selection pressures is human culture itself (Laland, Odling-Smee, and Myles 2010). Most of the genes at these sites are either related to digestion—reflecting a response to the changes that the culturally transmitted practices of animal husbandry, agriculture, and cooking have brought to our diet—or expressed in the brain, reflecting a response to the demands of the hyperinflation of our informational environment since humans became capable of transmitting and accumulating large amounts of culture.

9 Richerson and Boyd 2005, 115.

10 Samuels and Tyack 2000.

11 After Ken Norris's death in 1998, the scientific journal *Marine Mammal Science* dedicated an enlarged issue to his memory (edited by William Perrin, vol. 15, no. 4 [1999]).

12 Norris and Dohl 1980, 253.

13 Norris and Schilt (1988). We do not mean to imply that he would have endorsed all the views in this book. We wanted to point out that sharper minds than ours have long been thinking along these lines.

14 Part of this was a spinoff from growing prominence of the consideration of culture in the great apes (e.g., McGrew 1992; van Schaik et al. 2003; Whiten et al. 1999) and part was due to a number of papers specifically referring to the culture of whales and dolphins (Noad et al. 2000; Rendell and Whitehead 2001b, 2003; Yurk et al. 2002), as well as a well-attended workshop on marine mammal culture at the 2001 Biennial Conference on the Biology of Marine Mammals in Vancouver, British Columbia.

15 Rendell and Whitehead 2001b.
16 Premack and Hauser 2001; Fox 2001.
17 Reiss 1990.
18 Pearson 2011.
19 Lilly 1978, 1.

Chapter 2

1 Bonner 1980; Kroeber 1948, 8; Tylor 1871.
2 "The way we do things" is from McGrew (2003).
3 For example, Galef 1992.
4 Schultz and Lavenda 2009, 9, 18.
5 Laland, Kendal, and Kendal 2009.
6 Laland and Janik 2006, 542.
7 For more on types and grades of culture, see Slater (2001).
8 Richerson and Boyd 2005, 5.
9 Rendell et al. 2011a.
10 Richerson and Boyd 2005, 61.
11 Rendell and Whitehead (2001a), 364. Here we have here substituted "community" for the original "population or subpopulation" for clarity.
12 McGrew 2003.
13 Seppänen and Forsman 2007.
14 There are ways other than genetics and culture that information can move between animals. These include epigenetic effects (Tost 2008), such as maternal effects in which attributes of the mother are passed to the offspring through characteristics of the egg or her maternal care (Mousseau and Fox 1998).
15 How old is human culture? This is a big question, and frankly, we don't know. If the question is how long have humans had culture as defined in the broad way that we prefer, then the answer is almost certainly since before we actually became *Homo sapiens*. Early members of our genus, now extinct, seemed to be able to make what archaeologists regard as the first cultural artifacts—fine-edged stone tools shaped by splitting large pieces of flint (e.g., de la Torre 2011). However likely it may seem, though, there is little hard evidence that this skill was socially transmitted except for descriptions of archaeological sites where the skill was apparently practiced by novices in the presence of an expert (Fischer 1989). But how long have we had cumulative, symbolic, and normative culture mediated by language? That is a very open question, and one we return to in chap. 8.
16 On the inheritance of maternal effects, see Mousseau and Fox (1998).
17 Richerson and Boyd 2005, 50.
18 Box 1984, 231; Heyes 1994.
19 Aisner and Terkel 1992.
20 However, in the wild there may be very few rats whose only exposure to the technique is partially stripped cones (Aisner and Terkel 1992).
21 We still argue, e.g., over the very notion of cultural transmission by social learning. Take the archetypal case of human cultural transmission—the skill of making stone axes by

"knapping" flint. What exactly is being "transmitted" or "transferred" during the process through which a youth might acquire this skill? Obviously, there's no literal transfer, like the transfer of DNA from male to female via sperm. There is no physical entity we can point to that moves from demonstrator to observer. Or is there? Well, there might be—an appropriate hammer stone, e.g., might be passed from teacher to pupil. However, that is not sufficient. You still have to pass on the skill—the knowledge of how and where to strike the flint with the hammer stone, the foresight to plan how to fashion a usable ax, and the motor skills to follow this plan effectively. One school of thought argues that this can all be packaged whole into a conceptual unit—a meme—and transferred directly between individuals much like a gene (Dawkins 1976). Subconsciously acquiring a catchy tune seems to fit with this. Others disagree vociferously and argue that the process is much more active on the part of learners since they must literally reconstruct the knowledge themselves, and in doing so they will inevitably alter, adjust, or edit it according to their own needs, desires, or preferences (Sperber 2006). Again, intuitively, one can think of relevant examples—learning to play a complex piece of music from a score and a quick demonstration would seem to involve a healthy dose of active construction on the part of the learner. These are complex issues.

22 Hoppitt et al. 2008.
23 Rendell et al. 2011b.
24 Bikhchandani, Hirshleifer, and Welch 1992.
25 Rendell et al. 2011b.
26 From Goodall (1968). We have added "have the potential to interact" to include very large communities, such as nation states of humans or clans of sperm whales, in which quite a lot of the possible pairs of members have not actually interacted.
27 Wellman 2001.
28 For example, Connor 2007.
29 For example, Frantzis and Herzing 2002; Psarakos, Herzing, and Marten 2003.
30 For example, Galef and Laland 2005.
31 Cavalli-Sforza and Seielstad 2001.
32 See Cavalli-Sforza et al. 1982.
33 Our list of the forces of cultural evolution is from Richerson and Boyd (2005), 68.
34 Mesoudi, Whiten, and Laland (2006). Although initially used in ancient Greece, the very word "evolution" was first employed in modern scholarly discourse to describe the way that societies were supposed to progress inevitably from uncivilized "savagery" to the enlightened sophistication of mid-nineteenth-century Europe. Notwithstanding the fact that the prosperity of these societies rested substantially on the brutal enslavement and pillaging of their colonial empires, this ridiculous idea was put to rest by the barbarity shown by these same societies in World War I and their collective inability to prevent World War II. Anthropologists (the only scholars formally studying culture at that time) rightly and roundly rejected this linear perspective of cultural evolution (Carneiro 2003). Unfortunately, however, some have been unable or unwilling to recognize the difference between this idea of evolution and that represented by Darwin's incredibly powerful theory of evolution through natural selection. This was not helped by the rise of eugen-

ics, the pseudoscientific notion of "improving" races of humans by controlling breeding, abetted by some ill-informed evolutionary biology, in the 1930s. This distortion of Darwinism was also perceived to legitimize the most depraved barbarities of the Nazis. Evolutionary biology has rejected eugenics as roundly as anthropologists have rejected the linear improvement notion of "evolution," but suspicions linger. Feelings still run high, and the resultant clash between those who recognize the potential power of Darwinian theory to put the study of human cultural change on a scientific footing and those who smell a colonial or eugenic rat in any suggestion that cultures may evolve along similar principles to genes and species has been at times very heated (Ingold 2004; Mesoudi, Whiten, and Laland 2006). As biologists in everyday contact with Darwinian theory, and who see plenty of imperfections in our own societies, we smell no such rat. It would be a mistake to throw out Darwin's baby as we thoroughly dispose of the bathwater of eugenics.

35 Strimling, Enquist, and Eriksson 2009.
36 Harrison 2000.
37 Boyd and Richerson 1985, 82.
38 On female humpback whales' preference for rhyming by males, see Guinee and Payne (1988).
39 Strictly speaking, conformity in this sense is when individuals adopt the most common behavior at a higher rate than its actual prevalence in the population (Henrich 2004a).
40 You may not be surprised to learn that the role of conformity in human cultural affairs is contested by students of cultural evolution. Some believe it is critical to specific human features, such as the maintenance of cooperation (Richerson and Boyd 2005), while others question how it is we can have any cultural evolution at all if we are so conformist (Eriksson, Enquist, and Ghirlanda 2007). Regarding the idea that conformity may help stabilize cooperation, see Boyd and Richerson (1985) and Henrich and Boyd (1998).
41 Henrich and Henrich 2010.
42 Guided variation, biased transmission, and natural selection on cultural variants are sometimes called psychological selection to distinguish them from natural selection operating on gene frequencies (e.g., Mundinger 1980). On guided variation's lack of parallels in genetic evolution, see Richerson and Boyd (2005), 116.
43 Wilson (1994). These interactions are fascinating but understudied, perhaps because of their complexity. You can find out more in the relevant sections of the books by Coyne (2010) and Laland and Brown (2011).
44 For example, Enattah et al. 2002.
45 Tishkoff et al. 2007; Bersaglieri et al. 2004.
46 The details of this example can be found in Durham (1991) and Lindenbaum (2008).
47 Nicola Davies (not a blond) points out that the ideas that blonds are sexually attractive and less intelligent are themselves both cultural ideas, making this example even more complex.
48 Richerson and Boyd 2005, 162.
49 For the overall rejection of this theory by biologists, see, e.g., Trivers (1985). On the recent

theoretical comeback of genetic group selection in special circumstances, see Wilson and Dugatkin (1997).

50 Henrich 2004a.

51 Richerson and Boyd (2005), 162. However, yet again, this is not an unchallenged view. For a taste of the very technical debate, see Boyd, Richerson, and Henrich (2011) and Lehmann, Feldman, and Foster (2008).

52 See, e.g., Grillo (2003).

53 McGrew 2003.

54 Hill 2010, 324.

55 In reference to chimpanzees or bonobos being complex and unique species in their own right, see Boesch (2012).

56 McGrew 2009.

57 Grillo 2003, 158.

58 Consider the following definitions, which were gathered together by American anthropologist Lee Cronk (1999). The definitions are from Bodley (1994), Ember and Ember (1990), Moore (1992), and Peoples and Bailey (1997), respectively:

"The patterned and learned ways of life *and thought* shared by a *human* society."
"The learned set of behaviors, *beliefs, attitudes, values, or ideals* that are characteristic of a particular society or population."
"That complex of behavior *and beliefs* we learn from being members of our group."
"The socially transmitted knowledge and behavior shared by some group *of people*."

In the above definitions, we have highlighted those aspects that require us to know about the internal mental worlds of one's study subjects (e.g., beliefs, thoughts, attitudes, ideals) and, also, specific reference to humans. When these are removed, the definitions become:

"The patterned and learned ways of life . . . shared by a . . . society."
"The learned set of behaviors . . . that are characteristic of a particular society or population."
"That complex of behavior . . . [individuals] learn from being members of [their] group."
"The socially transmitted knowledge and behavior shared by some group . . ."

The similarity between these and our definition is obvious. The gap between what we and cultural anthropologists fundamentally consider to be culture is not as wide as it first seems; perhaps the major difference is the assumption of humanness understandably inherent in the ideas of cultural anthropologists.

59 Mesoudi, Whiten, and Laland 2006.

60 For the development of the modern synthesis, see Huxley (1942).

61 Tost 2008.

62 Richerson and Boyd 2005, 16–17.

63 Dawkins 1976, 92.

64 Despite capturing attention in the public sphere, the "meme" meme, has not generated

much in the way of a meaningful program of research, a failure exemplified by the closure of the *Journal of Memetics* in 2005 (Laland and Brown 2011). The term has, however, been adopted in the Internet age as a descriptor for encapsulated pieces of often hilarious silliness that spread and diversify at extraordinary speed across the web (e.g., the "Texts from Hillary" meme, accessed August 17, 2013, http://textsfromhillary.com/).

65 Boyd and Richerson 1985.

66 Richerson and Boyd 2005, 12–13.

67 For a comparison of the different approaches to understanding the evolution of human behavior, see Laland and Brown (2011).

68 Boyd and Richerson (1985); Richerson and Boyd (2005). Other proponents include, e.g., Cavalli-Sforza and Feldman (1981).

69 Wilson 1979, 167.

70 Richerson and Boyd 2005, 16.

71 Alison Greggor (2012) suggests a "functional paradigm" for research on nonhuman cultures with emphasis on how culture relates to social learning, ecology, social systems, and biology.

72 Popper (2002). Recognizing the testing of null hypotheses as sterile and flawed are Johnson (1999) and Garamszegi et al. (2009) and as dangerous, Ziliak and McCloskey (2008).

73 Akaike 1973; Burnham and Anderson 2002.

74 Bates and Byrne 2007.

75 However, the study of tool use in cetaceans is becoming more quantitative (e.g., Krützen et al. 2005; Martin, da Silva, and Rothery 2008).

76 For example, Payne 2000; Rendell and Whitehead 2003; Yurk et al. 2002.

77 One quantitative example in which movements are considered potentially to be cultural behavior is Whitehead and Rendell (2004).

78 Krützen et al. 2005; Mann et al. 2012; Sargeant et al. 2007.

79 Laland and Galef (2009b). Another very useful guide in our search for the key elements of human culture was Richerson and Boyd (2005).

80 Fitness can be quite a difficult technical concept. See Orr (2009).

81 For example, Chomsky 1965.

82 For example, Kirby, Cornish, and Smith 2008; Kirby, Dowman, and Griffiths 2007; Smith and Kirby 2008.

83 Tomasello 1994.

84 Technology can be defined as "the making, usage and knowledge of tools, techniques, crafts, systems or methods of organization in order to solve a problem or serve some purpose" ("Technology," Wikipedia, last modified on February 9, 2014, http://en.wikipedia.org/wiki/Technology).

85 There is an illuminating literature on "niche construction" in humans and other organisms (e.g., Laland, Odling-Smee, and Feldman 2000).

86 Tebbich et al. 2001.

87 Tomasello 2009.

88 For a book-length treatment of these arguments, see Tomasello (1999a).

89 Evidence for how humans compare to other primates in interacting with puzzles can be

found in Dean et al. (2012). For mathematical models of how cumulative culture might work, see, e.g., Lewis and Laland (2012).

90 A recent modeling study by Adam Powell and colleagues suggests that there is simply a threshold level of population density needed before even error-prone cultural transmission can generate cumulative adaptation (Powell, Shennan, and Thomas 2009). This is supported by data from traditional island communities in Oceania showing that islands with more people had bigger tool kits—they had more technological culture—than islands with fewer people (Kline and Boyd 2010). Others have suggested that what makes the difference between apes and humans might have something to do with apes being, with no political comment intended, more conservative (Whiten et al. 2009).

91 Broom 2003, 1.

92 Hamilton 1964.

93 In the unusual haplodiploid genetic inheritance of the social insects—ants, bees, etc.—males develop from unfertilized eggs and have only one set of chromosomes. This increases the relatedness among sisters, as they all receive the same genes from their father. This increased relatedness, it is postulated, leads to high levels of cooperation within large colonies of close kin.

94 Fehr and Fischbacher (2003). Once again, however, we've stumbled into an area of real controversy! There are, as ever, opposing viewpoints. Some scientists assert that, in fact, a general rule of thumb to be cooperative with those around you is all that needs to be explained and that this can be adequately done with kin selection theory (West, El Mouden, and Gardner 2011).

95 On the relationship between punishment and cooperation, see Boyd, Gintis, and Bowles (2010) and Fehr and Gachter (2002).

96 Orr 2009, 622–36.

97 Chudek and Henrich 2011; Gintis 2011.

98 Bernhard, Fischbacher, and Fehr 2006.

99 Hill et al. 2011.

100 Efferson, Lalive, and Fehr 2008.

101 Richerson and Boyd 2005, 195.

102 Richerson and Boyd 2005, 180–82.

103 McGrew 1987.

104 Henrich and Boyd 1998.

105 Hill 2009.

106 Bekoff and Pierce 2009; Broom 2003.

107 De Waal 1997.

108 The study documenting the poking of fingers up companions' noses is Perry and Manson (2003).

109 Perry 2009, 266.

110 Hill (2010). But remember that it is also somewhat controversial whether this even happens in humans.

111 Grant and Grant (1996). They document, e.g., a migrant bird arriving on their main study island and managing to breed with native females. The bird's attempts to learn the

local song were only partially successful, though, and he influenced the song of his offspring sufficiently so that they refused to breed with anyone but themselves, establishing a small, reproductively isolated population that the Grants recognized as an incipient species (Grant and Grant 2009).

112 Richerson and Boyd 2005, 144.

113 White 2007, 165.

114 Dawkins 1976; Wilson 1975.

115 On the controversial aspects of behavioral ecology being applied to humans, see Laland and Brown (2011).

116 Kuhn 1962.

117 Dawkins 1976; Lumsden and Wilson 1981.

118 See, e.g., William Hoppitt and Kevin Laland's recently published book, *Social Learning* (2013).

119 Among behavioral ecologists questioning the evidence and suggesting alternative explanations, see, e.g., Tyack (2001).

120 What follows is largely based on a section of Laland and Galef's introductory chapter to *The Question of Animal Culture* (2009a). The section is titled "A Brief History of the Animal Cultures Debate" (2–10).

121 Darwin 1874. On culture as an agent of trait inheritance, see Richerson and Boyd (2005).

122 In fact, some have even gone so far as to deny the importance of culture in human behavior. What we might describe as "hardcore" evolutionary psychology today admits a role for cultural transmission only insofar as it allows an individual to choose appropriate behavior from a genetically programmed and proscribed "jukebox" of alternative behavior patterns (Laland and Brown 2011). As biologists exploring the breadth and complexity of cultural transmission in nonhumans, this notion strikes us as absurd.

123 On elements of birdsong that are socially learned, see Thorpe (1961). The spread of milk bottle-opening techniques by blue tits was described by Fisher and Hinde (1949).

124 The recognition of primate cultures by Japanese scientists is described particularly well by Frans de Waal (2001). Sweet potato washing by monkeys on Koshima was described by Kawai (1965).

125 McGrew and Tutin 1978.

126 These conditions they termed "innovation," "dissemination," "standardization," "durability," "diffusion," "tradition," "nonsubsistence," and "natural adaptiveness," and it was on the last condition—that the population exhibiting the behavior had not been exposed to significant human impact—that they had their doubts, as the troop they studied was habituated through the prolonged presence of human researchers.

127 McGrew 1992.

128 Galef 1992.

129 McGrew 2003, 2004.

130 Whiten et al. 1999, 682.

131 Perry and Manson 2003; van Schaik et al. 2003.

132 Tomasello 2009.

133 Tomasello 2009, 218.

134 McGrew (2003). As we discuss in chap. 12, we feel this is unfair to birds and ornithologists alike. The cultural transmission of birdsong is fascinating and still has much to tell us about the evolutionary processes that promote learning from others. In recent years, thankfully, a series of studies on bird intelligence mean that "bird-brained" should now be viewed as much a compliment as an insult (e.g., Emery and Clayton 2004).

135 Rendell and Whitehead 2001b.

136 Boesch 2001.

137 Whiten 2001.

138 Ingold 2001.

139 Galef 2001.

140 Rendell and Whitehead 2001a.

141 Laland and Janik 2006.

142 This information is drawn from Thinh and colleagues (2011) and Geissmann (1984).

143 Kim Sterelny's chapter "Peacekeeping in the Culture Wars" in the book *The Question of Animal Culture* (2009) brought more fog than peace to the nonhuman culture conflict, in our opinion: he asserts that a criterion for "true" culture is that individuals "learn social facts socially" but fails to offer a convincing rationale for the assertion, as well as, in our eyes, an explanation of what this actually means such that it can exclude the learning of humpback song.

144 Laland and Hoppitt 2003; Laland and Janik 2006; Laland, Kendal, and Kendal 2009.

145 Laland, Kendal, and Kendal 2009.

146 Bateson and Martin 2000.

147 Hoppitt and Laland 2013; Laland, Kendal, and Kendal 2009; Whitehead 2009.

148 Henrich and Boyd 1998, 215.

Chapter 3

1 See McLaren and Smith (1985) regarding the fishing pigs of Tokelau.

2 Humans have had substantial impacts on coastal ecosystems, often through their hunting of marine mammals, for thousands of years (Jackson et al. 2001). The far-reaching effects that humans currently have on the oceans are addressed in many writings, e.g., Worm et al. (2005).

3 The name Carnivora is rather confusing because, while most Carnivora are carnivorous, some, such as the giant panda, are not, and there are many carnivorous mammals outside the Carnivora.

4 Sea turtles and seabirds share the "breed on land and feed in the ocean" lifestyle of the pinnipeds, and both have been successful. During their first twenty-three million years of aquatic life, the cetaceans went well beyond the pinnipeds in adapting to a marine existence—in fact, after twenty-three million years in the ocean, many whales were not too different from those around today. This may indicate that the pinnipeds have found a good between-two-worlds niche rather than being in transition toward becoming "full marine mammals."

5 For a summary of what is known of cetacean evolution, see Rice (2009) and Uhen (2010).

6 Uhen 2010.

7 Although cetaceans are large animals, little is known about many of them. Largely as a result of molecular-genetic studies, new species are discovered quite regularly (e.g., Dalebout et al. 2002), and familiar species, such as bottlenose dolphins or killer whales, may soon be split or merged. Thus we can only give an approximate number of species.

8 See, e.g., Au 1993.

9 On the power of the echolocation clicks of sperm whales, see Møhl et al. (2000).

10 For more on baleen, see Rice (2009).

11 Thanks to April Nason for this analogy.

12 For a detailed overview of the mammals and their evolution, see Feldhamer and colleagues (2007).

13 See Allman 2000, 99–101.

14 On the evolution of large-sized mammals, see Smith et al. (2010).

15 Huntley and Zhou 2004.

16 Culik 2001.

17 Tyack and Miller 2002.

18 Electrical sensing also works better in water than air (where it hardly works at all) but has a maximum range of just a few meters.

19 For instance, ecosystems form around deep-ocean hydrothermal vents based on chemosynthesis rather than photosynthesis.

20 Steele (1985). We will discuss red noise in more detail in chap. 9.

21 Inchausti and Halley 2002.

22 Sheldon, Prakash, and Sutcliffe (1972). Modern fishing techniques have changed this, preferentially removing the larger creatures in the ocean, thereby biasing the distribution of animal size in the ocean toward small animals, the terrestrial pattern (e.g., Levin et al. 2006).

23 For example, Estes et al. 1998; Myers et al. 2007; Worm and Myers 2003.

24 There remain in cetacean skeletons a few vestigial unattached bones where the pelvis used to be (which, for the creationists out there, can be explained only by an evolutionary history of terrestrial locomotion—or a pretty warped "creator").

25 Some pinnipeds, and especially sea otters, use specialized fur as insulation. However, in other respects, fur is a handicap when in the water.

26 Bats are most remarkable in this respect, with females of some of the smaller species giving birth to pups that are one-quarter of their own weight.

27 Whitehead and Mann 2000.

28 Allman 2000, 92.

29 Kooyman 1989.

30 Mirceta et al. 2013.

31 Schorr et al. 2014.

32 Connor 2000, 218.

33 Bowen 1997.

34 On vocalizations by manatees and seals, see, e.g., Nowacek et al. (2003), Schusterman (1978), and Schusterman, Balliet, and St. John (1970).

35 For instance, bowhead whales seem to use their sounds to navigate through pack ice (George et al. 1989).

36 See Tyack and Miller 2002.

37 Cranford 1999; Møhl et al. 2000.

38 Whitehead 2003a.

39 For very readable presentations of what is known of dolphin brains, see White (2007, 15–45) or Marino (2011).

40 In contrast, brain size variation within species is more related to body size than to cognitive ability.

41 Allman 2000, 160.

42 See Jerison (1973) for "encephalization quotient."

43 Byrne (1999); Deaner et al. (2007); Marino (2006). Absolute brain size may be a better general measure of cognitive ability, insofar as any such relatively crude measure can be related to complex cognition (Healy and Rowe 2007).

44 Northcutt 1977; Packard 1972.

45 Allman (2000), 86. Birds, the other warm-blooded group of animals, also have relatively large brains.

46 Allman 2000, 105.

47 Van Schaik 2006.

48 For a description of the centrality of calves in sperm whale social networks, see Gero, Engelhaupt, and Whitehead (2008) and Gero, Gordon, and Whitehead (2013). The other principal group of marine mammals—the pinnipeds (seals and their relatives)—appear much less social while in the ocean. They are found generally alone or in ephemeral aggregations where food is concentrated. Although the pinnipeds have developed other ways of reducing the impact of predators, such as sleeping at depth (Le Boeuf et al. 1986), they appear to have substantially higher mortality rates than the more social whales and dolphins: longevities are about thirteen to eighty-five years for the dolphins and porpoises, twenty-seventy to seventy-one for the beaked whales, sixty-five for the sperm whale, and forty-one to ninety-five for the baleen whales, as opposed to fourteen to forty-six for the pinnipeds (Whitehead and Mann 2000). This buttresses the argument that cetacean sociality developed to combat predation. But why are the pinnipeds not social at sea, if it is such an effective survival strategy? Nearly all pinniped mothers suckle their young on land or ice and leave the weaner to make its own way in the ocean. This lack of aquatic nursing of the young does not provide the foundation for advanced sociality. It is telling that the pinniped that does nurse at sea, and so develops a strong mother-offspring bond, the walrus, appears much more social in general than the other pinnipeds.

49 White 2007, 205.

50 Cetacean fathers seem to have no role in the care of their offspring. Presumably this is primarily because, within the ocean with its fluidity and three-dimensionality, males are never sufficiently certain of being the fathers of particular calves to make it worthwhile

to put in parental effort; see, e.g., Clutton-Brock (1989). However, in some cetacean societies, especially those of the large toothed whales, animals other than the mother may help raise the calf (see chap. 6).

51 An unusual and interesting exception is the case of the murre, an alcid seabird, in which the parents accompany the chick to sea (Paredes, Jones, and Boness 2006).

Chapter 4

1 Goldbogen, Pyenson, and Shadwick 2007.

2 See, e.g., Hain et al. 1982; Sharpe 2001.

3 Mayo and Marx 1990.

4 For more discussion, see Corkeron and Connor (1999).

5 Rasmussen et al. 2007.

6 Mikhalev 1997.

7 Stevick, McConnell, and Hammond (2002). For instance, in the North Atlantic the largest part of the humpback population feeds in the summer off Newfoundland and Labrador, with the great majority of the animals migrating to the Caribbean region to breed in the winter. Similarly, in the North Pacific, humpbacks move seasonally between southeast Alaska and Hawaii.

8 Burtenshaw et al. 2004.

9 Hucke-Gaete et al. 2004.

10 For example, Burtenshaw et al. 2004.

11 Kasuya 1995.

12 On lactation in whales, see Oftedal (1997); on fast growth of calves, see Whitehead and Mann (2000).

13 Whitehead and Mann 2000.

14 These are called "*r*-selected" because in traditional population models the parameter *r* usually represents the intrinsic growth rate of a population—i.e., the maximum rate at which the population can increase its numbers. Thus these species are selected to maximize their growth rate, so they might get hit hard in lean times but can bounce back quickly and at a faster rate than competitor species with lower intrinsic growth rates.

15 These are sometimes called "*K*-selected." Here, the parameter *K* usually represents the maximum population size a given environment can support—i.e., its carrying capacity. Populations of these species spend a large portion of their time at or near this capacity, so population growth potential is less important than ensuring smaller numbers of offspring really do make it to a successful adulthood.

16 Jefferson, Stacey, and Baird 1991.

17 See Ford et al. (2005) regarding the vulnerability of minke whales to killer whales.

18 George and Bockstoce 2008.

19 George et al. 1999; Nerini et al. 1984.

20 And, for the purist, the count should also include those genes that you share with relatives that are passed on in their own offspring. Thus it can pay to help your siblings and cousins reproduce. This total output is called an individual's inclusive fitness.

21 Off Newfoundland, groups of humpbacks feeding on schools of the fish capelin lunge synchronously, as do groups of finbacks (Whitehead 1983; Whitehead and Carlson 1988).
22 Würsig et al. 1985.
23 Clapham 2000; Weinrich 1991; Whitehead 1983; Whitehead and Carlson 1988.
24 "Humpback Whale Social Foraging and Its Implications on Community Structure," Alaska Whale Foundation, accessed July 12, 2011, http://www.alaskawhalefoundation.org/socialforaging.
25 Sharpe 2001.
26 Sharpe 2001.
27 Fred Sharpe, personal communication, December 5, 2011.
28 Sharpe (2001). The Chatham Strait animals sometimes follow herring runs into the Frederick Sound area.
29 Clapham 2000; Sharpe 2001.
30 See Würsig and colleagues (1985) for hints of social structure among foraging bowheads.
31 Payne and Webb 1971.
32 For example, Krause and Ruxton 2002.
33 Jefferson, Stacey, and Baird 1991.
34 Whitehead 1983.
35 Jefferson, Stacey, and Baird 1991; Whitehead and Glass 1985.
36 Ford et al. 2005.
37 Kraus and Hatch 2001; Payne and Dorsey 1983.
38 Payne and Dorsey 1983.
39 On mate-attracting sounds made by females, see Parks (2003).
40 Clapham et al. 1992; Tyack and Whitehead 1983.
41 An exception is the extraordinary case in which two right whale mothers seem to have switched calves soon after their birth in February 1987, probably accidentally, and gone on to raise each others' offspring (Frasier et al. 2010).
42 Amato (1993); see also, based on materials provided by Cornell University, "Secrets of Whales' Long-Distance Songs Are Being Unveiled," ScienceDaily, March 2, 2005, http://sciencedaily.kr/releases/2005/02/050223140605.htm. Scientists realized the possibility of this early on, speculating about the "acoustic herd" (Payne and Webb 1971), but collecting evidence on large-scale cooperation is a challenge.
43 Payne 1995, 357.
44 Payne 1995, 145.
45 Rothenberg 2010, 133–41.
46 Payne and McVay 1971.
47 "Songs of the Humpback Whale: The Original Album of Humpback Whale Songs, Produced by Roger Payne," Living Music Catalogue, accessed August 17, 2013, http://www.livingmusic.com/catalogue/albums/songshump.html. Perhaps in Winters's statement "our" should refer to all the inhabitants of the earth, not humanity. Humpback whale song is definitely culture, but not human culture.
48 The record has sold over thirty million copies and is still available. Rothenberg tells the

stories of the discovery of the whale songs, the making of the record, and its impact (Rothenberg 2010).

49 For example, Mackintosh 1965.

50 For instance, global shark populations are currently in a similar desperate position to that of the large whales in the 1960s (Baum et al. 2003), but a commercial moratorium seems unlikely. Sharks, as far as we know, don't sing.

51 Whaling in the 1960s had the populations of almost all the large whales heading directly toward extinction. The whales still face threats. In some cases, such as the North Atlantic right whales that are hit by ships and entangled in fishing gear, the threats are very serious and the survival of the species is uncertain. Also, in Japan and Norway, the two principal countries where commercial whaling continues (in the case of Japan, this is commercial whaling in the guise of science), whales are still regarded by many as resources ripe for exploitation.

52 Katherine Payne (2000) gives a clear description of the structure of the song of the humpback whale.

53 See Whitehead (2009) on the range of hearing humpback whale song.

54 Payne 2000. The longest continuous singing by an individual on record is twenty-one hours.

55 On note lengths, see Payne (2000).

56 Payne and McVay show why the humpback song is a "song" by any reasonable definition of the word, such as "a series of notes, generally of more than one type, uttered in succession and so related as to form a recognizable sequence or pattern in time" (as defined by W. B. Broughton in his essay, "Glossarial Index," in the book *Acoustic Behavior of Animals*, edited by R. Busnel in 1963 [quoted in Payne and McVay 1971]). See also, Spector (1994).

57 Rothenberg 2010, 137–39.

58 Guinee and Payne 1988.

59 Green et al. (2011); Suzuki, Buck, and Tyack (2006). These papers are indicative of some of the sophisticated quantitative analyses that have been applied to humpback songs.

60 Payne and McVay 1971, 597.

61 Hal was a member of the Paynes's laboratory in the late 1970s—an extraordinarily creative hub of whale science—but took no part in the analysis of the humpback songs.

62 Payne and Payne 1985; Payne, Tyack, and Payne 1983.

63 Payne 2000, 138.

64 Payne 2000.

65 Noad et al. 2000.

66 Jensen 2000.

67 Garland et al. 2011.

68 Garland et al. 2011.

69 Cerchio, Jacobsen, and Norris 2001.

70 The songs on each side of the Indian Ocean show little commonality in their evolution (Murray et al. 2012).

71 A few songs have been heard on the feeding grounds, in both summer and winter, but these are rare (McSweeney et al. 1989). Regarding the sex of the singers, see, e.g., Glock-

ner and Venus (1983). There is a rumor among scientists of one female singer off Hawaii, but we can find no published report of this.

72 Payne, Tyack, and Payne 1983.

73 The scientific controversy is well summarized by Rothenberg (2010), 161–68.

74 Tyack 1983.

75 Darling, Jones, and Nicklin 2006.

76 Smith et al. 2008.

77 These alternative explanations of humpback song include the conjectures that singers form a communal "lek" display that attracts females to the lek and then allows the females to make a choice among the displaying males; song warns other males of the presence of the singing whale and so might act as a spacing mechanism; song helps navigation or orientation by the singer (presumably listening for echoes off oceanic structures) or other whales who use the song as a beacon; or a type of sonar to locate females. This list is adapted from that in Darling Jones, and Nicklin (2006).

78 Wilson 1975, 108.

79 The quoted material in the text is from, respectively, Darling, Jones, and Nicklin (2006), 1051, and Rothenberg (2010), 166.

80 Payne, Tyack, and Payne 1983.

81 Catchpole and Slater 2008; Miller 2000.

82 Roger Payne notes eight striking similarities between human music and whale songs: rhythms, phrase length, song duration, frequency range, use of percussive sounds, overall song structure, note quality, and rhyme (1995, 146–47). This observation is placed in a more general context by Gray et al. (2001).

83 As originally proposed by Katherine Payne and colleagues (1983).

84 For example, Hekkert, Snelders, and Wieringen 2003.

85 Martindale 1990.

86 Everyone knows what birdsong is, right? You guessed it: scientists are involved, so it turns out everybody has a slightly different idea of what constitutes birdsong, and these varied ideas cannot easily be reconciled (Spector 1994). So all we can say is that, in general, song is a series of generally repeated vocalizations with a generally more elaborate form than other vocal outputs from the species in question and is generally used in the context of courtship and mating (Ball and Hulse 1998). Note that this means song is partly defined by its function, which, as we've seen, is tricky to pin down.

87 Cummings and Holliday 1987; Stafford et al. 2008; Würsig and Clark 1993.

88 Delarue, Laurinolli, and Martin 2009; Stafford et al. 2008; Tervo et al. 2011.

89 Tervo et al. 2011.

90 Gedamke, Costa, and Dunstan 2001.

91 Rankin and Barlow 2005; Tervo et al. 2011.

92 Oswald, Au, and Duennebier 2011.

93 Gedamke, Costa, and Dunstan 2001; Oswald, Au, and Duennebier 2011.

94 Watkins et al. 1987.

95 There is ongoing research into fin whale song, and, although we have summarized the picture in the current scientific literature (as of 2013), it looks as though fin whale song

is more complex and varied than we have sketched. These complexities and variation should become apparent in publications over the next few years.

96 Watkins et al. 1987.

97 Thompson, Findley, and Vidal 1992.

98 Payne and Webb 1971.

99 Croll et al. 2002; Thompson, Findley, and Vidal 1992; Watkins et al. 1987.

100 Thompson, Findley, and Vidal 1992.

101 Castellote, Clark, and Lammers 2011.

102 Stafford, Fox, and Clark 1998.

103 McDonald, Mesnick, and Hildebrand 2006.

104 Frank and Ferris 2011; McDonald, Mesnick, and Hildebrand 2006, 2009; Pangerc 2010.

105 McDonald et al. 2001.

106 McDonald, Mesnick, and Hildebrand (2006). We cannot quite hear blue whale song as it is sung. It is too low for human hearing. But if we speed up recordings, say eightfold, thus increasing pitch by three octaves, we can hear them clearly.

107 McDonald, Hildebrand, and Mesnick 2009.

108 Gavrilov, McCauley, and Gedamke 2012.

109 Other possible explanations considered by McDonald and colleagues (2009) include global warming and ocean acidification, as well as a general increase in blue whale size over the past thirty years.

110 Boyd and Richerson (1985), 82. On the apparent preference by females for lower sounds, see McDonald et al. (2001).

111 McDonald, Hildebrand, and Mesnick 2009.

112 "File:Hemline (Skirt Height) Overview Chart 1805–2005.svg," Wikimedia Commons, last modified on January 6, 2012, http://commons.wikimedia.org/wiki/File:Hemline_(skirt_height)_overview_chart_1805-2005.svg.

113 Personal communication from Mark McDonald, September 13, 2011.

114 McDonald, Mesnick, and Hildebrand 2006.

115 Smith et al. 2008.

116 Kraus and Hatch 2001; Payne and Dorsey 1983; Swartz 1986.

117 All the baleen whale species also make nonsong vocalizations, and these may be culture, too, but because they are less stereotyped and less studied, culture is much harder to infer in those instances.

118 Argentinean scientist Luciano Valenzuela and his colleagues recently reported that southern right whales also appear to inherit knowledge of summer feeding grounds culturally through the maternal line, judging from the fact that blubber samples from animals with different maternal lineages (assessed using DNA) had chemical signatures (specifically, the levels of the carbon-13 isotope) showing that the different lineages did their feeding in very different places (Valenzuela et al. 2009). For other baleen whales, we have an absence of evidence.

119 Baker et al. 1990; Palsbøll et al. 1995.

120 Herman 1979.

121 Herman 1979.
122 Weinrich et al. 2006.
123 Baker et al. 1994.
124 Weinrich et al. 2006.
125 See, e.g., Hewlett et al. (2011) on social learning mechanisms in human hunter-gatherers.
126 Sharpe 2001.
127 Ingebrigtsen 1929, 7.
128 Hain et al. 1982.
129 Hain et al. 1982.
130 There is evidence for this prey disorientation (Sharpe and Dill 1997).
131 Jurasz and Jurasz 1979; Sharpe 2001.
132 Sharpe 2001.
133 Hain et al. 1982; Weinrich, Schilling, and Belt 1992.
134 Weinrich, Schilling, and Belt 1992.
135 Weinrich, Schilling, and Belt 1992.
136 Allen et al. 2013.
137 To do this they used a statistical technique called "network-based diffusion analysis," which works on the straightforward notion that if social learning is important in how a behavior develops in an individual, then the rate at which a particular individual develops the behavior should be related to the amount of social contact that individual has with others who already know how to perform the behavior. In other words, the more you are exposed to the behavior, the more quickly you should learn it, if social learning is indeed involved. If there is no social influence, then it shouldn't matter a fig how often you see others do it. The calculated evidence ratios favored the models with a social learning effect by several orders of magnitude: it was literally tens of thousands of times more likely that social learning was involved in the development of lobtail feeding than it was not (Allen et al. 2013).
138 For instance, the data for Weinrich and colleagues' original study of the spread of lobtail feeding in the Gulf of Maine came from observations during 4,759 commercial whale-watch cruises (Weinrich, Schilling, and Belt 1992). For the quote in text, see Ingebrigtsen (1929), 7.
139 Hoelzel, Dorsey, and Stern 1989.
140 Payne 1976.
141 Payne 1995, 119.
142 Payne (1995), 119. There are speculations that right whale sailing may be involved in bottom feeding or thermoregulation, but we agree with Payne that play is much more likely.
143 Morete et al. 2003.
144 Morete et al. 2003.

Chapter 5

1 The classifications within the genus are now going through a period of flux as modern genome research provides a detailed window on evolution. There may be more than two

species—Australian scientists have suggested that a third species of *Tursiops*, *T. australis*, should be recognized with a small range off Southern Australia (Charlton-Robb et al. 2011).

2 For example, Hoelzel, Potter, and Best 1998; Torres et al. 2003.

3 Fearnbach et al. 2012; Parsons et al. 2006.

4 Leatherwood and Reeves 1990.

5 Parsons et al. 2006.

6 Urian et al. 2009; Wiszniewski et al. 2010.

7 Gero et al. (2005) look at preferred associates for particular activities within dolphin groups.

8 Frère et al. 2010b.

9 Evidence that fighting can lead to unconsciousness is found in Parsons, Durban, and Claridge (2003).

10 Documentation that males sometimes kill calves is in Patterson et al. (1998).

11 Wiszniewski et al. 2012.

12 "Meet the Dolphin Mafia," *Science*, March 27, 2012, http://news.sciencemag.org/sciencenow/2012/03/meet-the-dolphin-mafia.html.

13 Connor, Heithaus, and Barre 2001; Connor, Smolker, and Richards 1992; Moller et al. 2001; Parsons et al. 2003; Wells, Scott, and Irvine 1987.

14 "Meet the Dolphin Mafia"; Connor et al. 2010.

15 De Waal 1982.

16 Connor et al. 2010.

17 Connor 2007; Connor, Smolker, and Bejder 2006.

18 On training for synchronized behaviors, see, e.g., Herman (2002a).

19 Silva et al. 2003.

20 Andrews et al. 2010; Karczmarski et al. 2005.

21 Connor et al. 2000.

22 Bel'kovich et al. 1991.

23 Sargeant et al. 2005.

24 Silber and Fertl 1995.

25 Torres and Read 2009.

26 Gazda et al. 2005.

27 Finn, Tregenza, and Norman 2009, 1.

28 Finn, Tregenza, and Norman 2009, 3.

29 Sargeant and Mann 2009.

30 Mann and Sargeant 2003.

31 Mann and Sargeant 2003.

32 Krützen et al. 2005.

33 Mann et al. 2008.

34 Sargeant et al. 2007; Tyne et al. 2012.

35 On overlapping dolphin home ranges, see Frère et al. (2010a). Evidence of dolphins foraging in the same channels without carrying sponges can be found in Mann and Sargeant (2003).

36 Krützen et al. 2005.
37 Ackermann 2008.
38 Enquist et al. 2010.
39 Kopps and Sherwin 2012.
40 Mann et al. 2012.
41 Neil 2002.
42 Daura-Jorge et al. 2012; Simöes-Lopes, Fabián, and Menegheti 1998; Zappes et al. 2011.
43 Thanks to Maurico Cantor for providing these details.
44 For example, Zappes et al. 2011.
45 Confusingly, the "Ayeyarwady River" is the same as the "Irrawaddy River." "Irrawaddy" is the early transliteration into English of the Burmese name of the river and has become part of the English name of the dolphin species. "Ayeyarwady" is a later, more accurate, transliteration.
46 Special protection by the Burmese government: Chapter V, Protected Wildlife and Wild Plants, Article 15 of The Protection of Wildlife and Protected Areas Law (The State Law and Restoration Council Law No. 6/94); the Ministry of Livestock and Fisheries, Department of Fisheries (DoF), issued Notification No. 11/2005 on 28th. December 2005.
47 Tint Tun 2004, 9.
48 However, the catch per unit time was lower when fishing with dolphins because the fishers had to spend time queuing up for their chance to fish with the dolphins (Tint Tun 2005).
49 Robineau 1995.
50 Neil 2002.
51 Neil 2002.
52 On intentional signaling by honeyguides, see Isack and Reyer 1989.
53 Dean, Siegfried, and MacDonald 1990.
54 Daura-Jorge et al. 2012.
55 Mann et al. 2012.
56 Mann et al. 2012.
57 Chilvers and Corkeron 2001.
58 Chilvers and Corkeron 2001; Chilvers, Corkeron, and Puotinen 2003.
59 Chilvers and Corkeron 2001, 1904.
60 Ansmann et al. 2012.
61 On between-community mating, see Chilvers and Corkeron 2001.
62 Daura-Jorge et al. 2012; Mann et al. 2012.
63 Cantor and Whitehead 2013.
64 For example, Brotons, Grau, and Rendell 2008; Díaz López 2006; Gonzalvo et al. 2008.
65 Mann and Watson-Capps 2005.
66 Donaldson et al. 2012.
67 For more on the matrilines at Shark Bay, see "The Histories of the Beach Dolphins," Shark Bay Dolphin Project, accessed January 23, 2013, http://www.monkeymiadolphins.org/node/32.
68 "The Histories of the Beach Dolphins."

69 Holmes and Neil 2012, 397.
70 Neil 2002.
71 Hoelzel, Potter, and Best 1998, 1177.
72 For example, between the offshore and nearshore forms in the western North Atlantic (Hoelzel, Potter, and Best 1998).
73 Caldwell, Caldwell, and Tyack 1990; Janik, Sayigh, and Wells 2006; King et al. 2013.
74 King et al. 2013.
75 On doubts about signature whistles, see McCowan and Reiss (2001).
76 For the role of learning in the development of a signature whistle, see Sayigh et al. (2007).
77 Sayigh et al. 2007.
78 On the mimicking of other dolphins' signature whistles, see Sayigh et al. (1990) and Tyack (1986).
79 Janik and Slater 1998.
80 See Quick and Janik (2012) on the ways in which whistles are used; regarding recognition of individual whistles, see Janik, Sayigh, and Wells (2006).
81 King et al. 2013; King and Janik 2013.
82 Bruck 2013.
83 On brays, see Janik (2000a); on pops, see Connor and Smolker (1996).
84 For example, Hawkins 2010; Morisaka et al. 2005.
85 Lusseau 2007.
86 Play can be thought of as behavior that has no immediate function but may provide benefits later, e.g., in physical skills or when negotiating social interactions (Bekoff and Byers 1998). However, showing that some behavior has no immediate function is hard; uncovering later functions is even harder.
87 For more details, see "Billie the Dolphin—Her Story and Video," Adelaide Port River Dolphins, accessed August 12, 2013, http://www.portriverdolphins.com.au/billie-the-dolphin---her-story.html. Billie was originally thought to be a male, "Billy," but during her spell in captivity she was found to be female and called "Pat." Back in the wild, she became "Billie."
88 We thank Mike Bossley for sharing these details with us.
89 Paul Eccleston, "Wild Dolphins Learn to Walk on Water," *Telegraph*, August 19, 2008, http://www.telegraph.co.uk/earth/earthnews/3349840/Wild-dolphins-learn-to-walk-on-water.html.
90 Connor, Smolker, and Bejder 2006.
91 Connor et al. 2000.
92 Connor, Smolker, and Bejder 2006, 1377.
93 Connor, Smolker, and Bejder 2006.
94 For play as a way to rehearse and develop important skills, see Bekoff and Byers (1998).
95 Hagen and Bryant 2003.
96 There are multiple readily searchable videos of All-Black hakas on Youtube. Sadly, many of them are followed by Luke's adopted homeland, Wales, losing.
97 For example, Karczmarski 1999.

98 Herzing 2011.
99 On variation that might be cultural, consider, e.g., "Brush excitedly swam around squeaking and pec-rubbing her mom Paint. This was the Paint family tradition" (Herzing 2011, 241).
100 Gowans, Würsig, and Karczmarski 2007.
101 Würsig and Würsig 2010.
102 Dahooda and Benoit-Bird 2010.
103 Vaughn, Degrati, and McFadden 2010.
104 Pearson and Shelton 2010, 345.
105 Norris 1994, 303.
106 Norris and Schilt 1988.
107 Norris 1994, 304.
108 Berdahl et al. 2013.

Chapter 6

1 McComb et al. 2001, 2011.
2 The first detailed analysis of the social system of the "resident" killers was by Mike Bigg and colleagues (1990). An updated, authoritative, but less formal, description is by John Ford and colleagues (2000).
3 A detailed description of what we know of sperm whale society in the Pacific is in Whitehead (2003b). A recent update that contrasts Atlantic and Pacific populations is Whitehead et al. (2012).
4 Amos, Schlötterer, and Tautz 1993; Baird et al. 2008; Kasuya and Marsh 1984.
5 Gowans, Whitehead, and Hooker 2001.
6 "Killer Whale: The Top, Top Predator" (Pitman 2011c) is a collection of articles by many of the world's most influential killer whale scientists and is an excellent source of up-to-date information and speculation. There are also a number of good books about killer whales, such as Baird (2006), Ford and Ellis (1999), Ford, Ellis, and Balcomb (2000), and Hoyt (1991).
7 Pitman 2011b, 5.
8 Whitehead 2012.
9 Erich Hoyt captures this ambiguity about names in the title of his book *Orca, the Whale Called Killer* (1991).
10 Unlike the wild animals, captive killer whales have killed humans, including their trainers. This contrast is used as one of many arguments why these animals should not be kept in captivity (Marino and Frohoff 2011).
11 For example, Baird and Dill (1996). Studies of this kind are sometimes possible with terrestrial animals, such as lions or elephants, but nearly impossible for other cetaceans without sophisticated telemetry devices and some hefty assumptions.
12 Ford (2011). The names residents, transients, and offshores are artifacts of times when much less was known of these animals and are only weakly descriptive.
13 Foote et al. 2009.

14 De Bruyn, Tosh, and Terauds (2013) argue that, unlike the well-studied killer whales of the North Pacific, the division of Antarctic killer whales into ecotypes is premature; they should, for now, just be called morphotypes.
15 Pitman 2011a; Pitman and Durban 2010.
16 On wolves, see Koblmüller et al. (2009); on coyotes, see Sacks et al. (2008).
17 Stearns and Hoekstra 2000, 221–25.
18 Morin et al. 2010.
19 Barrett-Lennard 2011.
20 Barrett-Lennard and Heise 2011; Riesch et al. 2012.
21 Ford and Ellis 2006.
22 Pitman and Durban 2012.
23 On the preference for the lips and tongues of baleen whales by killer whales, see Jefferson, Stacey, and Baird (1991); on Gerlache killer whale food preferences, see Pitman and Durban (2010).
24 Barrett-Lennard 2011.
25 Ford and Ellis 1999, 83.
26 Ford and Ellis 1999, 83.
27 Barrett-Lennard 2011, 51.
28 Abbreviated and less complete versions of this argument have been made previously (Baird 2000; Boran and Heimlich 1999).
29 Riesch et al. (2012). Recent genetic analyses suggest that while some of the killer whale ecotypes originated in sympatry, i.e., together, others did not (Foote et al. 2011). For instance the Pacific transient and offshore ecotypes may have arisen in the Atlantic.
30 Baird 2011.
31 Estes et al. (1998). Estes and his colleagues have gone further, arguing that the switching of prey types by transient killer whales, following the destruction of baleen whale populations by commercial whaling, have led to the sequential decline of a number of other species of marine mammal (Springer et al. 2003), but this argument is extremely controversial (e.g., DeMaster et al. 2006; Whitehead and Reeves 2005).
32 Uhen 2010.
33 Uhen 2007.
34 McGowen, Spaulding, and Gatesy 2009.
35 Le Boeuf et al. (1988) discuss elephant seals resting at depth as a response to the threat posed by great white sharks, but killer whales could also be driving this behavior. On baleen whales migrating to avoid killer whales, see Corkeron and Connor (1999).
36 Connor 2000; Gowans, Würsig, and Karczmarski 2007.
37 Ford, Ellis, and Balcomb 2000.
38 This description of resident killer whale social structure is from Ford et al. (2000, 25).
39 Chadwick (2005). We are not sure who first used "momma's boys" to describe resident killer whale societies, but Chadwick's is the first written reference we have come across.
40 Barrett-Lennard 2000.
41 Baird and Whitehead 2000; Beck et al. 2011.
42 Bigg et al. 1990.

43 Riesch, Ford, and Thomsen 2006; Thomsen, Franck, and Ford 2002.
44 Ford 1991; Ford and Fisher 1983.
45 Deecke et al. 2010.
46 Yurk et al. 2002.
47 Deecke, Ford, and Spong 2000.
48 Deecke, Ford, and Spong 2000.
49 For example, Riesch et al. 2012; Strager 1995.
50 Deecke, Ford, and Slater 2005.
51 Ford, Ellis, and Balcomb (2000). The northern residents do occasionally perform the greeting ceremony, but it is very rare.
52 Pitman (2011b). Interestingly, the type with the most prominent eye patch, the Antarctic pack-ice killer whale, also shows the highest degree of coordination when hunting, synchronizing fluke movements (Pitman and Durban 2012).
53 Pitman (2011a), 42–43. For a more technical description, see Pitman and Durban (2012).
54 Similä and Ugarte 1993.
55 See Dial and colleagues (2008) for a discussion of how behavioral aptitudes scale with size.
56 Similä and Ugarte 1993.
57 Similä and Ugarte 1993, 1495.
58 Domenici et al. 2000.
59 Deecke et al. 2011; Simon et al. 2006.
60 Lopez and Lopez 1985, 182.
61 Guinet 1991.
62 Hoelzel 1991.
63 Guinet and Bouvier 1995; Lopez and Lopez 1985.
64 Visser 1999.
65 Duignan et al. 2000.
66 As described in Chadwick (2005).
67 Our description of the Twofold Bay killers is from Dakin (1934) and Wellings (1964). If you want to learn more about this cooperative, we recommend Mead's lively book *Killers of Eden* (1961) and the Australian Broadcasting Corporation's TV documentary *Killers in Eden* (produced and written by Klaus Toft [2005]), which contains original footage of the cooperative whaling.
68 Quoted from Dakin (1934), 156. Brierly also noted that the aboriginals in the area "regard the killers as incarnate spirits of their own departed ancestors and in this belief they go so far as to particularize and identify certain individual killer spirits."
69 Dakin 1934; Wellings 1964.
70 Baird and Whitehead 2000.
71 Matkin and Durban 2011.
72 For example, Pitman and Durban 2012.
73 Ford, Ellis, and Balcomb 2000.
74 Ford, Ellis, and Balcomb 2000.
75 Whitehead et al. 2004.

76 Baird 2011.
77 Baird 2011.
78 Durban and Pitman 2011.
79 Barrett-Lennard 2011.
80 There were perhaps 1,110,000 sperm whales before whaling (Whitehead 2002), with an average mass of about 19.5 metric tons each, which is equivalent to 433 million humans at fifty kilograms each. Human population reached this level in about 1300 A.D. ("World Population: Historical Estimates of World Population," U.S. Census Bureau, rev. December 19, 2013, http://www.census.gov/population/international/data/worldpop/table_history.php). Prewhaling biomass levels of fin and blues whales were similar to those of the sperms or, perhaps, a little larger (Pershing et al. 2010).
81 Clarke 1977.
82 There are a number of books about sperm whales, most famously *Moby Dick* (Melville [1851] 1972). The two most recent include most of what science knows of the species. *Sperm Whales: Social Evolution in the Ocean* (Whitehead 2003b) is written for scientists, and concentrates on their behavioral biology; *The Great Sperm Whale: A Natural History of the Ocean's Most Magnificent and Mysterious Creature* (Ellis 2011) is more recent, broader, and written for a wider audience. A very recent perspective on the sperm whale, written by scientists but for a general audience, is the 2012 special issue of the journal of the American Cetacean Society, the *Whalewatcher* (Whitehead 2012). It describes how we sperm whale scientists see the animals we study and why we are fixated on this strange creature.
83 Madsen (2012). We are beginning to get an idea of how the spermaceti organ works (Madsen 2002), but many of the details are not understood.
84 The blowhole on the left of the snout is seemingly a regression. While land mammals and seals have their nostrils on the snout, in other whales and dolphins it has migrated to the top of the head, allowing the animals to breathe more easily while traveling fast. In the sperm whale, much of the body has been rearranged around the needs of the sonar system, including the position of the blowhole. Consequently, sperm whales breathe at the surface in a rocking manner unlike that of the other whales.
85 Carl Linnaeus, the Swedish naturalist who was largely responsible for the system that scientists use to name species, thought that there were two species of sperm whale, *Physeter macrocephalus* and *Physeter catodon*. His confusion may have partly resulted from the differences between the sexes. The description of *Physeter macrocephalus* was based mainly on observations of males, while *Physeter catodon* was described as being smaller (Holthuis 1987; Husson and Holthuis 1974).
86 The males occasionally can move with sudden quickness, rather like sumo wrestlers, another set of very large males who look stiff and ungainly when not competing (Whitehead 2003b, 280–81).
87 Best 1979.
88 Assuming they make an average of one click per second over half of a sixty-year life span gives about 946,080,000 clicks.
89 Watwood et al. 2006.

90 Aoki et al. 2012; Madsen 2012; Watwood et al. 2006.
91 Whitehead 2003b, 167.
92 Whitehead 2003b, 168–84.
93 Gero (2012b) describes the social lives of the Group of Seven. See also Gero and colleagues (Gero et al. 2008; Gero, Gordon, and Whitehead 2013).
94 The Group of Seven is named after the Canadian landscape painters of the 1920s and 1930s. The family unit of sperm whales off Dominica has not always had seven members.
95 As we write, Fingers has given birth to another calf, and so will perhaps regain her central social position in the Group of Seven. See Gero (2012b).
96 Gero 2012b.
97 Gero 2012a.
98 Whitehead et al. 2012.
99 Whitehead 1999.
100 Rendell 2012; Watkins and Schevill 1977.
101 Madsen et al. 2002.
102 On the slow click/clang of breeding males, see Weilgart and Whitehead (1988).
103 Watkins and Schevill 1977.
104 Regarding the rarity of codas by males, see Marcoux, Whitehead, and Rendell (2006). However, males of the small and strange Mediterranean sperm whale population do routinely make codas (Frantzis and Alexiadou 2008). On the babbling of small calves, see Schulz et al. (2010).
105 Schulz et al. 2008.
106 Schulz et al. (2010); Rendell and Whitehead (2004). An exception is Fingers, the mother in the Group of Seven, who has a distinctive repertoire.
107 Antunes 2009.
108 Rendell 2012; Rendell and Whitehead 2003.
109 Rendell and Whitehead 2005.
110 Rendell and Whitehead 2003.
111 Whitehead and Rendell 2004.
112 Whitehead and Rendell 2004.
113 Whitehead 1996.
114 Whitehead (2003a), 305–6. The longest periods during an hour with no whales at the surface averaged about four minutes longer for Regular groups than for Plus-One groups.
115 Whitehead 2003a, 305.
116 Whitehead and Rendell 2004.
117 Marcoux, Rendell, and Whitehead 2007.
118 Rendell and Whitehead 2003.
119 Whitehead and Rendell 2004.
120 Rendell et al. 2012.
121 Whitehead 2003b, 300.
122 Antunes 2009.
123 Deecke, Ford, and Spong 2000.
124 Lyrholm et al. 1999.

125 Whitehead et al. 2012.

126 Antunes 2009.

127 Whitehead et al. 2012.

128 Whitehead et al. 2012.

129 Whitehead 2003b, 70–73.

130 Pitman et al. (2001); Whitehead et al. (2012). For an account of the most famous observation of killer whales attacking sperm whales, see Pitman and Chivers (1999).

131 On outward-facing formations, see Arnbom et al. (1987); on inward-facing ones, see Pitman et al. (2001). The latter is sometimes called the "marguerite" formation, after the shape of the flower of this name.

132 Rendell et al. 2012.

133 Zhou et al. 2005.

134 For hunter-gatherers the scaling ratio is about 4 (Hamilton et al. 2007). For baboons and elephants it is about 3 (Hill, Bentley, and Dunbar 2008).

135 Hill, Bentley, and Dunbar 2008.

136 Hill, Bentley, and Dunbar 2008; Zhou et al. 2005.

137 On clan sizes of roughly ten thousand, see Rendell and Whitehead (2003).

138 Richerson and Boyd 2005, 195.

139 Gero 2012b; Gero, Gordon, and Whitehead 2013.

140 Letteval et al. 2002.

141 On stealing fish from longlines, see Sigler et al. (2008); on entering shallow waters, see Scott and Sadove (1997).

142 See Whitehead 2003b, 277–83.

143 Weilgart and Whitehead 1988.

144 Beale 1839, 6.

145 The beluga and narwhal social systems are described as matrifocal by Palsbøll and colleagues (Palsbøll, Heide-Jørgensen, and Bérubé 2002; Palsbøll, Heide-J¢rgensen, Dietz 1997), rather than matrilineal. The killer whale is often included among the blackfish, although it is not of the Globicephalinae. Risso's dolphin may be a member of the Globicephalinae (McGowen, Spaulding, and Gatesy 2009), although it is not considered a blackfish, and there are no suggestions that it has a matrilineal social system (Hartman, Visser, and Hendriks 2008).

146 Aschettino et al. 2011; Baird et al. 2008; McSweeney et al. 2009.

147 They are often attracted to ships and boats but tend to follow rather than lead them, so "pilot" is perhaps not the most appropriate name.

148 Sergeant 1982.

149 Sergeant (1982). Twenty-eight percent were false killer whales.

150 Amos 1993; Amos, Schlötterer, and Tautz 1993.

151 Amos's analyses were for long-finned pilot whales. We do not have genetic data for short-finned pilot whale social systems, but Toshio Kasuya and Helene Marsh (1984) concluded that they, too, have a matrilineal social system based on the age distribution within driven schools off Japan; they, however, interpret the wide variation in the proportion of mature males among schools as suggesting that males move between schools.

152 Zachariassen 1993.
153 Ottensmeyer and Whitehead 2003.
154 De Stephanis et al. 2008b.
155 Group sizes off Tenerife can be found in Heimlich-Boran (1993), and off Japan, in Kasuya and Marsh (1984).
156 De Stephanis et al. (2008a).
157 Nemiroff 2009.
158 Baird et al. 2012.
159 Baird et al. 2008; McSweeney et al. 2009; Robin Baird personal communication, November 16, 2012.
160 Weller et al. 1996.
161 Palsbøll, Heide-Jørgensen, and Bérubé 2002.
162 Ridgway et al. 2012.
163 Colbeck et al. 2013.
164 Jefferson, Stacey, and Baird 1991; Norris and Schilt 1988.

Chapter 7

1 Marino and Frohoff 2011.
2 Kuczaj 2010; Marino and Frohoff 2011.
3 Madsen et al. 2003; Møhl et al. 2003.
4 Hoppitt and Laland 2008.
5 Richerson and Boyd 2005, 108; Tomasello 1999a.
6 See Laland, Kendal, and Kendal (2009) for a summary of the debate about the preeminence of imitation among social learning mechanisms.
7 Pryor 2001.
8 "Marine Mammal Program," U.S. Navy, http://www.public.navy.mil/spawar/Pacific/71500/Pages/default.aspx; William Gasperini, "Uncle Sam's Dolphins: In the Iraq War, Highly Trained Cetaceans Helped U.S. Forces Clear Mines in Umm Qasr's Harbor," *Smithsonian Magazine*, September 2003, http://www.smithsonianmag.com/science-nature/Uncle_Sams_Dolphins.html; John Pickrell, "Dolphins Deployed as Undersea Agents in Iraq," *National Geographic News*, March 28, 2003, http://news.nationalgeographic.com/news/2003/03/0328_030328_wardolphins.html.
9 Tayler and Saayman 1973.
10 Candland 1993.
11 Mary Vorsino, "Last Dolphin Dies at Marine Laboratory: A UH Official Says the Bottlenose Dolphin's Death Was a Surprise," *Honolulu Star-Bulletin*, February 26, 2004, http://archives.starbulletin.com/2004/02/26/news/story3.html.
12 Reported in Herman (2002a). To our knowledge, the work has not been published in the primary scientific literature, and we are grateful to Mark Xitco for helping us get the details right.
13 Herman 2002b, 281.
14 Bauer and Johnson 1994.
15 Bauer and Johnson 1994, 1312.

16 Mark Xitco, personal communication, November 20, 2012.
17 Call and Tomasello 1996.
18 For an insight into how this might happen in humans, recent work has shown how a stimulating environment in early childhood affects the size of certain brain areas in young adults (Avants et al. 2012).
19 Candland (1993); we will return to these children in the next chapter.
20 Jaakkola, Guarino, and Rodriguez 2010.
21 Jaakkola et al. 2013.
22 Jaakkola et al. 2013, 7.
23 Whiten 2001, 359.
24 Abramson et al. 2013.
25 Abramson et al. 2013, 3.
26 Abramson et al. 2013, 4.
27 Abramson et al. 2013, 7.
28 Janik and Slater 1997.
29 Janik and Slater 1997.
30 Shettleworth (1998)—"visual-tactile cross-modal performance" does, however, have the virtue of being a shorter way to describe this process.
31 See Rizzolatti and Craighero (2004) regarding the discovery of mirror neurons. For assumptions about the separation of the sensory systems, see Herman (2002a).
32 Catchpole and Slater 2008.
33 The vocal learning capabilities of spear-nosed bats are explored by Boughman (1998).
34 Richards 1986.
35 Reiss and McCowan 1993.
36 Ridgway et al. 2012.
37 Lilly 1965.
38 Miksis, Tyack, and Buck 2002.
39 Fripp et al. 2005; Sayigh et al. 1995.
40 Janik 2000b; King et al. 2013; King and Janik 2013; Quick and Janik 2012; Tyack 1986.
41 On the similarity of the whistles of adult male dolphins that have formed alliances, see Smolker and Pepper (1999) and Watwood, Tyack, and Wells (2004).
42 Miller et al. 2004.
43 Bain 1989.
44 Foote et al. 2006.
45 Csibra and Gergely 2009, 2011.
46 Byrne 1995.
47 Premack and Premack 1996.
48 Caro and Hauser 1992; Fogarty, Strimling, and Laland 2011; Hoppitt et al. 2008; Thornton and Raihani 2008.
49 Caro and Hauser 1992.
50 The definition in full reads: "An individual actor **A** can be said to teach if it modifies its behavior only in the presence of a naive observer, **B**, at some cost or at least without ob-

taining an immediate benefit for itself. **A**'s behavior thereby encourages or punishes **B**'s behavior, or provides **B** with experience, or sets an example for **B**. As a result, **B** acquires knowledge or learns a skill earlier in life or more rapidly or efficiently than it might otherwise do, or that it would not learn at all" (Caro and Hauser 1992, 153).

51 Byrne and Rapaport 2011.

52 Hoppitt et al. 2008.

53 Examples of teaching in animals are as follows: meerkats, Thornton and McAuliffe (2006); pied babblers, Raihani and Ridley (2008); and ants, Franks and Richardson (2006).

54 Fogarty, Strimling, and Laland 2011; Hoppitt et al. 2008.

55 Guinet 1991; Guinet and Bouvier 1995.

56 Hoelzel 1991; Lopez and Lopez 1985.

57 Thornton and McAuliffe 2006.

58 Herman and Pack 2001.

59 Bender, Herzing, and Bjorklund 2009.

60 Dolphins' understanding of the purpose of human's pointing is documented in Herman et al. (1999). Their use of body orientation to "point" is illustrated in Xitco, Gory, and Kuczaj (2001, 2004).

61 Byrne and Rapaport 2011.

62 Baker et al. 1998; Valenzuela et al. 2009.

63 Colbeck et al. 2013.

64 See Reader and Biro (2010).

65 Thornton and McAuliffe 2006.

66 Heyes 2012.

67 For social learning in octopuses, see Fiorito and Scotto (1992), and in tortoises, see Wilkinson et al. (2010).

Chapter 8

1 "Where's the Beef?" is the title of psychologist Jeff Galef's negative commentary on our 2001 review of culture in whales and dolphins (Galef 2001).

2 Byrne 2007.

3 Morgan 1894, 53.

4 Morgan 1903, 59.

5 Costall 1998; de Waal 2001; Fitzpatrick 2008; Sober 2005.

6 Fitzpatrick 2008.

7 *The Question of Animal Culture* is the title of the influential book edited by Kevin Laland and Jeff Galef (2009b).

8 Johnson 1999; Whitehead 2009.

9 Hoppitt and Laland 2013; Whitehead 2009.

10 Kasuya 2008, 768.

11 Laland and Janik 2006.

12 Allen et al. 2013.

13 Ford 1991; Yurk et al. 2002.

14 Deecke, Ford, and Spong 2000.

15 Bain 1989.

16 A preliminary analysis of autosomal genes, inherited through both parents, found no indication of differences between the clans of sperm whales with different dialects (Whitehead 2003b).

17 Bacher et al. 2010.

18 For example, Anderwald et al. 2011; Pastene et al. 2010.

19 Catchpole and Slater 2008.

20 Slagsvold and Wiebe 2011.

21 Bouchard et al. 1990.

22 For a detailed account of these examples and much more besides, see Douglas Candland's book *Feral Children and Clever Animals* (1993), from which our summaries are drawn.

23 Candland 1993, 18.

24 The details on Keiko that we present here are largely drawn from Simon et al. (2009).

25 Simon et al. 2009, 697.

26 These details are drawn from Francis and Hewlett (2007).

27 "Celebrating Springer—Orphan Orca That Overcame the Odds," NOAA Fisheries, July 12, 2012, http://www.nmfs.noaa.gov/pr/laws/mmpa/anniversary/celebrating_springer.html.

28 "Springer the Orphaned Killer Whale Spotted with Calf off B.C.'s North Coast," Keven Drews, *MacLean's*, posted July 8, 2013, http://www.macleans.ca/news/springer-the-orphaned-killer-whale-spotted-with-calf-off-b-c-s-north-coast/.

29 Wells, Bassos-Hull, and Norris 1998.

30 Watkins et al. 2004.

31 This epigraph is from Marks (2002), xvi. The phrase "perfectly absurd" is from anthropologist Tim Ingold's critical commentary on our 2001 paper (Ingold 2001, 337). The full sentence is: "Moreover, the idea that culture is a third determinant of behaviour, after allowance has been made for environmental and genetic determinants, is perfectly absurd."

32 For example, Hill 2009; Marks 2002.

33 This definition of meaning is from McGrew (1998).

34 Barkow 2001.

35 McGrew 1998; Tomasello 1999b.

36 Carrier et al. (1984); Liebenberg (2006); Wheeler (1991). For a more accessible account, see chap. 28 of Christopher McDougall's popular book *Born to Run* (2009).

37 Marks 2012, 155.

38 We owe this insight to conversations with the behavioral ecologist, and accomplished clarinet player, Luc-Alain Giraldeau.

39 Tool use by Galápagos finches is documented by Tebbich et al. (2001).

40 Tomasello 1999b.

41 Hill (2009), 283. The earliest and most tenuous of such evidence is in remains suggesting

the use of ochre pigment in South Africa some 164,000 years ago (Marean et al. 2007). By 100,000 years ago we have good evidence of relatively sophisticated processing of ochre (Henshilwood et al. 2011), and by 60,000 years ago there is much stronger evidence in the form of things like engraved eggshells (Texier et al. 2010).

42 Sterelny 2009.
43 Hofstede 1981.
44 Hill 2009, 275.
45 Perry 2009, 260.
46 On ethnic markers as indicators for cooperation, see Bowles and Gintis (2003) and McElreath, Boyd, and Richerson (2003).
47 Kelly 2002.
48 Cantor and Whitehead 2013.
49 Perry et al. 2003, 249.
50 Perry et al. 2003, 251.
51 Perry 2009, 262.
52 Perry 2009, 267.
53 That whistles may be used to refer to individuals is suggested in Janik (2000b). Evidence that this is the case in the wild can be found in King et al. (2013).
54 Nettle 1999.
55 Barrett-Lennard 2011.
56 Pitman et al. 2001.
57 Pitman and Chivers 1999.
58 Marks 2012.
59 Marks 2002, 2012.
60 Hill 2009, 271.

Chapter 9

1 Laland and Hoppitt 2003.
2 Richerson and Boyd 2005, 100.
3 When the conditions are right, natural selection can react to environmental change over just a few generations (e.g., Hairston et al. 2005; Svensson, Abbott, and Härdling 2005).
4 Auld, Agrawal, and Relyea 2010.
5 Agrawal 2001.
6 Auld, Agrawal, and Relyea 2010.
7 Thus some use the phrase "socially biased learning" rather than "social learning" (see also Fragaszy and Visalberghi 2001; Heyes 2012).
8 Rendell et al. 2011b; Rogers 1988.
9 Whitehead 2007.
10 This formulation is because energy is related to the inverse of frequency. See Halley (1996).
11 Theoretically we could also have "blue noise" in which the variation is greater over short scales than long ones, but real environments are unlikely to have this characteristic.

12 Whitehead 2007.
13 As explained in most introductory ecology texts such as Begon, Harper, and Townsend (1996).
14 Dommenget and Latif 2002; Hall and Manabe 1997; Leeuwenburgh and Stammer 2001; Pelletier 1997; Timmermann et al. 1998.
15 On the increasing redness of the variation for each step up the food chain, see Inchausti and Halley (2002).
16 Inchausti and Halley 2002.
17 Steele 1985.
18 Dommenget and Latif 2002; Hall and Manabe 1997; Leeuwenburgh and Stammer 2001; Timmermann et al. 1998.
19 Steele 1985.
20 Inchausti and Halley 2002.
21 For example, Gero 2012b.
22 Richerson and Boyd 2005, 131–36; 2013.
23 Learning in sharks is the topic of Guttridge et al. (2009).
24 Laland and Hoppitt 2003.
25 Rendell et al. 2011b.
26 An alternative explanation for the rarity of sophisticated multifaceted cultures is that, while many species can learn socially, only certain kinds of social learning, in particular imitation, leads to complex and cumulative cultures, and imitation is hard to evolve (Boyd and Richerson 1996; Tomasello 1994). However, more recently scholars have argued against a clear delineation of imitation among social learning mechanisms—in its nature, or our ability to categorize it, or in its effects (e.g., Heyes 2012; Laland, Kendal, and Kendal 2009).
27 Coussi-Korbel and Fragaszy 1995; Whitehead and Lusseau 2012.
28 De Waal 2001, 231.
29 De Waal and Bonnie 2009, 22.
30 De Waal and Bonnie 2009, 22–24.
31 De Waal 2001, 231.
32 McComb et al. 2001.
33 On social lives of sharks, see Mourier, Vercelloni, and Planes (2012).
34 Anderson 2004.
35 Anderson (2004), 65. However, sirenian social relationships are not totally random (Young Harper and Schulte 2005).
36 Reynolds and Wilcox 1986.
37 Sousa-Lima, Paglia, and Da Fonseca 2002.
38 Stewart and Fay 2001.
39 Sjare and Stirling 1996.
40 Estes et al. 2003.
41 Enquist et al. (2010). Brooke Sargeant and Janet Mann (2009) also discuss the relationship between individual specialization and culture.
42 Mithven 1999; Richerson and Boyd 2005, 141–43.

43 Richerson and Boyd 2005, 142.
44 Marino et al. 2003.
45 Marino, McShea, and Uhen 2004.
46 See Montgomery et al. 2013.
47 Personal communication from Lori Marino, March 15, 2012.
48 Ridgway and Au 1999.
49 Personal communication from Lori Marino, March 15, 2012.
50 Jerison 1986.
51 Marino et al. 2008, 4.
52 Thewissen 2009.
53 Thewissen 2009.
54 Würsig and Würsig 1980.
55 The most dramatic that we know of was the giant predatory sperm whale, Leviathan, of about twelve million years ago (Lambert et al. 2010).
56 Connor 2000.
57 The absence of vocal learning in other mammals is discussed by Janik and Slater (1997).
58 Moll and Tomasello 2007.
59 Vygotsky 1978.
60 The Machiavellian intelligence hypothesis was brought to prominence in 1976 by Nicholas Humphrey (1976) and then developed in an edited book published in 1988 (Byrne and Whiten 1988).
61 Moll and Tomasello 2007.
62 Marino, McShea, and Uhen 2004.
63 Marino, McShea, and Uhen 2004, figs. 2–4.
64 See Whitehead 2003b.
65 Horn and Rubinstein 1984.
66 Marino et al. (2008) discuss brain size and complexity of mysticetes.
67 Uhen 2010.
68 Personal communication from Fred Sharpe December 5, 2011.

Chapter 10

1 Maynard Smith and Szathmáry (1995). These same authors later wrote a less technical book on the same subject, aimed at a more general audience: *The Origins of Life: From the Birth of Life to the Origin of Language* (1999).
2 Maynard Smith and Szathmáry (1995, 6; 1999, 18). Back in 1999, when they wrote their second book, Maynard Smith and Szathmáry admitted that they had missed some important transitions, such as the evolution of the nervous system, the development of writing, and the current extraordinary transition to electronic means of transmitting, storing, and manipulating data: "Will our descendants live most of their lives in a virtual reality? Will some form of symbiosis between genetic and electronic storage evolve? Will electronic devices acquire the means for self-replication, and evolve to replace the primitive life forms that gave them birth?"(Maynard Smith and Szathmáry 1999, 170).
3 Richerson and Boyd 2005, 12–13.

4 Mithven 1999.
5 Email from Peter Richerson to Hal, January 11, 2001. For Richerson and Boyd's own thinking on this, see Richerson and Boyd (2013).
6 Maynard Smith and Szathmáry 1995.
7 Donald 1991.
8 Donald 1991; Dunbar 1996; Richerson and Boyd 2005.
9 On the need for a history of cultural evolution to generate language, see Kirby, Cornish, and Smith 2008; Kirby, Dowman, and Griffiths 2007; Smith and Kirby 2008.
10 For a summary of what was known of gene-culture coevolution in humans in 1996, see Feldman and Laland (1996).
11 Laland, Odling-Smee, and Myles 2010; Richerson, Boyd, and Henrich 2010.
12 Perry et al. 2007.
13 Cochran and Harpending 2009.
14 Morin et al. (2010).
15 Riesch et al. 2012.
16 Barrett-Lennard 2011.
17 Preliminary analysis by Charissa Fung (cited by Reeves et al. 2004) shows that mammal-eating transient killer whales have more robust mouth parts than the fish-eating residents.
18 The Loro Parque aquarium in Tenerife currently houses second-generation offspring from matings of Pacific mammal eaters and fish eaters with animals of unknown ecotype caught off Iceland (Kremers et al. 2012).
19 Foote et al. 2013a; Morin et al. 2010.
20 Foote et al. 2011.
21 Deecke et al. (2011) document whales that specialize on a particular food in the Atlantic, while Foote et al. (2009) explore the possibility of generalists being there as well.
22 Evidence for the evolution of separate ecotypes in the Atlantic being at an early stage is examined in Foote et al. (2013b).
23 Riesch et al. 2012.
24 Riesch et al. 2012.
25 The taxonomy of the Delphinidae has not been nailed down, so the number of species is uncertain. The pertinent information regarding having a living sister species joined by a common ancestor less than ten million years ago can be found in McGowen, Spaulding, and Gatesy (2009). Some of the nondolphin cetaceans are even more distinct than the killer whale. They include river dolphins, beaked whales, the pygmy right whale, and, of course, the sperm whale.
26 Riesch et al. 2012.
27 Traill, Bradshaw, and Brook 2007.
28 Matkin and Durban 2011.
29 Foote et al. (2009) describe the eastern North Atlantic generalists.
30 Whitehead 1998, 2003a.
31 See papers containing analyses of cetacean mitochondrial DNA diversity cited in Whitehead (1998, 2003a).

32 Lyrholm, Leimar, and Gyllensten 1996.
33 Whitehead 1998. However, the bottleneck theory has been bolstered for killer whales following a recent comprehensive study of both mitochondrial and nuclear genetic diversity finding signs of reductions in diversity during the last glaciation for some, but not all, populations (Moura et al. 2014).
34 Janik 2001.
35 For proposed demographic explanations, see Amos (1999); Siemann (1994); Tiedemann and Milinkovitch (1999).
36 Whitehead 2005.
37 Whitehead 1998.
38 Maynard Smith and Haigh 1974.
39 Whitehead and Rendell 2004.
40 Arnbom et al. 1987; Pitman et al. 2001.
41 Marcoux, Rendell, and Whitehead (2007). This study looked at mean rates of producing calves over a decade or so. For cultural hitchhiking to work, these differentials would have to persist over much longer periods.
42 Rendell et al. (2012) discuss the reduced diversity of mitochondrial genes in clans as compared with the population at large.
43 Deecke, Ford, and Spong 2000; Mesnick et al. 1999; Tiedemann and Milinkovitch 1999.
44 We know of one Pacific sperm whale, no. 236, who seems to have moved from the Regular clan to the Plus-One clan between 1985 and 1987 (Whitehead 2003b).
45 Whitehead (2005). The models showed that for cultural hitchhiking to reduce genetic diversity substantially there had to be a reasonably low rate of genetic mutation; that clans had to split fairly frequently when they had increased to constitute a substantial part of the population; that an average of fewer than ten animals move from any clan to other clans in any generation; that the culturally determined fitness of a clan changes by an average of more than about 0.005 percent per generation; and that the changes in a clan's fitness were more affected by cultural innovation within the clan than cultural assimilation from other clans.
46 Alexander et al. 2013.
47 Whitehead 2005.
48 Following the hypothesis that cultural hitchhiking is a potential explanation of the low mitochondrial diversity of the matrilineal whales, Hal Whitehead, Peter Richerson, and Robert Boyd (2002) suggested that cultural hitchhiking might also have occurred in another species, *Homo sapiens*. Humans have extraordinarily low diversity of the paternally inherited (i.e., inherited only from the father) Y-chromosome genes, and traditional human societies are largely patrilineal, with females transferring between tribes on marriage, so that cultural knowledge is largely transmitted through the male line. So the whale scenario is reversed. This is a very unusual situation in that a complex behavioral-genetic process first suggested for another species was then applied to humans.
49 Kopps et al. 2014.
50 For example, Foster and Ratnieks 2005.
51 Marsh and Kasuya 1984, 1986; Olesiuk, Bigg, and Ellis 1990.

52 Kasuya and Marsh 1984.

53 Marsh and Kasuya 1986.

54 Best, Canham, and Macleod (1984). Interestingly, six of the older nonreproductive female sperm whales were lactating, so they are contributing physiologically to new generations without giving birth themselves.

55 Hawkes et al. 1998; Kim, Coxworth, and Hawkes 2012.

56 Johnstone and Cant 2010.

57 Lahdenperä et al. 2004.

58 Foster et al. 2012.

59 Ford et al. 2011.

60 Hawkes et al. 1998.

61 Whitehead and Mann 2000.

62 Diamond 1998.

63 Diamond 1998, 123–24.

64 Arntz (1986) describes the effects of the super El Niño. Knowledge held by older whales and passed down to younger, less-experienced whales is discussed in McAuliffe and Whitehead (2005) and Rendell and Whitehead (2001b).

65 Headline in *Yes Magazine*, Vancouver, Canada, March 2001.

66 Butti et al. 2009; Hof and Van 2007.

67 Hakeem et al. 2009.

68 Boddy et al. 2012.

Chapter 11

1 Richerson and Boyd 2005, 12–13.

2 Estes et al. 1998.

3 Corkeron and Connor 1999; Whitehead et al. 2012.

4 *Nunatsiaq News*, as cited in the article, "Killer Whales, 'Wolves of the Sea,' Are Migrating North, Inuit Traditional Ecological Knowledge Reveals," Indian Country Today Media Network, http://indiancountrytodaymedianetwork.com/2012/02/02/killer-whales-wolves-sea-are-migrating-north-inuit-traditional-ecological-knowledge.

5 The ecological impact of sharks is explored by Stevens et al. (2000).

6 Although sharks may, on occasion, learn socially, too (Guttridge et al. 2012).

7 "Killer Whales, 'Wolves of the Sea,' Are Migrating North."

8 Ritchie and Johnson 2009.

9 "Top, top predator" is taken from an article title by Pitman (2011c).

10 The possible effects of sperm whales on the squid population are explored in Whitehead (2003b).

11 Sperm whales have a generally wider prey distribution than beaked whales or elephant seals (Whitehead, MacLeod, and Rodhouse 2003).

12 Although they rarely go into shallow waters, there are exceptions—for awhile sperm whales were sighted consistently in shallow waters off Long Island, New York (Scott and Sadove 1997).

13 Laws (1970); West, King, and White (2003). Elephants, starlings, and cetaceans are also adept at exploiting human-induced habitat change (Whitehead et al. 2004).
14 Laland, Odling-Smee, and Feldman 2000; Rendell, Fogarty, and Laland 2011.
15 Budnikova and Blokhin 2012.
16 Fearnbach et al. 2013.
17 Saulitis 2013.
18 Matkin and Durban 2011.
19 Saulitis 2013.
20 Matkin et al. 1997.
21 Mesnick et al. 2006.
22 Straley 2012, 41. See also Mathias et al. 2009.
23 Gazda et al. 2005.
24 Lusseau 2007.
25 Lusseau and Newman 2004.
26 Williams and Lusseau 2006.
27 Williams and Lusseau 2006.
28 Foley et al. 2008; McComb et al. 2001, 2011.
29 Fehring and Wells 1976.
30 Richerson and Boyd 2005, 12–13.
31 Brighton 2004.
32 "The Charge of the Light Brigade," Alfred, Lord Tennyson. 1854.
33 "The Charge of the Light Brigade."
34 Dixon 1994, 41.
35 Frantzis (1998); Simmonds (1997). There are also occasional "mass mortalities" in which the animals strand over large spatial extents and over long time periods of weeks or more. In these cases, the animals are already dead or sick. Disease and pollution are believed to be the causes (Simmonds 1991).
36 Weilgart 2007.
37 D'Amico et al. 2009.
38 Aristotle, *Historia Animalium* 1.5.
39 For example, Bradshaw, Evans, and Hindell 2006; Simmonds 1997.
40 Fehring and Wells 1976.
41 Bradshaw, Evans, and Hindell 2006; Simmonds 1997.
42 Robson 1988; Robson and van Bree 1971; Sergeant 1982.
43 Robson 1984.
44 Sergeant 1982.
45 Fehring and Wells 1976.
46 Kasuya 1975.
47 Diamond 2005.
48 Whitehead and Richerson 2009.
49 Kasuya 2008, 769.
50 Variants of the statement opening this section include, e.g., Louise Chilvers and Peter

Corkeron's discussion of the Moreton Bay "trawler" dolphins: "Should management strategies include considering maintaining these cultures? Assessing the consequences of human activities with relatively obvious effects on animal populations can be difficult. When mammals display complex social responses to activities not directed at them, the task of maintaining viable wildlife populations can become even more challenging" (2001, 1904).

51 For the official Japanese position on exploitation of whales, see, e.g., "Questions and Answers: Japan's Whale Research Programs (JARPN and JARPNII)," Institute for Cetacean Research, accessed August 14, 2013, http://www.icrwhale.org/QandA3.html.

52 McComb et al. 2001, 2011.

53 Chiquet et al. 2013.

54 Caldwell, Caldwell, and Rice 1966; Whitehead 2003b, 199.

55 Whitehead et al. 2004.

56 See Osborne (1999).

57 Ford and Ellis 2006.

58 Whitehead 2010.

59 Donaldson et al. 2012.

60 Read 2005; Saulitis 2013.

61 Sapolsky and Share 2004.

62 Hohn et al. 2006.

63 For formal definitions of ESUs and current perspectives, see Funk and colleagues (2012).

64 Whitehead et al. 2004.

65 The genetic differences between the sperm whale clans are discussed in Rendell et al. (2012) and Whitehead (2003b).

66 Our proposed definition for the ESU is: "A lineage demonstrating highly restricted flow of information that determines phenotypes from other such lineages within the higher organizational level (lineage) of the species" (Whitehead et al. 2004).

67 Ryan 2006.

68 Technically a "designatable unit" in Canada or a "distinct population segment" in the United States.

69 "Species at Risk Act Recovery Strategy Series: Recovery Strategy for the Northern and Southern Resident Killer Whales (*Orcinus orca*) in Canada," Species at Risk Public Registry, modified July 10, 2012, http://www.sararegistry.gc.ca/document/doc1341a/ind_e.cfm.

70 "Recovery Strategies: Recovery Strategy for the Northern and Southern Resident Killer Whales (*Orcinus orca*) in Canada," Species at Risk Public Registry, modified July 11, 2012, http://www.sararegistry.gc.ca/document/default_e.cfm?documentID=1341.

71 The ruling on the southern residents not forming a distinct population segment is examined in Teaney (2004).

72 Krahn et al. 2002.

73 The quote about the reversal of the U.S. government's decision is taken from "Endangered and Threatened Wildlife and Plants: Endangered Status for Southern Resident

Killer Whales," *Federal Register*, vol. 70, no. 222, November 18, 2005, Rules and Regulations, http://www.nmfs.noaa.gov/pr/pdfs/fr/fr70-69903.pdf, p. 69907.

74 "Endangered and Threatened Wildlife and Plants: Endangered Status for Southern Resident Killer Whales," pp. 69907–8.

75 Clapham, Young, and Brownell 1999.

76 Nei, Maruyama, and Chakraborty 1975.

77 Henrich 2004b.

78 Kraus and Rolland 2007a.

79 Hlista et al. 2009.

80 Reeves, Smith, and Josephson 2007.

81 Hlista et al. 2009; Mayo and Marx 1990.

82 Mate, Nieukirk, and Kraus 1997.

83 For a discussion of these ideas, see Kraus and Rolland (2007b).

84 Whitehead 2010.

85 Clapham, Aguilar, and Hatch 2008.

86 Clapham, Aguilar, and Hatch 2008.

Chapter 12

1 Kasuya 2008, 768.

2 The arguments why whales and dolphins should not be kept in captivity are laid out by Lori Marino and Toni Frohoff (2011). An alternative perspective is outlined by Stan Kuczaj (2010).

3 Sullivan 2013, 198.

4 Herzing 2011.

5 Marsh, Elfenbein, and Ambady 2007.

6 Johnson and Tyack 2003.

7 De Stephanis et al. 2008b; Ottensmeyer and Whitehead 2003.

8 Sapolsky and Share 2004.

9 Sapolsky and Share 2004.

10 Goodall 2005, 276.

11 Examples of cetaceans carrying their dead around can be found in, e.g., Caldwell and Caldwell (1966) and Ritter (2007).

12 Goodall 2005, 277.

13 Examples of when explorations of culture have turned into contests include Janik (2001) and Rendell and Whitehead (2001b).

14 Boesch 2012; Catchpole and Slater 2008; McGrew 2004.

15 Laland and Hoppitt 2003, 151.

16 Bonner 1980, 163.

17 Laland and Hoppitt 2003, 151.

18 Laland and Hoppitt 2003, 151.

19 Guttridge et al. 2012; Laland and Williams 1998.

20 Brown and Laland 2003.

21 Warner 1988.

22 Galef and Laland 2005.

23 Brown 2011; Galef and Laland 2005.

24 Aisner and Terkel 1992; Terkel 1996.

25 Janik and Slater (1997). That rodents and most other terrestrial mammals do not seem to learn vocalizations socially has, at least, been the prevalent view. Very recent evidence suggests that males of the humble dormouse may converge on a common frequency in their ultrasonic calls if housed together (Arriaga, Zhou, and Jarvis 2012), but this evidence is preliminary, and in any case, converging on a common frequency is a very long way from the kind of vocal learning we see in cetaceans or songbirds.

26 The lyrebird's remarkable vocal capabilities are described by David Attenborough in "Signals and Songs," episode 6 of *The Life of Birds*, the BBC nature documentary series written and presented by Attenborough (episode 6 originally aired November 25, 1998). A clip of Attenborough documenting the lyrebird can be seen at http://www.youtube.com/watch?v=VjE0Kdfos4Y, uploaded February 12, 2007. How lyrebirds and bowerbirds build their vocal repertoires by copying the sounds of their own species and others is the subject of Dalziell and Magrath (2012), Kelley and Healy (2010), and Putland et al. (2006).

27 Catchpole and Slater 2008.

28 Rothenberg 2010, 6.

29 Catchpole and Slater 2008; Kroodsma 2005.

30 Mann, Dingess, and Slater 2006; Templeton et al. 2013.

31 The fundamental frequency of three-wattled bellbird song dropped from 5.6 kilohertz in 1974 to 3.7 kilohertz in 2000, or about 1.6 percent per year (Kroodsma 2005). For another example, see Derryberry (2009).

32 Garland et al. 2011; Noad et al. 2000.

33 Hall and Schaller (1964) look at how sea otters use tools to break open shellfish; the phrase "one-trick ponies" comes from McGrew (2003).

34 Slagsvold and Wiebe 2011.

35 Ellis et al. 2003.

36 Cornell, Marzluff, and Pecoraro 2012; Madden et al. 2004; Wheatcroft and Price 2013.

37 Slagsvold and Wiebe 2011.

38 Fisher and Hinde 1949.

39 Hunt and Gray 2003.

40 Sherry and Galef 1984.

41 Bradbury 2003; Emery and Clayton 2004.

42 Catchpole and Slater 2008; Slagsvold and Wiebe 2011; Tebbich et al. 2001.

43 Emery et al. 2007.

44 Janik and Slater 1997.

45 This quandary is explored by Read (2011). Christophe Boesch (2012) has also written a book directly comparing chimpanzee and human culture.

46 Whiten et al. (1999). This paper has now been cited well over a thousand times.

47 Bloomsmith et al. 2006.

48 Kawai 1965.

49 These other foraging techniques by macaques are described in de Waal (2001) and Perry (2009).
50 Huffman 1984.
51 Huffman, Nahallage, and Leca 2008.
52 Huffman, Nahallage, and Leca 2008; Leca, Gunst, and Huffman 2007.
53 Huffman 1996.
54 Variations in stone handling are noted in Leca, Gunst, and Huffman (2007); evidence that it is socially learned can be found in Huffman, Nahallage, and Leca (2008).
55 Huffman 1996.
56 Perry 2009, 262.
57 Krützen, Willems, and van Schaik 2011.
58 Van Schaik et al. 2003.
59 Van Schaik et al. 2003.
60 McGrew (2003). In 2009, McGrew added a postscript titled "Revisiting the Battlefronts."
61 Whiten et al. 2001.
62 Huffman et al. 2010; Whiten et al. 2001.
63 Whiten et al. 2004.
64 Tomasello 2009.
65 Langergraber et al. 2011.
66 Experimental evidence of chimpanzees' capacity for social learning can be found in Whiten, Horner, and de Waal (2005) and Whiten et al. (2007).
67 Whiten 2011.
68 Thornton and McAuliffe 2006.
69 Patriquin (2012) considers the possibility of bat culture. Boughman (1998) and Boughman and Wilkinson (1998) have described shared calls among grouped spear-nosed bats.
70 Entering "Topic=(elephant*) AND Topic=(culture) AND Topic=(behavior OR behaviour)" in the Web of Knowledge database ("Web of Science," http://wokinfo.com/) produced no scientific papers on the culture of elephants. In contrast, when (elephant*) was replaced by (whale*), (dolphin*), or (chimpanzee*), the same search produced tens of papers on each of whales, dolphins, and chimpanzees.
71 Bradshaw 2009, 25.
72 Bradshaw 2009, 28; McComb et al. 2001; Poole et al. 2005.
73 Bates et al. 2008.
74 Bates et al. 2007.
75 McComb et al. 2014.
76 Bates et al. 2010.
77 Bates et al. 2010.
78 Foley, Pettorelli, and Foley 2009.
79 Adams (1979), 156. It was inevitable this quote would appear somewhere in the book!
80 Kako 1999.
81 Schulz et al. 2008.
82 Tomasello, Carpenter, and Liszkowski 2007.
83 Sayigh et al. 2007; Xitco, Gory, and Kuczaj 2004.

84 Christiansen and Kirby 2003.
85 Számadó and Szathmáry 2006.
86 Richerson and Boyd 2005, 110–11.
87 Weinrich, Schilling, and Belt 1992.
88 Lane et al. 2010; Wellman and Miller 2008.
89 De Waal 1997; Bekoff and Pierce 2009.
90 Rowlands (2012). Such views go back to Darwin himself—and beyond, to philosophers like David Hume.
91 Hamlin et al. 2011; Hamlin, Wynn, and Bloom 2007.
92 Monroe 1991.
93 As reported in "Is There a Hero Gene?" *The Big Issue*, February 13, 2013, http://www.bigissue.com/features/1951/there-hero-gene.
94 Rowlands 2012.
95 Broom 2003.
96 White 2007, 10.
97 Pitman and Durban 2009, 48.
98 Ray Lilley, "Dolphin Saves Stuck Whales, Guides Them Back to Sea," *National Geographic*, March 12, 2008, http://news.nationalgeographic.co.uk/news/2008/03/080312-AP-dolph-whal.html.
99 "Dolphins find missing sailor," *Cruising World*, cited in White (2007), 10.
100 See, e.g.: "Dolphins Rescuing Humans," Dolphins-World, accessed August 17, 2013, http://www.dolphins-world.com/Dolphins_Rescuing_Humans.html or "Dolphins Rescuing Humans," Save the Whales, accessed August 17, 2013, http://www.savethewhales.org/Dolphins_Rescuing_Humans.html.
101 While this point has doubtless been made many times before, we first encountered it in Lyons (2010).
102 White 2007, 161.
103 Pitman and Durban 2009, 48.
104 Norris 1991, 272.
105 On the frequency and power of sperm whale sonar, see Au (1993) and Møhl et al. (2000).
106 Whitehead 2011.
107 Cantor and Whitehead 2013.
108 The lack of evidence for symbolic marking in the apes and monkeys is discussed in Hill (2009) and Perry (2009).
109 Deecke, Ford, and Spong 2000.
110 Antunes 2009.
111 Marcoux, Rendell, and Whitehead (2007). However, the differences in reproductive rates between clans measured in different ways are not always consistent, so we cannot conclude irrefutably that the reproductive rates of the clans are different.
112 Marcoux, Whitehead, and Rendell 2007; Whitehead 2003b; Whitehead and Rendell 2004.
113 Mann et al. 2000.
114 Mann et al. 2008.
115 Harvey and Pagel 1991.

116 Dunbar 1998.

117 Richerson and Boyd 2005, 131.

118 The evolution of cetacean brains is examined in Marino, McShea, and Uhen (2004).

119 "Convention for the Safeguarding of the Intangible Cultural Heritage 2003," UNESCO, October 17, 2003, http://portal.unesco.org/en/ev.php-URL_ID=17716&URL_DO=DO_TOPIC&URL_SECTION=201.html.

120 Luck, Daily, and Ehrlich 2003.

121 "Questions and Answers: Whales/Whaling/Whale Research/International Whaling Commission," Japan Whaling Association, accessed August 17, 2013, http://www.whaling.jp/english/qa.html.

122 "Questions and Answers: Whales/Whaling/Whale Research/International Whaling Commission."

123 Condensed from the arguments in White (2007, 8–9).

124 As Thomas White (personal communication) puts it: "Complexities that go with being a person mean that the conditions that we need in order to grow, flourish or experience life in even a rudimentarily satisfying way are more complicated than the conditions non-persons need; it also means that these complexities make us vulnerable to harm in a way that nonpersons aren't."

125 List adapted from White (2007), 156–57. White also notes that these requirements for personhood come from the human experience, and if the concept of a "person" was developed from a cetacean's perspective there would be much greater emphasis on social intelligence, the whale or dolphin being generally a much more social being than the human (White 2007, 156–57).

126 White 2007.

127 Genesis 1:26 in the King James Bible reads: "And God said, Let us make man in our image, after our likeness: and let them have dominion over the fish of the sea, and over the fowl of the air, and over the cattle, and over all the earth, and over every creeping thing that creepeth upon the earth." Of course, there are also religions whose adherents would be less upset by the notion of nonhuman persons.

128 Though culture is not included directly in the usual requirements for personhood, the philosopher Michael Allen Fox (2001) makes the connection more directly from cetacean culture to cetacean rights in that the presence of culture increases cetaceans' similarity to humans and so deserving more of rights.

129 Gardner 2000, 34.

BIBLIOGRAPHY

Abramson, J. Z., V. Hernández-Lloreda, J. Call, and F. Colmenares. 2013. "Experimental evidence for action imitation in killer whales (*Orcinus orca*)." *Animal Cognition* 16:11–22.

Ackermann, C. 2008. "Contrasting vertical skill transmission patterns of a tool use behaviour in two groups of wild bottlenose dolphins (*Tursiops* sp.), as revealed by molecular genetic analysis." MSc thesis, University of Zurich.

Adams, D. 1979. *The Hitchhiker's Guide to the Galaxy*. New York: Pocket Books.

Agrawal, A. A. 2001. "Phenotypic plasticity in the interactions and evolution of species." *Science* 294:321–26.

Aisner, R., and J. Terkel. 1992. "Ontogeny of pine cone opening behaviour in the black rat, *Rattus rattus*." *Animal Behaviour* 44:327–36.

Akaike, H. 1973. "Information theory as an extension of the maximum likelihood principle." In *Second International Symposium on Information Theory*, edited by B. N. Petrov and F. Csaki, 267–81. Budapest: Akademiai Kiado.

Alexander, A., D. Steel, B. Slikas, K. Hoekzema, C. Carraher, M. Parks, R. Cronn, and C. S. Baker. 2013. "Low diversity in the mitogenome of sperm whales revealed by next-generation sequencing." *Genome Biology and Evolution* 5:113–29.

Allen, J., M. Weinrich, W. Hoppitt, and L. Rendell. 2013. "Network-based diffusion analysis reveals cultural transmission of lobtail feeding in humpback whales." *Science* 340:485–88.

Allen, S. J., L. Bejder, and M. Krützen. 2011. "Why do Indo-Pacific bottlenose dolphins (*Tursiops* sp.) carry conch shells (*Turbinella* sp.) in Shark Bay, Western Australia?" *Marine Mammal Science* 27:449–54.

Allman, J. M. 2000. *Evolving Brains*. New York: Scientific American Library.

Amato, I. 1993. "A sub surveillance network becomes a window on whales." *Science* 261:549–50.

Amos, B. 1993. "Use of molecular probes to analyse pilot whale pod structure: two novel analytical approaches." *Symposia of the Zoological Society, London* 66:33–48.

Amos, B., C. Schlötterer, and D. Tautz. 1993. "Social structure of pilot whales revealed by analytical DNA profiling." *Science* 260:670–72.

Amos, W. 1999. "Culture and genetic evolution in whales." *Science* 284:2055a.

Anderson, P. K. 2004. "Habitat, niche and evolution of sirenian mating systems." *Journal of Mammalian Evolution* 9:55–98.

Anderwald, P., A. K. Daníelsdóttir, T. Haug, F. Larsen, V. Lesage, R. J. Reid, G. A. Víkingsson, and A. R. Hoelzel. 2011. "Possible cryptic stock structure for minke whales in the North Atlantic: implications for conservation and management." *Biological Conservation* 144:2479–89.

Andrews, K. R., L. Karczmarski, W. W. L. Au, S. H. Rickards, C. A. Vanderlip, B. W. Bowen, E. G. Grau, and R. J. Toonen. 2010. "Rolling stones and stable homes: social structure, habitat diversity and population genetics of the Hawaiian spinner dolphin (*Stenella longirostris*)." *Molecular Ecology* 19:732–48.

Ansmann, I. C., G. P. Parra, B. L. Chilvers, and J. M. Lanyon. 2012. "Dolphins restructure social system after reduction of commercial fisheries." *Animal Behaviour* 84:575–81.

Antunes, R. N. C. 2009. "Variation in sperm whale (*Physeter macrocephalus*) coda vocalizations and social structure in the North Atlantic Ocean." PhD thesis, University of Saint Andrews.

Aoki, K., M. Amano, K. Mori, A. Kourogi, T. Kubodera, and N. Miyazaki. 2012. "Active hunting by deep-diving sperm whales: 3D dive profiles and maneuvers during bursts of speed." *Marine Ecology Progress Series* 444:289–322.

Arnbom, T., V. Papastavrou, L. S. Weilgart, and H. Whitehead. 1987. "Sperm whales react to an attack by killer whales." *Journal of Mammalogy* 68:450–53.

Arntz, W. E. 1986. "The two faces of El Niño, 1982–83." *Meeresforschung* 31:1–46.

Arriaga, G., E. P. Zhou, and E. D. Jarvis. 2012. "Of mice, birds, and men: the mouse ultrasonic song system has some features similar to humans and song-learning birds." *PLoS ONE* 7:e46610.

Aschettino, J. M., R. W. Baird, D. J. McSweeney, D. L. Webster, G. S. Schorr, J. L. Huggins, K. K. Martien, S. D. Mahaffy, and K. L. West. 2011. "Population structure of melon-headed whales (*Peponocephala electra*) in the Hawaiian Archipelago: evidence of multiple populations based on photo identification." *Marine Mammal Science* 28:666–89.

Au, W. W. L. 1993. *The Sonar of Dolphins*. New York: Springer Verlag.

Au, W. W. L., D. James, and K. Andrews. 2001. "High-frequency harmonics and source level of humpback whale songs." *Journal of the Acoustical Society of America* 110:2770.

Auld, J. R., A. A. Agrawal, and R. A. Relyea. 2010. "Re-evaluating the costs and limits of adaptive phenotypic plasticity." *Proceedings of the Royal Society of London*, ser. B 277: 503–11.

Avants, B., L. Betancourt, J. Giannetta, G. Lawson, J. Gee, M. Farah, and H. Hurt. 2012. "Early childhood home environment predicts frontal and temporal cortical thickness in the young adult brain." Abstract of paper presented at the Neuroscience Meeting, New Orleans, October 17, 2012, http://www.abstractsonline.com/Plan/ViewAbstract.aspx?sKey=734b1ccd-cfcf-4394-a945-083ca58f8033&cKey=7b3e8587-f590-4d94-ae3f-e050d52e8488.

Bacher, K., S. Allen, A. K. Lindholm, L. Bejder, and M. Krützen. 2010. "Genes or culture: are mitochondrial genes associated with tool use in bottlenose dolphins (*Tursiops* sp.)?" *Behavior Genetics* 40:706–14.

Bain, D. E. 1989. "An evaluation of evolutionary processes: studies of natural selection, dispersal, and cultural evolution in killer whales (*Orcinus orca*)." PhD thesis, University of California, Santa Cruz.

Baird, R. W. 2000. "The killer whale—foraging specializations and group hunting." In *Cetacean Societies*, edited by J. Mann, R. C. Connor, P. Tyack and H. Whitehead, 127–53. Chicago: University of Chicago Press.

———. 2006. *Killer Whales of the World: Natural History and Conservation*. Saint Paul: Voyageur Press.

———. 2011. "Predators, prey and play: killer whales and other marine mammals." *Journal of the American Cetacean Society* 40:54–57.

Baird, R. W., and L. M. Dill. 1995. "Occurrence and behavior of transient killer whales: seasonal and pod-specific variability, foraging behavior and prey handling." *Canadian Journal of Zoology* 73:1300–1311.

———. 1996. "Ecological and social determinants of group size in *transient* killer whales." *Behavioral Ecology* 7:408–16.

Baird, R. W., A. M. Gorgone, D. J. McSweeney, D. L. Webster, D. R. Salden, M. H. Deakos, A. D. Ligon, G. S. Schorr, J. Barlow, and S. D. Mahaffy. 2008. "False killer whales (*Pseudorca crassidens*) around the main Hawaiian Islands: long-term site fidelity, inter-island movements, and association patterns." *Marine Mammal Science* 24:591–612.

Baird, R. W., M. B. Hanson, G. S. Schorr, D. L. Webster, D. J. McSweeney, A. M. Gorgone, S. D. Mahaffy, D. M. Holzer, E. M. Oleson, and R. D. Andrews. 2012. "Range and primary habitats of Hawaiian insular false killer whales: informing determination of critical habitat." *Endangered Species Research* 18:47–61.

Baird, R. W., and H. Whitehead. 2000. "Social organization of mammal-eating killer whales: group stability and dispersal patterns." *Canadian Journal of Zoology* 78:2096–105.

Baker, C. S., L. Florez-Gonzalez, B. Abernethy, H. C. Rosenbaum, R. W. Slade, J. Capella, and J. L. Bannister. 1998. "Mitochondrial DNA variation and maternal gene flow among humpback whales of the southern hemisphere." *Marine Mammal Science* 14:721–37.

Baker, C. S., S. R. Palumbi, R. H. Lambertsen, M. T. Weinrich, J. Calambokidis, and S. J. O'Brien. 1990. "Influence of seasonal migration on geographic distribution of mitochondrial DNA haplotypes in humpback whales." *Nature* 344:238–40.

Baker, C. S., M. Weinrich, G. Early, and S. Palumbi. 1994. "Genetic impact of an unusual group mortality among humpback whales." *Journal of Heredity* 85:52–54.

Ball, G. F., and S. H. Hulse. 1998. "Birdsong." *American Psychologist* 53:37–58.

Barkow, J. H. 2001. "Culture and hyperculture: why can't a cetacean be more like a (hu)man?" *Behavioral and Brain Sciences* 24:324–25.

Barrett-Lennard, L. 2000. "Population structure and mating patterns of killer whales (*Orcinus orca*) as revealed by DNA analysis." PhD thesis, University of British Columbia.

———. 2011. "Killer whale evolution: populations, ecotypes, species, Oh my!" *Journal of the American Cetacean Society* 40:48–53.

Barrett-Lennard, L., and K. Heise. 2011. "Killer whale conservation: the perils of life at the top of the food chain." *Journal of the American Cetacean Society* 40:58–62.

Bates, L. A., and R. W. Byrne. 2007. "Creative or created: using anecdotes to investigate animal cognition." *Methods (San Diego)* 42:12–21.

Bates, L. A., R. Handford, P. C. Lee, N. Njiraini, J. H. Poole, K. Sayialel, S. Sayialel, C. J. Moss, and R. W. Byrne. 2010. "Why do African elephants (*Loxodonta africana*) simulate oestrus? an analysis of longitudinal data." *PLoS ONE* 5:e10052.

Bates, L. A., K. N. Sayialel, N. W. Njiraini, C. J. Moss, J. H. Poole, and R. W. Byrne. 2007.

"Elephants classify human ethnic groups by odor and garment color." *Current Biology* 17:1938–42.

Bates, L. A., K. N. Sayialel, N. W. Njiraini, J. H. Poole, C. J. Moss, and R. W. Byrne. 2008. "African elephants have expectations about the locations of out-of-sight family members." *Biology Letters* 4:34–36.

Bateson, P., and P. Martin. 2000. *Design for a Life*. New York: Simon & Schuster.

Bauer, G., and C. M. Johnson. 1994. "Trained motor imitation by bottlenose dolphins (*Tursiops truncatus*)." *Perceptual and Motor Skills* 79:1307–15.

Baum, J. K., R. A. Myers, D. G. Kehler, B. Worm, S. J. Harley, and P. A. Doherty. 2003. "Collapse and conservation of shark populations in the northwest Atlantic." *Science* 299:389–92.

Beale, T. 1839. *The Natural History of the Sperm Whale*. London: John van Voorst.

Beck, S., S. Kuningas, R. Esteban, and A. D. Foote. 2011. "The influence of ecology on sociality in the killer whale (*Orcinus orca*)." *Behavioral Ecology* 23:246–53.

Begon, M., J. L. Harper, and C. R. Townsend. 1996. *Ecology: Individuals, Populations and Communities*. Oxford: Blackwell Science.

Bekoff, M., and J. A. Byers. 1998. *Animal Play: Evolutionary, Comparative, and Ecological Approaches*. New York: Cambridge University Press.

Bekoff, M., and J. Pierce. 2009. *Wild Justice: The Moral Lives of Animals*. Chicago: University of Chicago Press.

Bel'kovich, V. M., A. V. Agafonov, O. V. Yefremenkova, L. B. Kozarovitsky, and S. P. Kharitonov. 1991. "Herd structure, hunting, and play: bottlenose dolphins in the Black Sea." In *Dolphin Societies: Discoveries and Puzzles*, edited by K. S. Norris and K. Pryor, 17–77. Berkeley: University of California Press.

Bender, C., D. Herzing, and D. Bjorklund. 2009. "Evidence of teaching in Atlantic spotted dolphins (*Stenella frontalis*) by mother dolphins foraging in the presence of their calves." *Animal Cognition* 12:43–53.

Berdahl, A., C. J. Torney, C. C. Ioannou, J. J. Faria, and I. D. Couzin. 2013. "Emergent sensing of complex environments by mobile animal groups." *Science* 339:574–76.

Bernhard, H., U. Fischbacher, and E. Fehr. 2006. "Parochial altruism in humans." *Nature* 442:912–15.

Bersaglieri, T., P. C. Sabeti, N. Patterson, T. Vanderploeg, S. F. Schaffner, J. A. Drake, M. Rhodes, D. E. Reich, and J. N. Hirschhorn. 2004. "Genetic signatures of strong recent positive selection at the lactase gene." *American Journal of Human Genetics* 74:1111–20.

Best, P. B. 1979. "Social organization in sperm whales, *Physeter macrocephalus*." In *Behavior of Marine Animals*, edited by H. E. Winn and B. L. Olla, 3:227–89. New York: Plenum.

Best, P. B., P. A. S. Canham, and N. Macleod. 1984. "Patterns of reproduction in sperm whales, *Physeter macrocephalus*." *Reports of the International Whaling Commission (Special Issue)* 6:51–79.

Bigg, M. A., P. F. Olesiuk, G. M. Ellis, J. K. B. Ford, and K. C. Balcomb. 1990. "Social organization and genealogy of resident killer whales (*Orcinus orca*) in the coastal waters of British Columbia and Washington State." *Reports of the International Whaling Commission (Special Issue)* 12:383–405.

Bikhchandani, S., D. Hirshleifer, and I. Welch. 1992. "A theory of fads, fashion, custom, and cultural change as informational cascades." *Journal of Political Economy* 100:992–1026.

Bloomsmith, M. A., K. C. Baker, S. Ross, and S. Lambeth. 2006. "Early rearing conditions and captive chimpanzee behavior: some surprising findings." In *Nursery Rearing of Nonhuman Primates in the 21st Century*, edited by G. P. Sackett, G. C. Ruppentahal, and K. Elias, 289–312. New York: Springer US.

Boddy, A. M., M. R. McGowen, C. C. Sherwood, L. I. Grossman, M. Goodman, and D. E. Wildman. 2012. "Comparative analysis of encephalization in mammals reveals relaxed constraints on anthropoid primate and cetacean brain scaling." *Journal of Evolutionary Biology* 25:981–94.

Bodley, J. H. 1994. *Cultural Anthropology: Tribes, States and the Global System*. Mountain View, CA: Mayfield Publishing.

Boesch, C. 2001. "Sacrileges are welcome in science! opening a discussion about culture in animals." *Behavioral and Brain Sciences* 24:327–28.

———. 2012. *Wild Cultures: A Comparison between Chimpanzee and Human Cultures*. Cambridge: Cambridge University Press.

Bonner, J. T. 1980. *The Evolution of Culture in Animals*. Princeton, NJ: Princeton University Press.

Boran, J. R., and S. L. Heimlich. 1999. "Social learning in cetaceans: hunting, hearing and hierarchies." *Symposia of the Zoological Society, London* 73:282–307.

Bouchard, T., D. Lykken, M. McGue, N. Segal, and A. Tellegen. 1990. "Sources of human psychological differences: the Minnesota study of twins reared apart." *Science* 250:223–28.

Boughman, J. W. 1998. "Vocal learning by greater spear-nosed bats." *Proceedings of the Royal Society of London*, ser. B 265:227–33.

Boughman, J. W., and G. S. Wilkinson. 1998. "Greater spear-nosed bats discriminate group mates by vocalizations." *Animal Behaviour* 55:1717–32.

Bowen, W. D. 1997. "Role of marine mammals in aquatic ecosystems." *Marine Ecology Progress Series* 158:267–74.

Bowles, S., and H. Gintis. 2003. "Origins of human cooperation." In *Genetic and Cultural Evolution of Cooperation*, edited by P. Hammerstein, 429–43. Cambridge, MA: MIT Press.

Box, H. 1984. *Primate Behaviour and Social Ecology*. London: Chapman & Hall.

Boyd, R., H. Gintis, and S. Bowles. 2010. "Coordinated punishment of defectors sustains cooperation and can proliferate when rare." *Science* 328:617–20.

Boyd, R., and P. Richerson. 1985. *Culture and the Evolutionary Process*. Chicago: Chicago University Press.

———. 1996. "Why culture is common, but cultural evolution is rare." *Proceedings of the British Academy* 88:77–93.

Boyd, R., P. J. Richerson, and J. Henrich. 2011. "Rapid cultural adaptation can facilitate the evolution of large-scale cooperation." *Behavioral Ecology and Sociobiology* 65:431–44.

Bradbury, J. W. 2003. "Vocal communication in wild parrots." In *Animal Social Complexity: Intelligence, Culture and Individualized Societies*, edited by F. B. M. de Waal and P. L. Tyack, 293–316. Cambridge, MA: Harvard University Press.

Bradshaw, C. J. A., K. Evans, and M. A. Hindell. 2006. "Mass cetacean strandings—a plea for empiricism." *Conservation Biology* 20:584–86.

Bradshaw, G. A. 2009. *Elephants on the Edge: What Animals Teach Us about Humanity*. New Haven, CT: Yale University Press.

Brighton, T. 2004. *Hell Riders: The True Story of the Charge of the Light Brigade*. New York: Viking.

Broom, D. 2003. *The Evolution of Morality and Religion*. Cambridge: Cambridge University Press.

Brotons, J. M., A. M. Grau, and L. Rendell. 2008. "Estimating the impact of interactions between bottlenose dolphins and artisanal fisheries around the Balearic Islands." *Marine Mammal Science* 24:112–27.

Brown, C., and K. N. Laland. 2003. "Social learning in fishes: a review." *Fish and Fisheries* 4:280–88.

Brown, M. F. 2011. "Social influences on rat spatial choice." *Comparative Cognition and Behavior Reviews* 6:5–23.

Bruck, J. N. 2013. "Decades-long social memory in bottlenose dolphins." *Proceedings of the Royal Society of London*, ser. B 280:20131726.

Budnikova, L. L., and S. A. Blokhin. 2012. "Food contents of the eastern gray whale *Eschrichtius robustus* Lilljeborg, 1861 in the Mechigmensky Bay of the Bering Sea." *Russian Journal of Marine Biology* 38:149–55.

Burnham, K. P., and D. R. Anderson. 2002. *Model Selection and Multimodel Inference: A Practical Information-Theoretic Approach*. New York: Springer-Verlag.

Burtenshaw, J. C., E. M. Oleson, J. A. Hildebrand, M. A. McDonald, R. K. Andrew, B. M. Howe, and J. A. Mercer. 2004. "Acoustic and satellite remote sensing of blue whale seasonality and habitat in the Northeast Pacific." *Deep Sea Research Part 2: Topical Studies in Oceanography* 51:967–86.

Butti, C., C. C. Sherwood, A. Y. Hakeem, J. M. Allman, and P. R. Hof. 2009. "Total number and volume of Von Economo neurons in the cerebral cortex of cetaceans." *Journal of Comparative Neurology* 515:243–59.

Byrne, R. W. 1995. *The Thinking Ape: Evolutionary Origins of Intelligence*. Oxford: Oxford University Press.

———. 1999. "Human cognitive evolution." In *The Descent of Mind*, edited by M. C. Corballis and S. E. G. Lea, 71–87. Oxford: Oxford University Press.

———. 2007. "Culture in great apes: using intricate complexity in feeding skills to trace the evolutionary origin of human technical prowess." *Philosophical Transactions of the Royal Society of London*, ser. B 362:577–85.

Byrne, R. W., and L. G. Rapaport. 2011. "What are we learning from teaching?" *Animal Behaviour* 82:1207–11.

Byrne, R. W., and A. Whiten. 1988. *Machiavellian Intelligence*. Oxford: Clarendon.

Caldwell, M. C., and D. K. Caldwell. 1966. "Epimeletic (care-giving) behavior in Cetacea." In *Whales, Dolphins, and Porpoises*, edited by K. S. Norris, 755–89. Berkeley: University of California Press.

Caldwell, D. K., M. C. Caldwell, and D. W. Rice. 1966. "Behavior of the sperm whale *Physeter catodon* L." In *Whales, Dolphins, and Porpoises*, edited by K. S. Norris, 677–717. Berkeley: University of California Press.

Caldwell, M. C., D. K. Caldwell, and P. L. Tyack. 1990. "A review of the signature whistle hypothesis for the Atlantic bottlenose dolphin, *Tursiops truncatus*." In *The Bottlenose Dolphin: Recent Progress in Research*, edited by S. Leatherwood and R. R. Reeves, 199–234. San Diego: Academic Press.

Call, J., and M. Tomasello. 1996. "The effect of humans on the cognitive development of apes." In *Reaching into Thought*, edited by A. E. Russon, K. A. Bard, and S. T. Parker, 371–403. New York: Cambridge University Press.

Candland, D. K. 1993. *Feral Children and Clever Animals: Reflections on Human Nature*. Oxford: Oxford University Press.

Cantor, M., and H. Whitehead. 2013. "The interplay between social networks and culture: theoretically and among whales and dolphins." *Philosophical Transactions of the Royal Society of London*, ser. B 368:149–65.

Carneiro, R. L. 2003. *Evolutionism in Cultural Anthropology: A Critical History*. Boulder, CO: Westview Press.

Caro, T. M., and M. D. Hauser. 1992. "Is there teaching in nonhuman animals?" *Quarterly Review of Biology* 67:151–74.

Carrier, D. R., A. K. Kapoor, T. Kimura, M. K. Nickels, Satwanti, E. C. Scott, J. K. So, and E. Trinkaus. 1984. "The energetic paradox of human running and hominid evolution." *Current Anthropology* 25:483–95.

Carter, R. 2006. "Boat remains and maritime trade in the Persian Gulf during sixth and fifth millennia BC." *Antiquity* 80:52–63.

Castellote, M., C. W. Clark, and M. O. Lammers. 2011. "Fin whale (*Balaenoptera physalus*) population identity in the western Mediterranean Sea." *Marine Mammal Science* 28:325–44.

Catchpole, C. K., and P. J. B. Slater. 2008. *Bird Song: Biological Themes and Variations*. 2nd ed. Cambridge: Cambridge University Press.

Cavalli-Sforza, L. L., and M. W. Feldman. 1981. *Cultural Transmission and Evolution: A Quantitative Approach*. Princeton, NJ: Princeton University Press.

Cavalli-Sforza, L. L., M. W. Feldman, K. H. Chen, and S. M. Dornbusch. 1982. "Theory and observation in cultural transmission." *Science* 218:19–27.

Cavalli-Sforza, L. L., and M. Seielstad. 2001. *Genes, Peoples and Languages*. New York: Penguin Press.

Cerchio, S., J. K. Jacobsen, and T. F. Norris. 2001. "Temporal and geographical variation in songs of humpback whales, *Megaptera novaeangliae*: synchronous change in Hawaiian and Mexican breeding assemblages." *Animal Behaviour* 62:313–29.

Chadwick, D. 2005. "Investigating a killer." *National Geographic* 207:86–105.

Charlton-Robb, K., L. A. Gershwin, R. Thompson, J. Austin, K. Owen, and S. McKechnie. 2011. "A new dolphin species, the Burrunan Dolphin *Tursiops australis* sp. nov., endemic to southern Australian coastal waters." *PloS One* 6:e24047.

Chilvers, B. L., and P. J. Corkeron. 2001. "Trawling and bottlenose dolphins' social structure." *Proceedings of the Royal Society of London*, ser. B 268:1901–5.

Chilvers B.L., Corkeron P.J., and Puotinen M.L. 2003. "Influence of trawling on the behaviour and spatial distribution of Indo-Pacific bottlenose dolphins (*Tursiops aduncus*) in Moreton Bay, Australia." *Canadian Journal of Zoology* 81:1947–55.
Chiquet, R. A., B. Ma, A. S. Ackleh, N. Pal, and N. Sidorovskaia. 2013. "Demographic analysis of sperm whales using matrix population models." *Ecological Modeling* 248:71–79.
Chomsky, N. 1965. *Aspects of the Theory of Syntax*. Massachusetts Institute of Technology, Research Laboratory of Electronics, Special Technical Report 11. Cambridge, MA: MIT Press.
Christiansen, M., and S. Kirby. 2003. "Language evolution: the hardest problem in science?" In *Language Evolution*, edited by M. Christiansen and S. Kirby, 1–15. Oxford: Oxford University Press.
Chudek, M., and J. Henrich. 2011. "Culture-gene coevolution, norm-psychology and the emergence of human prosociality." *Trends in Cognitive Sciences* 15:218–26.
Clapham, P. J. 2000. "The humpback whale: seasonal breeding and feeding in a baleen whale." In *Cetacean Societies*, edited by J. Mann, R. C. Connor, P. L. Tyack, and H. Whitehead, 173–96. Chicago: University of Chicago Press.
Clapham, P. J., A. Aguilar, and L. T. Hatch. 2008. "Determining spatial and temporal scales for management: lessons from whaling." *Marine Mammal Science* 24:183–201.
Clapham, P. J., P. J. Palsbøll, D. K. Matilla, and O. Vasquez. 1992. "Composition and dynamics of humpback whale competitive groups in the West Indies." *Behaviour* 122:182–94.
Clapham, P. J., S. B. Young, and R. L. Brownell. 1999. "Baleen whales: conservation issues and the status of the most endangered populations." *Mammal Review* 29:35–60.
Clarke, M. R. 1977. "Beaks, nets and numbers." *Symposia of the Zoological Society, London* 38:89–126.
Clutton-Brock, T. H. 1989. "Mammalian mating systems." *Proceedings of the Royal Society of London*, ser. B 236:339–72.
Cochran, G., and H. Harpending. 2009. *The 10,000 Year Explosion: How Civilization Accelerated Human Evolution*. New York: Basic Books.
Colbeck, G. J., P. Duchesne, L. D. Postma, V. Lesage, M. O. Hammill, and J. Turgeon. 2013. "Groups of related belugas (*Delphinapterus leucas*) travel together during their seasonal migrations in and around Hudson Bay." *Proceedings of the Royal Society of London*, ser. B280: 20122552.
Connor, R. C. 2000. "Group living in whales and dolphins." In *Cetacean Societies*, edited by J. Mann, R. C. Connor, P. L. Tyack, and H. Whitehead, 199–218. Chicago: University of Chicago Press.
———. 2007. "Dolphin social intelligence: complex alliance relationships in bottlenose dolphins and a consideration of selective environments for extreme brain size evolution in mammals." *Philosophical Transactions of the Royal Society of London*, ser. B 362:587–602.
Connor, R. C., M. R. Heithaus, and L. M. Barre. 2001. "Complex social structure, alliance stability and mating access in a bottlenose dolphin 'super-alliance.'" *Proceedings of the Royal Society of London*, ser. B 268:263–67.

Connor, R. C., M. R. Heithaus, P. Berggen, and J. L. Miksis. 2000. "'Kerplunking': surface fluke-splashes during shallow-water bottom foraging by bottlenose dolphins." *Marine Mammal Science* 16:646–53.

Connor, R. C., and R. A. Smolker. 1996. "'Pop' goes the dolphin: a vocalization male bottlenose dolphins produce during consortships." *Behaviour* 133:643–62.

Connor, R. C., R. Smolker, and L. Bejder. 2006. "Synchrony, social behaviour and alliance affiliation in Indian Ocean bottlenose dolphins, *Tursiops truncatus*." *Animal Behaviour* 72:1371–78.

Connor, R. C., R. A. Smolker, and A. F. Richards. 1992. "Two levels of alliance formation among male bottlenose dolphins (*Tursiops* sp.)." *Proceedings of the National Academy of Sciences of the United States of America* 89:987–90.

Connor, R. C., J. J. Watson-Capps, W. B. Sherwin, and M. Krützen. 2010. "A new level of complexity in the male alliance networks of Indian Ocean bottlenose dolphins (*Tursiops* sp.)." *Biology Letters* 7:623–26.

Connor, R. C., R. S. Wells, J. Mann, and A. J. Read. 2000. "The bottlenose dolphin: social relationships in a fission-fusion society." In *Cetacean Societies*, edited by J. Mann, R. C. Connor, P. L. Tyack, and H. Whitehead, 91–126. Chicago: University of Chicago Press.

Corkeron, P. J., and R. C. Connor. 1999. "Why do baleen whales migrate?" *Marine Mammal Science* 15:1228–45.

Cornell, H. N., J. M. Marzluff, and S. Pecoraro. 2012. "Social learning spreads knowledge about dangerous humans among American crows." *Proceedings of the Royal Society of London*, ser. B 279:499–508.

Costall, A. 1998. "Lloyd Morgan, and the rise and fall of 'Animal Psychology.'" *Society and Animals* 6:13–29.

Coussi-Korbel, S., and D. M. Fragaszy. 1995. "On the relation between social dynamics and social learning." *Animal Behaviour* 50:1441–53.

Coyne, J. A. 2010. *Why Evolution Is True*. Oxford: Oxford University Press.

Cranford, T. W. 1999. "The sperm whale's nose: sexual selection on a grand scale?" *Marine Mammal Science* 15:1133–57.

Croll, D. A., C. W. Clark, A. Acevedo, B. Tershy, S. Flores, J. Gedamke, and J. Urban. 2002. "Only male fin whales sing loud songs." *Nature* 417:809.

Cronk, L. 1999. *That Complex Whole: Culture and the Evolution of Human Behavior*. Boulder, CO: Westview Press.

Csibra, G., and G. Gergely. 2009. "Natural pedagogy." *Trends in Cognitive Sciences* 13:148–53.

———. 2011. "Natural pedagogy as evolutionary adaptation." *Philosophical Transactions of the Royal Society of London*, ser. B 366:1149–57.

Culik, B. 2001. "Finding food in the open ocean: foraging strategies in Humboldt penguins." *Zoology* 104:327–38.

Cummings, W., and D. Holliday. 1987. "Sounds and source levels from bowhead whales off Pt. Barrow, Alaska." *Journal of the Acoustical Society of America* 82:814–21.

Dahooda, A. D., and K. J. Benoit-Bird. 2010. "Dusky dolphins foraging at night." In *The Dusky Dolphin: Master Acrobat off Different Shores*, edited by B. G. Würsig and M. Würsig, 99–114. Amsterdam: Elsevier/Academic Press.

Dakin, W. J. 1934. *Whalemen Adventurers*. Sydney: Angus and Robertson Ltd.

Dalebout, M. L., J. G. Mead, C. S. Baker, A. N. Baker, and A. L. van Helden. 2002. "A new species of beaked whale *Mesoplodon perrini* sp. n. (Cetacea: Ziphiidae) discovered through phylogenetic analyses of mitochondrial DNA sequences." *Marine Mammal Science* 18:577–608.

Dalla Rosa, L., and E. R. Secchi. 2007. "Killer whale (*Orcinus orca*) interactions with the tuna and swordfish longline fishery off southern and south-eastern Brazil: a comparison with shark interactions." *Journal of the Marine Biological Association of the United Kingdom* 87:135–40.

Dalziell, A. H., and R. D. Magrath. 2012. "Fooling the experts: accurate vocal mimicry in the song of the superb lyrebird, *Menura novaehollandiae*." *Animal Behaviour* 83:1401–10.

D'Amico, A., R. C. Gisiner, D. R. Ketten, J. A. Hammock, C. Johnson, P. L. Tyack, and J. Mead. 2009. "Beaked whale strandings and naval exercises." *Aquatic Mammals* 35:452–72.

Darling, J. D., M. E. Jones, and C. P. Nicklin. 2006. "Humpback whale songs: do they organize males during the breeding season?" *Behaviour* 143:1051–1101.

Darwin, C. 1874. *The Descent of Man, and Selection in Relation to Sex*. London: John Murray.

Daura-Jorge, F. G., M. Cantor, S. N. Ingram, D. Lusseau, and P. C. Simões-Lopes. 2012. "The structure of a bottlenose dolphin society is coupled to a unique foraging cooperation with artisanal fishermen." *Biology Letters* 8:702–5.

Dawkins, R. 1976. *The Selfish Gene*. Oxford: Oxford University Press.

Dean, L. G., R. L. Kendal, S. J. Schapiro, B. Thierry, and K. N. Laland. 2012. "Identification of the social and cognitive processes underlying human cumulative culture." *Science* 335:1114–18.

Dean, W. R. J., W. R. Siegfried, and I. A. W. MacDonald. 1990. "The fallacy, fact, and fate of guiding behavior in the greater honeyguide." *Conservation Biology* 4:99–101.

Deaner, R. O., K. Isler, J. Burkart, and C. van Schaik. 2007. "Overall brain size, and not encephalization quotient, best predicts cognitive ability across non-human primates." *Brain, Behavior and Evolution* 70:115–24.

de Bruyn, P. J. N., C. A. Tosh, and A. Terauds. 2013. "Killer whale ecotypes: is there a global model?" *Biological Reviews* 88:62–80.

Deecke, V., L. Barrett-Lennard, P. Spong, and J. Ford. 2010. "The structure of stereotyped calls reflects kinship and social affiliation in resident killer whales (*Orcinus orca*)." *Die Naturwissenschaften* 97:513–18.

Deecke, V. B., J. K. B. Ford, and P. J. B. Slater. 2005. "The vocal behaviour of mammal-eating killer whales: communicating with costly calls." *Animal Behaviour* 69:395–405.

Deecke, V. B., J. K. B. Ford, and P. Spong. 2000. "Dialect change in resident killer whales: implications for vocal learning and cultural transmission." *Animal Behaviour* 40:629–38.

Deecke, V. B., M. Nykänen, A. D. Foote, and V. M. Janik. 2011. "Vocal behaviour and feeding ecology of killer whales *Orcinus orca* around Shetland, U.K." *Aquatic Biology* 13:79–88.

Delarue, J., M. Laurinolli, and B. Martin. 2009. "Bowhead whale (*Balaena mysticetus*) songs in the Chukchi Sea between October 2007 and May 2008." *Journal of the Acoustical Society of America* 126:3319–28.

de la Torre, I. 2011. "The origins of stone tool technology in Africa: a historical perspective." *Philosophical Transactions of the Royal Society of London*, ser. B 366:1028–37.

DeMaster, D. P., A. W. Trites, P. Clapham, S. Mizroch, P. Wade, R. J. Small, and J. V. Hoef. 2006. "The sequential megafaunal collapse hypothesis: testing with existing data." *Progress in Oceanography* 68:329–42.

Derryberry, E. 2009. "Ecology shapes birdsong evolution: variation in morphology and habitat explains variation in white-crowned sparrow song." *American Naturalist* 174:24–33.

De Stephanis, R., S. García-Tiscar, P. Verborgh, R. Esteban-Pavo, S. Pérez, L. Minvielle-Sebastia, and C. Guinet. 2008a. "Diet of the social groups of long-finned pilot whales (*Globicephala melas*) in the Strait of Gibraltar." *Marine Biology* 154:603–12.

De Stephanis, R., P. Verborgh, S. Pérez, R. Esteban, L. Minvielle-Sebastia, and C. Guinet. 2008b. "Long-term social structure of long-finned pilot whales (*Globicephala melas*) in the Strait of Gibraltar." *Acta Ethologica* 11:81–94.

de Waal, F. B. M. 1982. *Chimpanzee Politics: Power and Sex among Apes*. Baltimore: Johns Hopkins University Press.

———. 1997. *Good Natured: The Origins of Right and Wrong in Humans and Other Animals*. Cambridge, MA: Harvard University Press.

———. 2001. *The Ape and the Sushi Master*. New York: Basic Books.

de Waal, F. B. M., and K. E. Bonnie. 2009. "In tune with others: the social side of primate culture." In *The Question of Animal Culture*, edited by K. N. Laland and B. G. Galef, 19–39. Cambridge, MA: Harvard University Press.

Dial, K. P., E. Greene, and D. J. Irschick. 2008. "Allometry of behavior." *Trends in Ecology and Evolution* 23:394–401.

Diamond, J. 1998. *Why Is Sex Fun? The Evolution of Human Sexuality*. New York: Basic Books.

———. 2005. *Collapse: How Societies Choose to Fail or Succeed*. New York: Penguin.

Díaz López, B. 2006. "Interactions between Mediterranean bottlenose dolphins (*Tursiops truncatus*) and gillnets off Sardinia, Italy." *ICES Journal of Marine Science: Journal du Conseil* 63:946–51.

Dixon, N. F. 1994. *On the Psychology of Military Incompetence*. London: Random House.

Domenici, P., R. S. Batty, T. Simila, and E. Ogam. 2000. "Killer whales (*Orcinus orca*) feeding on schooling herring (*Clupea harengus*) using underwater tail-slaps: kinematic analyses of field observations." *Journal of Experimental Biology* 203:283–94.

Dommenget, D., and M. Latif. 2002. "Analysis of observed and simulated SST spectra in the midlatitudes." *Climate Dynamics* 19:277–88.

Donald, M. 1991. *Origins of the Modern Mind: Three Stages in the Evolution of Culture and Cognition*. Cambridge, MA: Harvard University Press.

Donaldson, R., H. Finn, L. Bejder, D. Lusseau, and M. Calver. 2012. "The social side of human-wildlife interaction: wildlife can learn harmful behaviours from each other." *Animal Conservation* 15:427–35.

Douglas-Hamilton, I., R. F. W. Barnes, H. Shoshani, A. C. Williams, and A. J. T. Johnsingh. 2001. "Elephants." In *The New Encyclopedia of Mammals*, edited by D. Macdonald, 436–45. Oxford: Oxford University Press.

Duignan, P. J., J. E. B. Hunter, I. N. Visser, A. Jones, and A. Nutman. 2000. "Stingray spines: a potential cause of killer whale mortality in New Zealand." *Aquatic Mammals* 26:143–47.

Dunbar, R. I. M. 1996. *Grooming, Gossip, and the Evolution of Language*. Cambridge, MA: Harvard University Press.

———. 1998. "The social brain hypothesis." *Evolutionary Anthropology* 6:178–90.

Durban, J. W., and R. L. Pitman. 2011. "Antarctic killer whales make rapid, round-trip movements to subtropical waters: evidence for physiological maintenance migrations?" *Biology Letters* 8:274–77.

Durham, W. H. 1991. *Coevolution: Genes, Culture, and Human Diversity*. Stanford, CA: Stanford University Press.

Efferson, C., R. Lalive, and E. Fehr. 2008. "The coevolution of cultural groups and ingroup favoritism." *Science* 321:1844–49.

Ellis, D. H., W. J. L. Sladen, W. A. Lishman, K. R. Clegg, J. W. Duff, G. F. Gee, and J. C. Lewis. 2003. "Motorized migrations: the future or mere fantasy?" *Bioscience* 53:260–64.

Ellis, R. 2011. *The Great Sperm Whale: A Natural History of the Ocean's Most Magnificent and Mysterious Creature*. Lawrence: University of Kansas Press.

Ember, M., and C. R. Ember. 1990. *Anthropology*. 6th ed. Englewood Cliffs, NJ: Prentice Hall.

Emery, N. J., and N. S. Clayton. 2004. "The mentality of crows: convergent evolution of intelligence in corvids and apes." *Science* 306:1903–7.

Emery, N. J., A. M. Seed, A. M. P. von Bayern, and N. S. Clayton. 2007. "Cognitive adaptations of social bonding in birds." *Philosophical Transactions of the Royal Society of London*, ser. B 362:489–505.

Enattah, N. S., T. Sahi, E. Savilahti, J. D. Terwilliger, L. Peltonen, and I. Jarvela. 2002. "Identification of a variant associated with adult-type hypolactasia." *Nature Genetics* 30:233–37.

Enquist, M., P. Strimling, K. Eriksson, K. Laland, and J. Sjostrand. 2010. "One cultural parent makes no culture." *Animal Behaviour* 79:1353–62.

Eriksson, K., M. Enquist, and S. Ghirlanda. 2007. "Critical points in current theory of conformist social learning." *Journal of Evolutionary Psychology* 5:67–87.

Estes, J. A., M. L. Riedman, M. M. Staedler, M. T. Tinker, and B. E. Lyon. 2003. "Individual variation in prey selection by sea otters: patterns, causes and implications." *Journal of Animal Ecology* 72:144–55.

Estes, J. A., M. T. Tinker, T. M. Williams, and D. F. Doak. 1998. "Killer whale predation on sea otters linking oceanic and nearshore ecosystems." *Science* 282:473–74.

Fearnbach, H., J. W. Durban, D. K. Ellifrit, J. W. Waite, C. O. Matkin, C. R. Lunsford, M. J. Peterson, J. Barlow, and P. R. Wade. 2013. "Spatial and social connectivity of fish-eating 'resident' killer whales (*Orcinus orca*) in the far North Pacific." *Marine Biology*.161:459–72.

Fearnbach, H., J. Durban, K. Parsons, and D. Claridge. 2012. "Seasonality of calving and predation risk in bottlenose dolphins on Little Bahama Bank." *Marine Mammal Science* 28:402–11.

Fehr, E., and U. Fischbacher. 2003. "The nature of human altruism." *Nature* 425:785–91.

Fehr, E., and S. Gachter. 2002. "Altruistic punishment in humans." *Nature* 415:137–40.

Fehring, W. K., and R. S. Wells. 1976. "A series of strandings by a single herd of pilot whales on the west coast of Florida." *Journal of Mammalogy* 57:191–94.

Feldhamer, G. A., L. C. Drickhamer, S. H. Vessey, J. F. Merritt, and C. Krajewski. 2007. *Mammalogy: Adaptation, Diversity, Ecology*. 3rd ed. Baltimore: Johns Hopkins University Press.

Feldman, M. W., and K. N. Laland. 1996. "Gene-culture coevolutionary theory." *Trends in Ecology and Evolution* 11:453–57.

Finn, J., T. Tregenza, and M. Norman. 2009. "Preparing the perfect cuttlefish meal: complex prey handling by dolphins." *PLoS ONE* 4:e4217.

Fiorito, G., and P. Scotto. 1992. "Observational learning in *Octopus vulgaris*." *Science* 256:545–47.

Fischer, A. 1989. "A late palaeolithic 'school' of flint-knapping at Trollesgave, Denmark: results from refitting." *Acta Archaeologica* 60:33–49.

Fisher, J., and R. A. Hinde. 1949. "The opening of milk bottles by birds." *British Birds* 42:347–57.

Fitzpatrick, S. 2008. "Doing away with Morgan's Canon." *Mind and Language* 23:224–46.

Fogarty, L., P. Strimling, and K. Laland. 2011. "The evolution of teaching." *Evolution* 65:2760–70.

Foley, C. N. Pettorelli, and L. Foley. 2009. "Severe drought and calf survival in elephants." *Biology Letters* 5:541–44.

Foote, A. D., R. M. Griffin, D. Howitt, L. Larsson, P. J. O. Miller, and A. Rus Hoelzel. 2006. "Killer whales are capable of vocal learning." *Biology Letters* 2:509–12.

Foote, A. D., P. A. Morin, J. W. Durban, E. Willerslev, L. Orlando, and M. T. Gilbert. 2011. "Out of the Pacific and back again: insights into the matrilineal history of Pacific killer whale ecotypes." *PLoS ONE* 6:e24980.

Foote, A. D., P. A. Morin, R. L. Pitman, M. C. Ávila-Arcos, J. W. Durban, A. Helden, M. S. Sinding, and M. T. Gilbert. 2013a. "Mitogenomic insights into a recently described and rarely observed killer whale morphotype." *Polar Biology* 36:1519–23.

Foote, A. D., J. Newton, M. C. Ávila-Arcos, M. Kampmann, J. A. Samaniego, K. Post, A. Rosing-Asvid, M. S. Sinding, and M. T. P. Gilbert. 2013b. "Tracking niche variation over millennial timescales in sympatric killer whale lineages." *Proceedings of the Royal Society of London*, ser. B, vol. 280:20131481.

Foote, A. D., J. Newton, S. B. Piertney, E. Willerslev, and M. T. P. Gilbert. 2009. "Ecological, morphological and genetic divergence of sympatric North Atlantic killer whale populations." *Molecular Ecology* 18:5207–17.

Ford, J. K. B. 1991. "Vocal traditions among resident killer whales (*Orcinus orca*) in coastal waters of British Columbia." *Canadian Journal of Zoology* 69:1454–83.

———. 2011. "Killer whales of the Pacific northwest coast: from pest to paragon." *Journal of the American Cetacean Society* 40:15–23.

Ford, J. K. B., and G. M. Ellis. 1999. *Transients: Mammal-Hunting Killer Whales*. Vancouver: University of British Columbia Press.

———. 2006. "Selective foraging by fish-eating killer whales *Orcinus orca* in British Columbia." *Marine Ecology Progress Series* 316:185–99.

Ford, J. K. B., G. M. Ellis, and K. C. Balcomb. 2000. *Killer Whales*. Vancouver: University of British Columbia Press.

Ford, J. K. B., G. M. Ellis, D. R. Matkin, K. C. Balcomb, D. Briggs, and A. B. Morton. 2005. "Killer whale attacks on minke whales: prey capture and antipredator tactics." *Marine Mammal Science* 21:603–18.

Ford, J. K. B., and H. D. Fisher. 1983. "Group-specific dialects of killer whales (*Orcinus orca*) in British Columbia." In *Communication and Behavior of Whales*, edited by R. Payne, 129–61. Boulder, CO: Westview Press.

Ford, M. J., M. B. Hanson, J. A. Hempelmann, K. L. Ayres, C. K. Emmons, G. S. Schorr, R. W. Baird, et al. 2011. "Inferred paternity and male reproductive success in a killer whale (*Orcinus orca*) population." *Journal of Heredity* 102:537–53.

Foster, E. A., D. W. Franks, S. Mazzi, S. K. Darden, K. C. Balcomb, J. K. B. Ford, and D. P. Croft. 2012. "Adaptive prolonged postreproductive life span in killer whales." *Science* 337:1313.

Foster, K. R., and F. L. W. Ratnieks. 2005. "A new social vertebrate?" *Trends in Ecology and Evolution* 20:363–64.

Fox, M. A. 2001. "Cetacean culture: philosophical implications." *Behavioral and Brain Sciences* 24:333–34.

Fragaszy, D., and E. Visalberghi. 2001. "Recognizing a swan: socially-biased learning." *Psychologia* 44:82–98.

Francis, D., and G. Hewlett. 2007. *Operation Orca: Springer, Luna and the Struggle to Save West Coast Killer Whales*. Madeira Park, BC: Harbour Publishing.

Frank, S. D., and A. N. Ferris. 2011. "Analysis and localization of blue whale vocalizations in the Solomon Sea using waveform amplitude data." *Journal of the Acoustical Society of America* 130:731–36.

Franks, N. R., and T. Richardson. 2006. "Teaching in tandem-running ants." *Nature* 439: 153.

Frantzis, A. 1998. "Does acoustic testing strand whales?" *Nature* 392:29.

Frantzis, A., and P. Alexiadou. 2008. "Male sperm whale (*Physeter macrocephalus*) coda production and coda-type usage depend on the presence of conspecifics and the behavioural context." *Canadian Journal of Zoology* 86:62–75.

Frantzis, A., and D. L. Herzing. 2002. "Mixed-species associations of striped dolphins (*Stenella coeruleoalba*), short-beaked common dolphins (*Delphinus delphis*), and Risso's dolphins (*Grampus griseus*) in the Gulf of Corinth (Greece, Mediterranean Sea)." *Aquatic Mammals* 28:188–97.

Frasier, T. R., P. K. Hamilton, M. W. Brown, S. D. Kraus, and B. N. White. 2010. "Reciprocal exchange and subsequent adoption of calves by two North Atlantic right whales (*Eubalaena glacialis*)." *Aquatic Mammals* 36:115–20.

Frère, C. H., M. Krützen, J. Mann, R. C. Connor, L. Bejder, and W. B. Sherwin. 2010b. "Social and genetic interactions drive fitness variation in a free-living dolphin population." *Proceedings of the National Academy of Sciences of the United States of America* 107:19949–54.

Frère, C. H., M. Krützen, J. Mann, J. Watson-Capps, Y. J. Tsai, E. M. Patterson, R. Connor,

L. Bejder, and W. B. Sherwin. 2010a. "Home range overlap, matrilineal and biparental kinship drive female associations in bottlenose dolphins." *Animal Behaviour* 80:481–86.

Fripp, D., C. Owen, E. Quintana-Rizzo, A. Shapiro, K. Buckstaff, K. Jankowski, R. Wells, and P. Tyack. 2005. "Bottlenose dolphin (*Tursiops truncatus*) calves appear to model their signature whistles on the signature whistles of community members." *Animal Cognition* 8:17–26.

Funk, W. C., J. K. McKay, P. A. Hohenlohe, and F. W. Allendorf. 2012. "Harnessing genomics for delineating conservation units." *Trends in Ecology and Evolution* 27:489–96.

Galef, B. G. 1992. "The question of animal culture." *Human Nature* 3:157–78.

———. 2001. "Where's the beef? evidence of culture, imitation, and teaching, in cetaceans?" *Behavioral and Brain Sciences* 24:335.

Galef, B. G., and K. N. Laland. 2005. "Social learning in animals: empirical studies and theoretical models." *Bioscience* 55:489–99.

Garamszegi, L. Z., S. Calhim, N. Dochtermann, G. Hegyi, P. L. Hurd, C. Jørgensen, N. Kutsukake, M. J. Lajeunesse, K. A. Pollard, and H. Schielzeth. 2009. "Changing philosophies and tools for statistical inferences in behavioral ecology." *Behavioral Ecology* 20:1363–75.

Gardner, H. E. 2000. *Intelligence Reframed: Multiple Intelligences for the 21st Century*. New York: Basic Books.

Garland, E. C., A. W. Goldizen, M. L. Rekdahl, R. Constantine, C. Garrigue, N. D. Hauser, M. M. Poole, J. Robbins, and M. J. Noad. 2011. "Dynamic horizontal cultural transmission of humpback whale song at the ocean basin scale." *Current Biology* 21:687–91.

Gavrilov, A. N., R. D. McCauley, and J. Gedamke. 2012. "Steady inter and intra-annual decrease in the vocalization frequency of Antarctic blue whales." *Journal of the Acoustical Society of America* 131:4476–80.

Gazda, S. K., R. C. Connor, R. K. Edgar, and F. Cox. 2005. "A division of labour with role specialization in group-hunting bottlenose dolphins (*Tursiops truncatus*) off Cedar Key, Florida." *Proceedings of the Royal Society of London*, ser. B 272:135–40.

Gedamke, J., D. P. Costa, and A. Dunstan. 2001. "Localization and visual verification of a complex minke whale vocalization." *Journal of the Acoustical Society of America* 109:3038–47.

Geissmann, T. 1984. "Inheritance of song parameters in the gibbon song, analyzed in 2 hybrid gibbons (*Hylobates pileatus* x *H. lar*)." *Folia Primatologica* 42:216–35.

George, J. C., J. Bada, J. Zeh, L. Scott, S. E. Brown, T. O'Hara, and R. Suydam. 1999. "Age and growth estimates of bowhead whales (*Balaena mysticetus*) via aspartic acid racemization." *Canadian Journal of Zoology* 77:571–80.

George, J. C., and J. Bockstoce. 2008. "Two historical weapon fragments as an aid to estimating the longevity and movements of bowhead whales." *Polar Biology* 31:751–54.

George, J. C., C. Clark, G. M. Carroll, and W. T. Ellison. 1989. "Observations on the ice-breaking and ice navigation behavior of migrating bowhead whales (*Balaena mysticetus*) near Point Barrow, Alaska, spring 1985." *Arctic* 42:24–30.

Gero, S. 2012a. "The dynamics of social relationships and vocal communication between individual and social units of sperm whales." PhD thesis, Dalhousie University.

———. 2012b. "The surprisingly familiar family lives of sperm whales." *Journal of the American Cetacean Society* 42:16–20.

Gero, S., L. Bejder, H. Whitehead, J. Mann, and R. C. Connor. 2005. "Behaviourally specific preferred associations in bottlenose dolphins, *Tursiops* sp." *Canadian Journal of Zoology* 83:1566–73.

Gero, S., D. Engelhaupt, and H. Whitehead. 2008. "Heterogeneous social associations within a sperm whale, *Physeter macrocephalus*, unit reflect pairwise relatedness." *Behavioral Ecology and Sociobiology* 63:143–51.

Gero, S., J. Gordon, and H. Whitehead. 2013. "Calves as social hubs: dynamics of the social network within sperm whale units." *Proceedings of the Royal Society of London*, ser. B, 280: 20131113.

Gintis, H. 2011. "Gene-culture coevolution and the nature of human sociality." *Philosophical Transactions of the Royal Society of London*, ser. B 366:878–88.

Glockner, D. A., and S. Venus. 1983. "Determining the sex of humpback whales (*Megaptera novaeangliae*) in their natural environment." In *Communication and Behavior of Whales*, edited by R. Payne, 447–64. Boulder, CO: Westview Press.

Goldbogen, J. A., N. D. Pyenson, and R. E. Shadwick. 2007. "Big gulps require high drag for fin whale lunge feeding." *Marine Ecology Progress Series* 349:289–301.

Gonzalvo, J., M. Valls, L. Cardona, and A. Aguilar. 2008. "Factors determining the interaction between common bottlenose dolphins and bottom trawlers off the Balearic Archipelago (western Mediterranean Sea)." *Journal of Experimental Marine Biology and Ecology* 367:47–52.

Goodall, J. 1968. "Behaviour of free-living chimpanzees of the Gombe Stream Reserve." *Animal Behaviour Monograph* 1:163–311.

———. 2005. "Do chimpanzees have souls?" In *Spiritual Information: 100 Perspectives on Science and Religion*, edited by C. L. Harper, 275–78. West Conshohocken, PA: Templeton Press.

Gowans, S., H. Whitehead, and S. K. Hooker. 2001. "Social organization in northern bottlenose whales (*Hyperoodon ampullatus*): not driven by deep water foraging?" *Animal Behaviour* 62:369–77.

Gowans, S., B. Würsig, and L. Karczmarski. 2007. "The social structure and strategies of delphinids: predictions based on an ecological framework." *Advances in Marine Biology* 53:195–294.

Grant, B. R., and P. R. Grant. 1996. "Cultural inheritance of song and its role in the evolution of Darwin's finches." *Evolution* 50:2471–87.

Grant, P. R., and B. R. Grant. 2009. "The secondary contact phase of allopatric speciation in Darwin's finches." *Proceedings of the National Academy of Sciences of the United States of America* 106:20141–48.

Gray, P. M., B. Krause, J. Atema, R. Payne, C. Krumhansl, and L. Baptista. 2001. "The music of nature and the nature of music." *Science* 291:52–54.

Green, S. R., E. Mercado III, A. A. Pack, and L. M. Herman. 2011. "Recurring patterns in the songs of humpback whales (*Megaptera novaeangliae*)." *Behavioural Processes* 86:284–94.

Greggor, A. L. 2012. "A functional paradigm for evaluating culture: an example with cetaceans." *Functional Zoology* 58:271–86.

Grillo, R. D. 2003. "Cultural essentialism and cultural anxiety." *Anthropological Theory* 3:157–73.

Guinee, L. N., and K. B. Payne. 1988. "Rhyme-like repetitions in songs of humpback whales." *Ethology* 79:295–306.

Guinet, C. 1991. "Intentional stranding apprenticeship and social play in killer whales (*Orcinus orca*)." *Canadian Journal of Zoology* 69:2712–16.

Guinet, C., L. G. Barrett-Lennard, and B. Loyer. 2000. "Co-ordinated attack behavior and prey sharing by killer whales at Crozet Archipelago: strategies for feeding on negatively-buoyant prey." *Marine Mammal Science* 16:829–34.

Guinet, C., and J. Bouvier. 1995. "Development of intentional stranding hunting techniques in killer whale (*Orcinus orca*) calves at Crozet Archipelago." *Canadian Journal of Zoology* 73:27–33.

Guinet, C., P. Domenici, R. De Stephanis, L. Barrett-Lennard, J. K. B. Ford, and P. Verborgh. 2007. "Killer whale predation on bluefin tuna: exploring the hypothesis of the endurance-exhaustion." *Marine Ecology Progress Series* 347:111–19.

Guttridge, T., A. Myrberg, I. Porcher, D. Sims, and J. Krause. 2009. "The role of learning in shark behaviour." *Fish and Fisheries* 10:450–69.

Guttridge, T., S. van Dijk, E. Stamhuis, J. Krause, S. Gruber, and C. Brown. 2012. "Social learning in juvenile lemon sharks, *Negaprion brevirostris*." *Animal Cognition* 16:55–64.

Hagen, E., and G. Bryant. 2003. "Music and dance as a coalition signaling system." *Human Nature* 14:21–51.

Hain, J. H. W., G. Carter, S. Kraus, C. Mayo, and H. Winn. 1982. "Feeding behavior of the humpback whale in the Western North Atlantic." *Fishery Bulletin US* 80:259–68.

Hairston, N. G., S. P. Ellner, M. A. Geber, T. Yoshida, and J. A. Fox. 2005. "Rapid evolution and the convergence of ecological and evolutionary time." *Ecology Letters* 8:1114–27.

Hakeem, A. Y., C. C. Sherwood, C. J. Bonar, C. Butti, P. R. Hof, and J. M. Allman. 2009. "Von Economo neurons in the elephant brain." *Anatomical Record: Advances in Integrative Anatomy and Evolutionary Biology* 292:242–48.

Hall, A., and S. Manabe. 1997. "Can local linear stochastic theory explain sea surface temperature and salinity variability?" *Climate Dynamics* 13:167–80.

Hall, K. R. L., and G. B. Schaller. 1964. "Tool-using behavior of the California sea otter." *Journal of Mammalogy* 45:287–98.

Halley, J. M. 1996. "Ecology, evolution and 1/*f* noise." *Trends in Ecology and Evolution* 11:33–37.

Hamilton, M. J., B. T. Milne, R. S. Walker, O. Burger, and J. H. Brown. 2007. "The complex structure of hunter-gatherer social networks." *Proceedings of the Royal Society of London*, ser. B 274:2195–2203.

Hamilton, W. D. 1964. "The genetical evolution of social behaviour." *Journal of Theoretical Biology* 7:1–52.

Hamlin, J. K., K. Wynn, and P. Bloom. 2007. "Social evaluation by preverbal infants." *Nature* 450:557–59.

Hamlin, J. K., K. Wynn, P. Bloom, and N. Mahajan. 2011. "How infants and toddlers react to antisocial others." *Proceedings of the National Academy of Sciences of the United States of America* 108:19931–36.

Harrison, R. K. 2000. *Introduction to the Old Testament*. Grand Rapids, MI: William B Eerdmans.

Hartman, K. L., F. Visser, and A. J. E. Hendriks. 2008. "Social structure of Risso's dolphins (*Grampus griseus*) at the Azores: a stratified community based on highly associated social units." *Canadian Journal of Zoology* 86:294–306.

Harvey, P. H., and M. D. Pagel. 1991. *The Comparative Method in Evolutionary Biology*. Oxford: Oxford University Press.

Hawkes, K., J. F. O'Connell, N. G. Blurton Jones, H. Alvarez, and E. L. Charnov. 1998. "Grandmothering, menopause, and the evolution of human life histories." *Proceedings of the National Academy of Sciences of the United States of America* 95:1336–39.

Hawkins, E. R. 2010. "Geographic variations in the whistles of bottlenose dolphins (*Tursiops aduncus*) along the east and west coasts of Australia." *Journal of the Acoustical Society of America* 128:924–35.

Healy, S. D., and C. Rowe. 2007. "A critique of comparative studies of brain size." *Proceedings of the Royal Society of London*, ser. B 274:453–64.

Heimlich-Boran, J. R. 1993. "Social organization of the short-finned pilot whale *Globicephala macrorhynchus*, with special reference to the comparative social ecology of delphinids." PhD thesis, Cambridge University.

Hekkert, P., D. Snelders, and P. C. W. Wieringen. 2003. "'Most advanced, yet acceptable': Typicality and novelty as joint predictors of aesthetic preference in industrial design." *British Journal of Psychology* 94:111–24.

Henrich, J. 2004a. "Cultural group selection, coevolutionary processes and large-scale cooperation." *Journal of Economic Behavior and Organization* 53:3–35.

———. 2004b. "Demography and cultural evolution: how adaptive cultural processes can produce maladaptive losses—the Tasmanian case." *American Antiquity* 69:197–214.

Henrich, J., and R. Boyd. 1998. "The evolution of conformist transmission and the emergence of between-group differences." *Evolution and Human Behavior* 19:215–41.

Henrich, J., and N. Henrich. 2010. "The evolution of cultural adaptations: Fijian food taboos protect against dangerous marine toxins." *Proceedings of the Royal Society of London*, ser. B 277:3715–24.

Henshilwood, C. S., F. d'Errico, K. L. van Niekerk, Y. Coquinot, Z. Jacobs, S. Lauritzen, M. Menu, and R. García-Moreno. 2011. "A 100,000-year-old ochre-processing workshop at Blombos Cave, South Africa." *Science* 334:219–22.

Herman, L. M. 1979. "Humpback whales in Hawaiian waters: a study in historical ecology." *Pacific Science* 33:1–15.

———. 2002a. "Vocal, social, and self-imitation by bottlenosed dolphins." In *Imitation in Animals and Artifacts*, edited by K. Dautenhahn and C. L. Nehaniv, 63–108. Cambridge, MA: MIT Press.

———. 2002b. "Exploring the cognitive world of the bottlenosed dolphin." In *The Cognitive*

Animal: Empirical and Theoretical Perspectives in Animal Cognition, edited by M. Bekoff, C. Allen, and G. M. Burghardt, 275–83. Cambridge, MA: MIT Press.

Herman, L. M., S. L. Abichandani, A. N. Elhajj, E. Y. Herman, J. L. Sanchez, and A. A. Pack. 1999. "Dolphins (*Tursiops truncatus*) comprehend the referential character of the human pointing gesture." *Journal of Comparative Psychology* 113:347–64.

Herman, L. M., and A. A. Pack. 2001. "Laboratory evidence for cultural transmission mechanisms." *Behavioral and Brain Sciences* 24:335–36.

Herzing, D. L. 2011. *Dolphin Diaries: My 25 Years with Spotted Dolphins in the Bahamas*. New York: St. Martin's Press.

Hewlett, B. S., H. N. Fouts, A. H. Boyette, and B. L. Hewlett. 2011. "Social learning among Congo Basin hunter-gatherers." *Philosophical Transactions of the Royal Society of London*, ser B 366:1168–78.

Heyes, C. M. 1994. "Social learning in animals: categories and mechanisms." *Biological Reviews* 69:207–31.

———. 2012. "What's social about social learning?" *Journal of Comparative Psychology* 126:193–202.

Hill, K. 2009. "Animal 'culture.'" In *The Question of Animal Culture*, edited by K. N. Laland and B. G. Galef Jr., 269–87. Cambridge, MA: Harvard University Press.

———. 2010. "Experimental studies of animal social learning in the wild: trying to untangle the mystery of human culture." *Learning and Behavior* 38:319–28.

Hill, K. R., R. S. Walker, M. Božičević, J. Eder, T. Headland, B. Hewlett, A. M. Hurtado, F. Marlowe, P. Wiessner, and B. Wood. 2011. "Co-residence patterns in hunter-gatherer societies show unique human social structure." *Science* 331:1286–89.

Hill, R. A., R. A. Bentley, and R. I. M. Dunbar. 2008. "Network scaling reveals consistent fractal pattern in hierarchical mammalian societies." *Biology Letters* 4:748–51.

Hlista, B., H. Sosik, L. Traykovski, R. Kenney, and M. Moore. 2009. "Seasonal and interannual correlations between right-whale distribution and calving success and chlorophyll concentrations in the Gulf of Maine, USA." *Marine Ecology Progress Series* 394:289–302.

Hoelzel, A. R. 1991. "Killer whale predation on marine mammals at Punta Norte, Argentina: food sharing, provisioning and foraging strategy." *Behavioral Ecology and Sociobiology* 29:197–204.

Hoelzel, A. R., E. M. Dorsey, and S. J. Stern. 1989. "The foraging specializations of individual minke whales." *Animal Behaviour* 38:786–94.

Hoelzel, A. R., C. W. Potter, and P. B. Best. 1998. "Genetic differentiation between parapatric 'nearshore' and 'offshore' populations of the bottlenose dolphin." *Proceedings of the Royal Society of London*, ser. B 265:1177–83.

Hof, P. R., and D. G. Van. 2007. "Structure of the cerebral cortex of the humpback whale, *Megaptera novaeangliae* (Cetacea, Mysticeti, Balaenopteridae)." *Anatomical Record: Advances in Integrative Anatomy and Evolutionary Biology* 290:1–31.

Hofstede, G. 1981. "Culture and organizations." *International Studies of Management and Organization* 10:15–41.

Hohn, A. A., D. S. Rotstein, C. A. Harms, and B. L. Southall. 2006. *Report on Marine Mammal Unusual Mortality Event UMESE0501Sp: Multispecies Mass Stranding of Pilot Whales* (Globicephala macrorhynchus), *Minke Whale* (Balaenoptera acutorostrata), *and Dwarf Sperm Whales* (Kogia sima) *in North Carolina on 15–16 January 2005*. NOAA Technical Memorandum NMFS-SEFSC-53716217. Miami: U.S. Dept. of Commerce, National Oceanic and Atmospheric Administration, National Marine Fisheries Service, Southeast Fisheries Science Center.

Holmes, B. J., and D. T. Neil. 2012. "Gift giving by wild bottlenose dolphins (*Tursiops* sp.) to humans at a wild dolphin provisioning program, Tangalooma, Australia." *Anthrozoos* 25:397–413.

Holthuis, L. B. 1987. "The scientific name of the sperm whale." *Marine Mammal Science* 3:87–89.

Hoppitt, W., and K. N. Laland. 2008. "Social processes influencing learning in animals: a review of the evidence." *Advances in the Study of Behavior* 38:105–65.

———. 2013. *Social Learning: An Introduction to Mechanisms, Methods, and Models*. Princeton, NJ: Princeton University Press.

Hoppitt, W. J. E., G. R. Brown, R. Kendal, L. Rendell, A. Thornton, M. M. Webster, and K. N. Laland. 2008. "Lessons from animal teaching." *Trends in Ecology and Evolution* 23:486–93.

Horn, H. S., and D. I. Rubenstein. 1984. "Behavioural adaptations and life history." In *Behavioural Ecology: An Evolutionary Approach*, edited by J. R. Krebs, and N. B. Davies, 279–98. 2nd ed. Oxford: Blackwell Science Publications.

Hoyt, E. 1991. *Orca, the Whale Called Killer*. 3rd ed. Richmond Hill, ON: Firefly Books.

Hucke-Gaete, R., L. P. Osman, C. A. Moreno, K. P. Findlay, and D. K. Ljungblad. 2004. "Discovery of a blue whale feeding and nursing ground in southern Chile." *Proceedings of the Royal Society of London*, ser. B 271:S170–S173.

Huffman, M. A. 1984. "Stone-play of *Macaca fuscata* in Arashiyama B troop: transmission of a non-adaptive behavior." *Journal of Human Evolution* 13:725–35.

———. 1996. "Acquisition of innovative cultural behaviors in nonhuman primates: a case study of stone handling, a socially transmitted behavior in Japanese macaques." In *Social Learning in Animals*, edited by C. M. Heyes and B. G. Galef, 267–89. San Diego: Academic Press.

Huffman, M. A., C. A. D. Nahallage, and J. Leca. 2008. "Cultured monkeys: social learning cast in stones." *Current Directions in Psychological Science* 17:410–14.

Huffman, M. A., C. Spiezio, A. Sgaravatti, and J. Leca. 2010. "Leaf swallowing behavior in chimpanzees (*Pan troglodytes*): biased learning and the emergence of group level cultural differences." *Animal Cognition* 13:871–80.

Humphrey, N. K. 1976. "The social function of intellect." In *Growing Points in Ethology*," edited by P. P. G. Bateson and R. A. Hinde, 303–17. Cambridge: Cambridge University Press.

Hunt, G. R., and R. D. Gray. 2003. "Diversification and cumulative evolution in New Caledonian crow tool manufacture." *Proceedings of the Royal Society of London*, ser. B 270:867–74.

Huntley, M. E., and M. Zhou. 2004. “Influence of animals on turbulence in the sea.” *Marine Ecology Progress Series* 273:65–79.

Husson, A. M., and L. B. Holthuis. 1974. “*Physeter macrocephalus* Linnaeus, 1758, the valid name for the sperm whale.” *Zoologische Mededelingen* 48:205–17.

Huxley, J. S. 1942. *Evolution: The Modern Synthesis*. London: Allen & Unwin.

Inchausti, P., and J. Halley. 2002. “The long-term temporal variability and spectral colour of animal populations.” *Evolutionary Ecology Research* 4:1033–48.

Ingebrigtsen, A. 1929. “Whales caught in the North Atlantic and other areas.” *Conseil Permanent International pour l’Exploration de la Mer, Rapports et Proces-Verbaux des Reunions* 55:1–26.

Ingold, T. 2001. “The use and abuse of ethnography.” *Behavioral and Brain Sciences* 24:337.

———. 2004. “Beyond biology and culture: the meaning of evolution in a relational world.” *Social Anthropology* 12:209–21.

Isack, H. A., and H. U. Reyer. 1989. “Honeyguides and honey gatherers: interspecific communication in a symbiotic relationship.” *Science* 243:1343–46.

Jaakkola, K., E. Guarino, and M. Rodriguez. 2010. “Blindfolded imitation in a bottlenose dolphin (*Tursiops truncatus*).” *International Journal of Comparative Psychology* 23:671–88.

Jaakkola, K., E. Guarino, M. Rodriguez, and J. Hecksher. 2013. “Switching strategies: a dolphin’s use of passive and active acoustics to imitate motor actions.” *Animal Cognition* 16:701–9.

Jackson, J. B. C., M. X. Kirby, W. H. Berger, K. A. Bjorndal, L. W. Botsford, B. J. Bourque, R. H. Bradbury, R. Cooke, J. Erlandson, and J. A. Estes. 2001. “Historical overfishing and the recent collapse of coastal ecosystems.” *Science* 293:629–38.

Janik, V. M. 2000a. “Food-related bray calls in wild bottlenose dolphins (*Tursiops truncatus*).” *Proceedings of the Royal Society of London*, ser. B 267:923–27.

———. 2000b. “Whistle matching in wild bottlenose dolphins (*Tursiops truncatus*).” *Science* 289:1355–57.

———. 2001. “Is cetacean social learning unique?” *Behavioral and Brain Sciences* 24:337–38.

Janik, V. M., L. S. Sayigh, and R. S. Wells. 2006. “Signature whistle shape conveys identity information to bottlenose dolphins.” *Proceedings of the National Academy of Sciences of the United States of America* 103:8293–97.

Janik, V. M., and P. J. B. Slater. 1997. “Vocal learning in mammals.” *Advances in the Study of Behavior* 26:59–99.

———. 1998. “Context-specific use suggests that bottlenose dolphin signature whistles are cohesion calls.” *Animal Behaviour* 56:829–38.

Jefferson, T. A., P. J. Stacey, and R. W. Baird. 1991. “A review of killer whale interactions with other marine mammals: predation to co-existence.” *Mammal Review* 4:151–80.

Jensen, M. N. “Whales’ cultural revolution.” *Science NOW*, November 29, http://news.sciencemag.org/2000/11/whales-cultural-revolution.

Jerison, H. J. 1973. *Evolution of the Brain and Intelligence*. New York: Academic Press.

———. 1986. “The perceptual world of dolphins.” In *Dolphin Cognition and Behavior: A Comparative View*, edited by R. J. Schusterman, J. A. Thomas, and F. G. Wood, 141–66. Hillsdale, NJ: Lawrence Erlbaum Associates.

Johnson, D. H. 1999. "The insignificance of statistical significance testing." *Journal of Wildlife Management* 63:763–72.

Johnson, M. P., and P. L. Tyack. 2003. "A digital acoustic recording tag for measuring the response of wild marine mammals to sound." *IEEE Journal of Ocean Engineering* 28:3–12.

Johnstone, R. A., and M. A. Cant. 2010. "The evolution of menopause in cetaceans and humans: the role of demography." *Proceedings of the Royal Society of London*, ser. B 277:3765–71.

Jurasz, C., and V. Jurasz. 1979. "Feeding modes of the humpback whale, *Megaptera novaeangliae*, in southeast Alaska." *Scientific Reports of the Whales Research Institute of Tokyo* 31:69–83.

Kako, E. 1999. "Elements of syntax in the systems of three language-trained animals." *Animal Learning and Behavior* 27:1–14.

Karczmarski, L. 1999. "Group dynamics of humpback dolphins (*Sousa chinensis*) in the Algoa Bay region, South Africa. *Journal of Zoology* 249:283–93.

Karczmarski, L., B. Würsig, G. Gailey, K. W. Larson, and C. Vanderlip. 2005. "Spinner dolphins in a remote Hawaiian atoll: social grouping and population structure." *Behavioral Ecology* 16:675–85.

Kasuya, T. 1975. "Past occurrence of *Globicephala melaena* in the western North Pacific." *Scientific Reports of the Whales Research Institute* 27:95–110.

———. 1995. "Overview of cetacean life histories: an essay in their evolution." In *Whales, Seals, Fish and Man*, edited by A. S. Blix, L. Walløe, and O. Ulltang, 481–97. Amsterdam: Elsevier Science.

———. 2008. "The Kenneth S. Norris lifetime achievement award lecture: presented on 29 November 2007 Cape Town, South Africa." *Marine Mammal Science* 24:749–73.

Kasuya, T., and H. Marsh. 1984. "Life history and reproductive biology of the short-finned pilot whale, *Globicephala macrorhynchus*, off the Pacific coast of Japan." *Reports of the International Whaling Commission (Special Issue)* 6:259–310.

Kawai, M. 1965. "Newly acquired pre-cultural behavior of the natural troop of Japanese monkeys on Koshima Inlet." *Primates* 2:1–30.

Kelley, L. A., and S. D. Healy. 2010. "Vocal mimicry in male bowerbirds: who learns from whom?" *Biology Letters* 6:626–29.

Kelly, E. 2002. "Hate crime: the struggle for justice in Scotland." *Criminal Justice Matters* 48:16–17.

Kim, P. S., J. E. Coxworth, and K. Hawkes. 2012. "Increased longevity evolves from grandmothering." *Proceedings of the Royal Society of London*, ser. B 279:4880–84.

King, S. L., and V. M. Janik. 2013. "Bottlenose dolphins can use learned vocal labels to address each other." *Proceedings of the National Academy of Sciences* 110:13216–21.

King, S. L., L. S. Sayigh, R. S. Wells, W. Fellner, and V. M. Janik. 2013. "Vocal copying of individually distinctive signature whistles in bottlenose dolphins." *Proceedings of the Royal Society of London*, ser. B 280:20130053.

Kirby, S., H. Cornish, and K. Smith. 2008. "Cumulative cultural evolution in the laboratory: an experimental approach to the origins of structure in human language." *Proceedings of the National Academy of Sciences of the United States of America* 105:10681–86.

Kirby, S., M. Dowman, and T. L. Griffiths. 2007. "Innateness and culture in the evolution of language." *Proceedings of the National Academy of Sciences of the United States of America* 104:5241–45.

Klatsky, L. J., R. S. Wells, and J. C. Sweeney. 2007. "Offshore bottlenose dolphins (*Tursiops truncatus*): movement and dive behavior near the Bermuda Pedestal." *Journal of Mammalogy* 88:59–66.

Kline, M. A., and R. Boyd. 2010. "Population size predicts technological complexity in Oceania." *Proceedings of the Royal Society of London*, ser. B 277:2559–64.

Koblmüller, S., M. Nord, R. K. Wayne, and J. A. Leonard. 2009. "Origin and status of the Great Lakes wolf." *Molecular Ecology* 18:2313–26.

Kooyman, G. L. 1989. *Diverse Divers*. Berlin: Springer-Verlag.

Kopps, A. M., C. Y. Ackermann, W. B. Sherwin, S. J. Allen, L. Bejder, and M. Krützen. 2014. "Cultural transmission of tool use combined with habitat specializations leads to fine-scale genetic structure in bottlenose dolphins." *Proceedings of the Royal Society of London*, ser. B 281:1782 20133245.

Kopps, A. M., and W. B. Sherwin. 2012. "Modelling the emergence and stability of a vertically transmitted cultural trait in bottlenose dolphins." *Animal Behaviour* 84:1347–62.

Krahn, M. M., P. R. Wade, S. T. Kalinowski, M. E. Dahlheim, B. L. Taylor, M. B. Hanson, G. M. Ylitalo, R. P. Angliss, J. E. Stein, and R. S. Waples. 2002. *Status Review of Southern Resident Killer Whales* (Orcinus orca) *under the Endangered Species Act.* NOAA Technical Memorandum NMFS-NWFSC-5416219. [Seattle]: U.S. Dept. of Commerce, National Oceanic and Atmospheric Administration, National Marine Fisheries Service [Northwest Fisheries Science Center].

Kraus, S. D., and J. J. Hatch. 2001. "Mating strategies in the North Atlantic right whale (*Eubalaena glacialis*)." *Journal of Cetacean Research and Management Special Issue* 2:237–44.

Kraus, S. D., and R. M. Rolland. 2007a. "Right whales in an urban ocean." In *The Urban Whale: North Atlantic Right Whales at the Crossroads*, edited by S. D. Kraus and R. M. Rolland, 1–38. Cambridge, MA: Harvard University Press.

———. 2007b. "The urban whale syndrome." In *The Urban Whale: North Atlantic Right Whales at the Crossroads*, edited by S. D. Kraus and R. M. Rolland, 488–513. Cambridge, MA: Harvard University Press.

Krause, J., and G. Ruxton. 2002. *Living in Groups*. Oxford: Oxford University Press.

Kremers, D., A. Lemasson, J. Almunia, and R. Wanker. 2012. "Vocal sharing and individual acoustic distinctiveness within a group of captive orcas (*Orcinus orca*)." *Journal of Comparative Psychology* 126:433–45.

Kroeber, A. L. 1948. *Anthropology*. New York: Harcourt, Brace.

Kroodsma, D. E. 2005. *The Singing Life of Birds: The Art and Science of Listening to Birdsong*. Vol. 1. Boston: Houghton Mifflin Harcourt.

Krützen, M., J. Mann, M. R. Heithaus, R. C. Connor, L. Bejder, and W. B. Sherwin. 2005. "Cultural transmission of tool use in bottlenose dolphins." *Proceedings of the National Academy of Sciences of the United States of America* 102:8939–43.

Krützen, M., E. Willems, and C. van Schaik. 2011. "Culture and geographic variation in orangutan behavior." *Current Biology* 21:1808–12.

Kuczaj, S. A. 2010. "Research with captive marine mammals is important: an introduction to the special issue." *International Journal of Comparative Psychology* 23:225–26.

Kuhn, T. S. 1962. *The Structure of Scientific Revolutions*. Chicago: University of Chicago Press.

Lahdenperä, M., V. Lummaa, S. Helle, M. Tremblay, and A. F. Russell. 2004. "Fitness benefits of prolonged post-reproductive lifespan in women." *Nature* 428:178–81.

Laland, K. N., and G. R. Brown. 2011. *Sense and Nonsense: Evolutionary Perspectives on Human Behaviour*. 2nd ed. Oxford: Oxford University Press.

Laland, K. N., and B. G. Galef. 2009a. "Introduction." In *The Question of Animal Culture*, edited by K. N. Laland and B. G. Galef Jr., 1–18. Cambridge, MA: Harvard University Press.

———. 2009b. *The Question of Animal Culture*. Cambridge, MA: Harvard University Press.

Laland, K. N., and W. Hoppitt. 2003. "Do animals have culture?" *Evolutionary Anthropology* 12:150–59.

Laland, K. N., and V. M. Janik. 2006. "The animal cultures debate." *Trends in Ecology and Evolution* 21:542–47.

Laland, K. N., J. R. Kendal, and R. L. Kendal. 2009. "Animal culture: problems and solutions." In *The Question of Animal Culture*, edited by K. N. Laland and B. G. Galef Jr., 174–97. Cambridge, MA: Harvard University Press.

Laland, K. N., J. Odling-Smee, and M. W. Feldman. 2000. "Niche construction, biological evolution and cultural change." *Behavioral and Brain Sciences* 23:131–75.

Laland, K. N., J. Odling-Smee, and S. Myles. 2010. "How culture shaped the human genome: bringing genetics and the human sciences together." *Nature Reviews Genetics* 11:137–48.

Laland, K. N., and K. Williams. 1998. "Social transmission of maladaptive information in the guppy." *Behavioral Ecology* 9:493–99.

Lambert, O., G. Bianucci, K. Post, C. de Muizon, R. Salas-Gismondi, M. Urbina, and J. Reumer. 2010. "The giant bite of a new raptorial sperm whale from the Miocene epoch of Peru." *Nature* 466:105–8.

Lane, J. D., H. M. Wellman, S. L. Olson, J. LaBounty, and D. C. R. Kerr. 2010. "Theory of mind and emotion understanding predict moral development in early childhood." *British Journal of Developmental Psychology* 28:871–89.

Langergraber, K. E., C. Boesch, E. Inoue, M. Inoue-Murayama, J. C. Mitani, T. Nishida, A. Pusey, et al. 2011. "Genetic and 'cultural' similarity in wild chimpanzees." *Proceedings of the Royal Society of London*, ser. B 278:408–16.

Laws, R. M. 1970. "Elephants as agents of habitat and landscape change in East Africa." *Oikos* 21:1–15.

Leatherwood, S., and R. R. Reeves. 1990. *The Bottlenose Dolphin*. San Diego: Academic Press.

Le Boeuf, B. J., D. P. Costa, A. C. Huntley, and S. D. Feldkamp. 1988. "Continuous, deep diving in female northern elephant seals, *Mirounga angustirostris*." *Canadian Journal of Zoology* 66:446–58.

Le Boeuf, B. J., D. P. Costa, A. C. Huntley, G. L. Kooyman, and R. W. Davis. 1986. "Pattern and depth of dives in northern elephant seals, *Mirounga angustirostris*." *Journal of Zoology* 208:1–7.

Leca, J., N. Gunst, and M. A. Huffman. 2007. "Japanese macaque cultures: inter- and intra-troop behavioural variability of stone handling patterns across 10 troops." *Behaviour* 144:251–81.

Leeuwenburgh, O., and D. Stammer. 2001. "The effect of ocean currents on sea surface temperature anomalies." *Journal of Physical Oceanography* 31:2340–58.

Lehmann, L., M. W. Feldman, and K. R. Foster. 2008. "Cultural transmission can inhibit the evolution of altruistic helping." *American Naturalist* 172:12–24.

Letteval, E., C. Richter, N. Jaquet, E. Slooten, S. Dawson, H. Whitehead, J. Christal, and P. McCall Howard. 2002. "Social structure and residency in aggregations of male sperm whales." *Canadian Journal of Zoology* 80:1189–96.

Levin, P. S., E. E. Holmes, K. R. Piner, and C. J. Harvey. 2006. "Shifts in a Pacific Ocean fish assemblage: the potential influence of exploitation." *Conservation Biology* 20:1181–90.

Lewis, H. M., and K. N. Laland. 2012. "Transmission fidelity is the key to the build-up of cumulative culture." *Philosophical Transactions of the Royal Society of London*, ser. B 367:2171–80.

Lewis, J. S., and W. W. Schroeder. 2003. "Mud plume feeding, a unique foraging behavior of the bottlenose dolphin in the Florida Keys." *Gulf of Mexico Science* 21:92–97.

Liebenberg, L. 2006. "Persistence hunting by modern hunter gatherers." *Current Anthropology* 47:1017–26.

Lilly, J. C. 1965. "Vocal mimicry in *Tursiops:* ability to match numbers and durations of human vocal bursts." *Science* 147:300–301.

———. 1978. *Communication between Man and Dolphin: The Possibilities of Talking with Other Species*. New York: Crown.

Lindenbaum, S. 2008. "Review: Understanding kuru: the contribution of anthropology and medicine." *Philosophical Transactions of the Royal Society of London*, ser. B 363:3715–20.

Lopez, B. D. 2012. "Bottlenose dolphins and aquaculture: interaction and site fidelity on the north-eastern coast of Sardinia (Italy)." *Marine Biology* 159:2161–72.

Lopez, J. C., and D. Lopez. 1985. "Killer whales (*Orcinus orca*) of Patagonia, and their behavior of intentional stranding while hunting nearshore." *Journal of Mammalogy* 66:181–83.

Luck, G. W., G. C. Daily, and P. R. Ehrlich. 2003. "Population diversity and ecosystem services." *Trends in Ecology and Evolution* 18:331–36.

Lumsden, C. J., and E. O. Wilson. 1981. *Genes, Mind, and Culture: The Coevolutionary Process*. Cambridge, MA: Harvard University Press.

Lusseau, D. 2007. "Evidence for social role in a dolphin social network." *Evolutionary Ecology* 21:357–66.

Lusseau, D., and M. E. J. Newman. 2004. "Identifying the role that animals play in social networks." *Proceedings of the Royal Society of London*, ser. B 271: S477–481.

Lyons, S. 2010. Review of *Wild Justice*, by Mark Bekoff and Jessica Pierce. *Philosophy Now* 79:36–37.

Lyrholm, T., O. Leimar, and U. Gyllensten. 1996. "Low diversity and biased substitution patterns in the mitochondrial DNA control region of sperm whales: implications for estimates of time since common ancestry." *Molecular Biology and Evolution* 13:1318–26.

Lyrholm, T., O. Leimar, B. Johanneson, and U. Gyllensten. 1999. "Sex-biased dispersal in sperm whales: contrasting mitochondrial and nuclear genetic structure of global populations." *Proceedings of the Royal Society of London*, ser. B 266:347–54.

Mackintosh, N. A. 1965. *The Stocks of Whales*. London: Fishing News (Books) Ltd.

Madden, J. R., T. J. Lowe, H. V. Fuller, K. K. Dasmahapatra, and R. L. Coe. 2004. "Local traditions of bower decoration by spotted bowerbirds in a single population." *Animal Behaviour* 68:759–65.

Madsen, P. T. 2002. "Sperm whale sound production—in the acoustic realm of the biggest nose on record." In "Sperm Whale Sound Production," 1–39. PhD Thesis, University of Aarhus.

———. 2012. "Foraging with the biggest nose on record." *Journal of the American Cetacean Society* 41:9–15.

Madsen, P. T., D. A. Carder, W. W. L. Au, P. E. Nachtigall, B. Mohl, and S. H. Ridgway. 2003. "Sound production in neonate sperm whales (L)." *Journal of the Acoustical Society of America* 113:2988–91.

Madsen, P., R. Payne, N. Kristiansen, M. Wahlberg, I. Kerr, and B. Møhl. 2002. "Sperm whale sound production studied with ultrasound time/depth-recording tags." *Journal of Experimental Biology* 205:1899–1906.

Mann, J., R. C. Connor, L. M. Barre, and M. R. Heithaus. 2000. "Female reproductive success in bottlenose dolphins (*Tursiops* sp.): life history, habitat, provisioning, and group size effects." *Behavioral Ecology* 11:210–19.

Mann, J., and B. Sargeant. 2003. "Like mother, like calf: the ontogeny of foraging traditions in wild Indian Ocean bottlenose dolphins (*Tursiops* sp.)." In *The Biology of Traditions: Models and Evidence*, edited by D. M. Fragaszy and S. Perry, 236–66. Cambridge: Cambridge University Press.

Mann, J., B. L. Sargeant, J. Watson-Capps, Q. A. Gibson, M. R. Heithaus, R. C. Connor, and E. Patterson. 2008. "Why do dolphins carry sponges?" *PLoS ONE* 3:e3868.

Mann, J., M. Stanton, E. M. Patterson, E. J. Bienenstock, and L. O. Singh. 2012. "Social networks reveal cultural behaviour in tool-using using dolphins." *Nature Communications* 3:980.

Mann, J., and J. Watson-Capps. 2005. "Surviving at sea: ecological and behavioural predictors of calf mortality in Indian Ocean bottlenose dolphins, *Tursiops* sp." *Animal Behaviour* 69:899–909.

Mann, N. I., K. A. Dingess, and P. J. B. Slater. 2006. "Antiphonal four-part synchronized chorusing in a Neotropical wren." *Biology Letters* 2:1–4.

Marcoux, M., L. Rendell, and H. Whitehead. 2007. "Indications of fitness differences among vocal clans of sperm whales." *Behavioural Ecology and Sociobiology* 61:1093–98.

Marcoux, M., H. Whitehead, and L. Rendell. 2006. "Coda vocalizations recorded in breeding areas are almost entirely produced by mature female sperm whales (*Physeter macrocephalus*)." *Canadian Journal of Zoology* 84:609–14.

———. 2007. "Sperm whale feeding variation by location, year, social group and clan: evidence from stable isotopes." *Marine Ecology Progress Series* 333:309–14.

Marean, C. W., M. Bar-Matthews, J. Bernatchez, E. Fisher, P. Goldberg, A. I. R. Herries,

Z. Jacobs, et al. 2007. "Early human use of marine resources and pigment in South Africa during the Middle Pleistocene." *Nature* 449:905–8.

Marino, L. 1998. "A comparison of encephalization between odontocete cetaceans and anthropoid primates." *Brain, Behaviour and Evolution* 51:230–38.

———. 2006. "Absolute brain size: did we throw the baby out with the bathwater?" *Proceedings of the National Academy of Sciences of the United States of America* 103:13563–64.

———. 2011. "Brain structure and intelligence in cetaceans." In *Whales and Dolphins: Cognition, Culture, Conservation and Human Perceptions*, edited by P. Brakes and M. P. Simmonds, 115–28. London: Earthscan.

Marino, L., C. Butti, R. C. Connor, R. E. Fordyce, L. M. Herman, P. R. Hof, L. Lefebvre, D. Lusseau, B. McCowan, and E. A. Nimchinsky. 2008. "A claim in search of evidence: reply to Manger's thermogenesis hypothesis of cetacean brain structure." *Biological Reviews* 83:417–40.

Marino, L., and T. Frohoff. 2011. "Towards a new paradigm of non-captive research on cetacean cognition." *PLoS ONE* 6:e24121.

Marino, L., D. W. McShea, and M. D. Uhen. 2004. "Origin and evolution of large brains in toothed whales." *Anatomical Record: Advances in Integrative Anatomy and Evolutionary Biology* 281A:1247–55.

Marino, L., M. D. Uhen, N. D. Pyenson, and B. Frohlich. 2003. "Reconstructing cetacean brain evolution using computed tomography." *Anatomical Record: Advances in Integrative Anatomy and Evolutionary Biology* 272B:107–17.

Marks, J. 2002. *What It Means to Be 98% Chimpanzee: Apes, People, and Their Genes*. Berkeley: University of California Press.

———. 2012. "The biological myth of human evolution." *Contemporary Social Science* 7:139–57.

Marsh, A. A., H. A. Elfenbein, and N. Ambady. 2007. "Separated by a common language: nonverbal accents and cultural stereotypes about Americans and Australians." *Journal of Cross-Cultural Psychology* 38:284–301.

Marsh, H., and T. Kasuya. 1984. "Changes in the ovaries of the short-finned pilot whale, *Globicephala macrorhynchus*, with age and reproductive activity." *Reports of the International Whaling Commission (Special Issue)* 6:311–35.

———. 1986. "Evidence for reproductive senescence in female cetaceans." *Reports of the International Whaling Commission (Special Issue)* 8:57–74.

Martin, A. R., V. M. F. da Silva, and P. Rothery. 2008. "Object carrying as sociosexual display in an aquatic mammal." *Biology Letters* 4:243–45.

Martindale, C. 1990. *The Clockwork Muse: The Predictability of Artistic Change*. New York: Basic Books.

Mate, B. R., S. L. Nieukirk, and S. D. Kraus. 1997. "Satellite-monitored movements of the northern right whale." *Journal of Wildlife Management* 61:1393–1405.

Mathias, D., A. Thode, J. Straley, and K. Folkert. 2009. "Relationship between sperm whale (*Physeter macrocephalus*) click structure and size derived from videocamera images of a depredating whale." *Journal of Acoustical Society of America* 125:3444–53.

Matkin, C., and J. Durban. 2011. "Killer whales in Alaskan waters." *Journal of the American Cetacean Society* 40:24–29.

Matkin, C. O., D. R. Matkin, G. M. Ellis, E. Saulitis, and D. McSweeney. 1997. "Movements of resident killer whales in southeastern Alaska and Prince William Sound, Alaska." *Marine Mammal Science* 13:469–75.

Maynard Smith, J. 1989. *Evolutionary Genetics*. Oxford: Oxford University Press.

Maynard Smith, J., and J. Haigh. 1974. "The hitch-hiking effect of a favourable gene." *Genetics Research* 23:23–35.

Maynard Smith, J., and E. Szathmáry. 1995. *The Major Transitions in Evolution*. Oxford: Oxford University Press.

———. 1999. *The Origins of Life: From the Birth of Life to the Origin of Language*. Oxford: Oxford University Press.

Mayo, C. A., and M. K. Marx. 1990. "Surface foraging behavior of the North Atlantic right whale and associated plankton characteristics." *Canadian Journal of Zoology* 68:2214–20.

McAuliffe, K., and H. Whitehead. 2005. "Eusociality, menopause and information in matrilineal whales." *Trends in Ecology and Evolution* 20:650.

McComb, K., G. Shannon, S. M. Durant, K. Sayialel, R. Slotow, J. Poole, and C. Moss. 2011. "Leadership in elephants: the adaptive value of age." *Proceedings of the Royal Society of London*, ser. B 278:3270–76.

McComb, K., G. Shannon, K. Sayialel, and C. Moss. 2014. "Elephants can determine ethnicity, gender, and age from acoustic cues in human voices." *Proceedings of the National Academy of Sciences of the United States of America*

McComb, K., C. Moss, S. M. Durant, L. Baker, and S. Sayialel. 2001. "Matriarchs as repositories of social knowledge in African elephants." *Science* 292:491–94.

McCowan, B., and D. Reiss. 2001. "The fallacy of 'signature whistles' in bottlenose dolphins: a comparative perspective of 'signature information' in animal vocalizations." *Animal Behaviour* 62:1151–62.

McDonald, M. A., J. Calambokidis, A. M. Teranishi, and J. A. Hildebrand. 2001. "The acoustic calls of blue whales off California with gender data." *Journal of the Acoustical Society of America* 109:1728–35.

McDonald, M. A., J. A. Hildebrand, and S. Mesnick. 2009. "Worldwide decline in tonal frequencies of blue whale songs." *Endangered Species Research* 9:13–21.

McDonald, M. A., S. L. Mesnick, and J. A. Hildebrand. 2006. "Biogeographic characterization of blue whale song worldwide: using song to identify populations." *Journal of Cetacean Research and Management* 8:55–65.

McDougall, C. 2009. *Born to Run: A Hidden Tribe, Superathletes, and the Greatest Race the World Has Never Seen*. New York: Knopf Doubleday.

McElreath, R., R. Boyd, and P. Richerson. 2003. "Shared norms and the evolution of ethnic markers." *Current Anthropology* 44:122–30.

McGowen, M. R., M. Spaulding, and J. Gatesy. 2009. "Divergence date estimation and a comprehensive molecular tree of extant cetaceans." *Molecular Phylogenetics and Evolution* 53:891–906.

McGrew, W. C. 1987. "Tools to get food—the subsistants of Tasmanian aborigines and Tanzanian chimpanzees compared." *Journal of Anthropological Research* 43:247–58.

———. 1992. *Chimpanzee Material Culture: Implications for Human Evolution.* Cambridge: Cambridge University Press.

———. 1998. "Culture in nonhuman primates?" *Annual Review of Anthropology* 27:301–28.

———. 2003. "Ten dispatches from the chimpanzee culture wars." In *Animal Social Complexity: Intelligence, Culture, and Individualized Societies*, edited by F. B. M. de Waal and P. L. Tyack, 419–39. Cambridge, MA: Harvard University Press.

———. 2004. *The Cultured Chimpanzee: Reflections on Cultural Primatology.* Cambridge: Cambridge University Press.

———. 2009. "Ten dispatches from the chimpanzee culture wars, plus postscript (revisiting the battlefronts)." In *The Question of Animal Culture*, edited by K. N. Laland and B. G. Galef, 41–69. Cambridge, MA: Harvard University Press.

McGrew, W. C., and C. E. G. Tutin. 1978. "Evidence for a social custom in wild chimpanzees?" *Man* 13:234–51.

McLaren, I. A., and T. G. Smith. 1985. "Population ecology of seals: retrospective and prospective views." *Marine Mammal Science* 1:54–83.

McSweeney, D. J., R. W. Baird, S. D. Mahaffy, D. L. Webster, and G. S. Schorr. 2009. "Site fidelity and association patterns of a rare species: pygmy killer whales (*Feresa attenuata*) in the main Hawaiian Islands." *Marine Mammal Science* 25:557–72.

McSweeney, D. J., K. C. Chu, W. F. Dolphin, and L. N. Guinee. 1989. "North Pacific humpback whale songs: a comparison of southeast Alaskan feeding ground songs with Hawaiian wintering ground songs." *Marine Mammal Science* 5:139–48.

Mead, T. 1961. *Killers of Eden: The Story of the Killer Whales of Twofold Bay.* Sydney: Angus and Robertsony.

Melville, H. (1851) 1972. *Moby Dick; or, The Whale.* London: Penguin.

Mesnick, S., N. Warner, J. Straley, V. O'Connell, M. Purves, C. Guinet , J. E. Dyb, C. Lunsford, C. Roche, and N. Gasco. 2006. "Global sperm whale (*Physeter macrocephalus*) depredation of demersal longlines." Abstract. Paper presented at Symposium on Fisheries Depredation by Killer and Sperm Whales: Behavioural Insights, Behavioural Solutions, Pender Island, British Columbia, October 2–5.

Mesnick, S. L., B. L. Taylor, R. G. Le Duc, S. E. Treviño, G. M. O'Corry-Crowe, and A. E. Dizon. 1999. "Culture and genetic evolution in whales." *Science* 284:2055a.

Mesoudi, A., A. Whiten, and K. N. Laland. 2006. "Towards a unified science of cultural evolution." *Behavioral and Brain Sciences* 29:329–46.

Mikhalev, Y. A. 1997. "Humpback whales *Megaptera novaeangliae* in the Arabian Sea." *Marine Ecology Progress Series* 149:13–21.

Miksis, J. L., P. L. Tyack, and J. R. Buck. 2002. "Captive dolphins, *Tursiops truncatus*, develop signature whistles that match acoustic features of human-made model sounds." *Journal of the Acoustical Society of America* 112:728–39.

Miller, G. F. 2000. "Evolution of human music through sexual selection." In *The Origins of Music*, edited by N. L. Wallin, B. Merker, and S. Brown, 360. Cambridge, MA: MIT Press.

Miller, P. J. O., A. D. Shapiro, P. L. Tyack, and A. R. Solow. 2004. "Call-type matching in vocal exchanges of free-ranging resident killer whales, *Orcinus orca*." *Animal Behaviour* 67:1099–1107.

Mirceta, S., A. V. Signore, J. M. Burns, A. R. Cossins, K. L. Campbell, and M. Berenbrink. 2013. "Evolution of mammalian diving capacity traced by myoglobin net surface charge." *Science* 340:1234192.

Mithven, S. 1999. "Imitation and cultural change: a view from the Stone Age, with specific reference to the manufacture of handaxes." *Symposia of the Zoological Society, London* 72:389–99.

Møhl, B., P. T. Madsen, M. Wahlberg, W. W. L. Au, P. E. Nachtigall, and S. H. Ridgway. 2003. "Sound transmission in the spermaceti complex of a recently expired sperm whale calf." *Acoustics Research Letters Online* 4:19–24.

Møhl, B., M. Wahlberg, P. T. Madsen, L. A. Miller, and A. Surlykke. 2000. "Sperm whale clicks: directionality and source level revisited." *Journal of the Acoustical Society of America* 107:638–48.

Moll, H., and M. Tomasello. 2007. "Cooperation and human cognition: the Vygotskian intelligence hypothesis." *Philosophical Transactions of the Royal Society of London*, ser. B 362:639–48.

Moller, L. M., L. B. Beheregaray, R. G. Harcourt, and M. Krutzen. 2001. "Alliance membership and kinship in wild male bottlenose dolphins (*Tursiops aduncus*) of southeastern Australia." *Proceedings of the Royal Society of London*, ser. B 268:1941–47.

Monroe, K. R. 1991. "John Donne's people: explaining differences between rational actors and altruists through cognitive frameworks." *Journal of Politics* 53:394–433.

Montgomery, S. H., J. H. Geisler, M. R. McGowen, C. Fox, L. Marino, and J. Gatesy. 2013. "The evolutionary history of cetacean brain and body size." *Evolution* 67:3339–53.

Moore, A. 1992. *Cultural Anthropology: The Field Study of Human Beings*. San Diego: Collegiate Press.

Morete, M., A. Freitas, M. Engel, R. M. Pace III, and P. Clapham. 2003. "A novel behavior observed in humpback whales on wintering grounds at Abrolhos Bank (Brazil)." *Marine Mammal Science* 19:694–707.

Morgan, C. L. 1894. *An Introduction to Comparative Psychology*. London: Walter Scott.

———. 1903. *An Introduction to Comparative Psychology*. 2nd ed. London: Walter Scott.

Morin, P. A., F. I. Archer, A. D. Foote, J. Vilstrup, E. E. Allen, P. Wade, J. Durban, K. Parsons, R. Pitman, and L. Li. 2010. "Complete mitochondrial genome phylogeographic analysis of killer whales (*Orcinus orca*) indicates multiple species." *Genome Research* 20:908–16.

Morisaka, T., M. Shinohara, F. Nakahara, and T. Akamatsu. 2005. "Geographic variations in the whistles among three Indo-Pacific bottlenose dolphin *Tursiops aduncus* populations in Japan." *Fisheries Science* 71:568–76.

Moura, A. E., C. J. van Rensburg, M. Pilot, A. Tehrani, P. B. Best, M. Thornton, S. Plön, P. J. N. de Bruyn, K. C. Worley, R. A. Gibbs, M. E. Dahlheim, and A. R. Hoelzel. 2014. "Killer whale nuclear genome and mtDNA reveal widespread population bottleneck during the last glacial maximum." *Molecular Biology and Evolution*. Published electronically February 4, 2014. doi: 10.1093/molbev/msu058.

Mourier, J., J. Vercelloni, and S. Planes. 2012. "Evidence of social communities in a spatially structured network of a free-ranging shark species." *Animal Behaviour* 83:389–401.

Mousseau, T. A., and C. W. Fox. 1998. "The adaptive significance of maternal effects." *Trends in Ecology and Evolution* 13:403–7.

Mundinger, P. C. 1980. "Animal cultures and a general theory of cultural evolution." *Ethology and Sociobiology* 1:183–223.

Murray, A., S. Cerchio, R. McCauley, C. S. Jenner, Y. Razafindrakoto, D. Coughran, S. McKay, and H. Rosenbaum. 2012. "Minimal similarity in songs suggests limited exchange between humpback whales (*Megaptera novaeangliae*) in the southern Indian Ocean." *Marine Mammal Science* 28: E41–E57.

Myers, R. A., J. K. Baum, T. D. Shepherd, S. P. Powers, and C. H. Peterson. 2007. "Cascading effects of the loss of apex predatory sharks from a coastal ocean." *Science* 315:1846–50.

Nei, M., T. Maruyama, and R. Chakraborty. 1975. "The bottleneck effect and genetic variability in populations." *Evolution* 29:1–10.

Neil, D. T. 2002. "Cooperative fishing interactions between Aboriginal Australians and dolphins in eastern Australia." *Anthrozoos* 15:3–18.

Nemiroff, L. 2009. "Structural variation and communicative functions of long-finned pilot whale (*Globicephala melas*) pulsed calls and complex whistles." MSc thesis, Dalhousie University.

Nerini, M. K., H. W. Braham, W. M. Marquette, and D. J. Rugh. 1984. "Life history of the bowhead whale, *Balaena mysticetus* (Mammalia: Cetacea)." *Journal of Zoology* 204:443–68.

Nettle, D. 1999. "Language variation and the evolution of societies." In *The Evolution of Culture*, edited by R. I. M. Dunbar, C. Knight, and C. Power, 214–27. Piscataway, NJ: Rutgers University Press.

Noad, M. J., D. H. Cato, M. M. Bryden, M. N. Jenner, and K. C. S. Jenner. 2000. "Cultural revolution in whale songs." *Nature* 408:537.

Norris, K. S. 1991. *Dolphin Days: The Life and Times of the Spinner Dolphin*. New York: Norton.

———. 1994. "Comparative view of cetacean social ecology, culture, and evolution." In *The Hawaiian Spinner Dolphin*, edited by K. S. Norris, B. Würsig, R. S. Wells, and M. Würsig, 301–44. Berkeley: University of California Press.

Norris, K. S., and T. P. Dohl. 1980. "The structure and functions of cetacean schools." In *Cetacean Behavior: Mechanisms and Functions*, edited by L. M. Herman, 211–61. New York: Wiley-Interscience.

Norris, K. S., and C. R. Schilt. 1988. "Cooperative societies in three-dimensional space: on the origins of aggregations, flocks and schools, with special reference to dolphins and fish." *Ethology and Sociobiology* 9:149–79.

Northcutt, R. G. 1977. "Elasmobranch central nervous system organization and its possible evolutionary significance." *American Zoologist* 17:411–29.

Nowacek, D. P., B. M. Casper, R. S. Wells, S. M. Nowacek, and D. A. Mann. 2003. "Intraspecific and geographic variation of West Indian manatee (*Trichechus manatus* spp.) vocalizations (L)." *Journal of the Acoustical Society of America* 114:66–93.

Oftedal, O. T. 1997. "Lactation in whales and dolphins: evidence of divergence between baleen- and toothed-species." *Journal of Mammary Gland Biology and Neoplasia* 2:205–30.

Olesiuk, P., M. A. Bigg, and G. M. Ellis. 1990. "Life history and population dynamics of resident killer whales (*Orcinus orca*) in the coastal waters of British Columbia and Washington State." *Reports of the International Whaling Commission (Special Issue)* 12:209–43.

Orr, H. A. 2009. "Fitness and its role in evolutionary genetics." *Nature Review Genetics* 10:531–39.

Osborne, R. W. 1999. "A historical ecology of Salish Sea 'resident' killer whales (*Orcinus orca*): with implications for management." PhD thesis, University of Victoria.

Oswald, J. N., W. W. L. Au, and F. Duennebier. 2011. "Minke whale (*Balaenoptera acutorostrata*) boings detected at the Station ALOHA Cabled Observatory." *Journal of the Acoustical Society of America* 129:3353–60.

Ottensmeyer, C. A., and H. Whitehead. 2003. "Behavioural evidence for social units in long-finned pilot whales." *Canadian Journal of Zoology* 81:1327–38.

Packard, A. 1972. "Cephalopods and fish: the limits of convergence." *Biological Reviews* 47:241–307.

Palsbøll, P. J., P. J. Clapham, D. K. Matilla, F. Larsen, R. Sears, H. R. Siegismund, J. Sigurjónsson, O. Vasquez, and P. Arctander. 1995. "Distribution of mtDNA haplotypes in North Atlantic humpback whales: the influence of behaviour on population structure." *Marine Ecology Progress Series* 116:1–10.

Palsbøll, P. J., M. P. Heide-Jørgensen, and M. Bérubé. 2002. "Analysis of mitochondrial control region nucleotide sequences from Baffin Bay belugas (*Delphinapterus leucas*): detecting pods or sub-populations?" In *Belugas in the North Atlantic and Russian Arctic*, edited by M. P. Heide-Jørgensen and Ø. Wiig, 39–50. NAMMCO Scientific Publication 4. Tromsø, Norway: North Atlantic Marine Mammal Commission.

Palsbøll, P. J., M. P. Heide-Jørgensen, and R. Dietz. 1997. "Population structure and seasonal movements of narwhals, *Monodon monoceros*, determined from mtDNA analysis." *Heredity* 78:284–92.

Pangerc, T. 2010. "Baleen whale acoustic presence around South Georgia." Ph.D. thesis, University of East Anglia.

Paredes, R., I. L. Jones, and D. J. Boness. 2006. "Parental roles of male and female thick-billed murres and razorbills at the Gannet Islands, Labrador." *Behaviour* 143:451–81.

Parks, S. E. 2003. "Response of North Atlantic right whales (*Eubalaena glacialis*) to playback of calls recorded from surface active groups in both the North and South Atlantic." *Marine Mammal Science* 19:563–80.

Parsons, K. M., J. W. Durban, and D. E. Claridge. 2003. "Male-male aggression renders bottlenose dolphin (*Tursiops truncatus*) unconscious." *Aquatic Mammals* 29:360–62.

Parsons, K. M., J. W. Durban, D. E. Claridge, K. C. Balcomb, L. R. Noble, and P. M. Thompson. 2003. "Kinship as a basis for alliance formation between male bottlenose dolphins, *Tursiops truncatus*, in the Bahamas." *Animal Behaviour* 66:185–94.

Parsons, K. M., J. W. Durban, D. E. Claridge, D. L. Herzing, K. C. Balcomb, and L. R. Noble.

2006. "Population genetic structure of coastal bottlenose dolphins (*Tursiops truncatus*) in the northern Bahamas." *Marine Mammal Science* 22:276–98.

Pastene, L. A., J. Acevedo, M. Goto, A. N. Zerbini, P. Acuna, and A. Aguayo-Lobo. 2010. "Population structure and possible migratory links of common minke whales, *Balaenoptera acutorostrata*, in the Southern Hemisphere." *Conservation Genetics* 11:1553–58.

Patriquin, K. P. 2012. "The causes and consequences of fission-fusion dynamics in female northern long-eared bats (*Myotis septentrionalis*)." PhD thesis, Dalhousie University.

Patterson, I. A. P., R. J. Reid, B. Wilson, K. Grellier, H. M. Ross, and P. M. Thompson. 1998. "Evidence for infanticide in bottlenose dolphins: an explanation for violent interactions with harbour porpoises?" *Proceedings of the Royal Society of London*, ser. B 256:1167–70.

Payne, K. 2000. "The progressively changing songs of humpback whales: a window on the creative process in a wild animal." In *The Origins of Music*, edited by N. L. Wallin, B. Merker, and S. Brown, 135–50. Cambridge, MA: MIT Press.

Payne, K., and R. S. Payne. 1985. "Large-scale changes over 17 years in songs of humpback whales in Bermuda." *Zeitschrift Fur Tierpsychologie* 68:89–114.

Payne, K., P. Tyack, and R. Payne. 1983. "Progressive changes in the songs of humpback whales (*Megaptera novaeangliae*): a detailed analysis of two seasons in Hawaii." In *Communication and Behavior of Whales*, edited by R. Payne, 9–57. Boulder, CO: Westview Press.

Payne, R. 1976. "At home with right whales." *National Geographic* 149:322–41.

———. 1995. *Among Whales*. New York: Simon and Schuster.

Payne, R., and E. M. Dorsey. 1983. "Sexual dimorphism and aggressive use of callosities in right whales (*Eubalaena australis*)." In *Communication and Behavior of Whales*, edited by R. Payne, 295–329. Boulder, CO: Westview Press.

Payne, R., and S. McVay. 1971. "Songs of humpback whales." *Science* 173:587–97.

Payne, R. S., and D. Webb. 1971. "Orientation by means of long-range acoustic signaling in baleen whales." *Annals of the New York Academy of Sciences* 188:110–42.

Pearson, H. C. 2011. "Sociability of female bottlenose dolphins (*Tursiops* spp.) and chimpanzees (*Pan troglodytes*): understanding evolutionary pathways toward social convergence." *Evolutionary Anthropology: Issues, News, and Reviews* 20:85–95.

Pearson, H. C., and D. E. Shelton. 2010. "A large-brained social animal." In *The Dusky Dolphin: Master Acrobat off Different Shores*, edited by B. Würsig, and M. Würsig, 333–53. Amsterdam: Elsevier/Academic Press.

Pelletier, J. 1997. "Analysis and modeling of the natural variability of climate." *Journal of Climate* 10:1331–42.

Peoples, J., and G. Bailey. 1997. *Humanity: An Introduction to Cultural Anthropology*. Belmont, CA: West/Wadsworth.

Perry, G. H., N. J. Dominy, K. G. Claw, A. S. Lee, H. Fiegler, R. Redon, J. Werner, et al. 2007. "Diet and the evolution of human amylase gene copy number variation." *Nature Genetics* 39:1256–60.

Perry, P., M. Baker, L. Fedigan, J. Gros-Louis, K. Jack, K. C. MacKinnon, J. H. Manson,

M. Panger, K. Pyle, and L. Rose. 2003. "Social conventions in wild white-faced capuchin monkeys." *Current Anthropology* 44:241–68.

Perry, S. 2009. "Are non-human primates likely to exhibit cultural capacities like those of humans?" In *The Question of Animal Culture*, edited by K. N. Laland and B. G. Galef Jr., 247–68. Cambridge, MA: Harvard University Press.

Perry, S., and J. H. Manson. 2003. "Traditions in monkeys." *Evolutionary Anthropology* 12:71–81.

Pershing, A. J., L. B. Christensen, N. R. Record, G. D. Sherwood, and P. B. Stetson. 2010. "The impact of whaling on the ocean carbon cycle: why bigger was better." *PLoS ONE* 5:e12444.

Pinker, S. 1994. *The Language Instinct*. New York: William Morrow.

Pitman, R. L. 2011a. "Antarctic killer whales: top of the food chain at the bottom of the world." *Journal of the American Cetacean Society* 40:39–45.

———. 2011b. "An introduction to the world's premier predator." *Journal of the American Cetacean Society* 40:2–5.

———, ed. 2011c. "Killer Whale: The Top, Top Predator." Special issue, *Whalewatcher*, vol. 40, no. 1.

Pitman, R. L., L. T. Balance, S. L. Mesnick, and S. Chivers. 2001. "Killer whale predation on sperm whales: observations and implications." *Marine Mammal Science* 17:494–507.

Pitman, R. L., and S. J. Chivers. 1999. "Terror in black and white." *Natural History* 107:26–29.

Pitman, R. L., and J. W. Durban. 2009. "Save the seal! whales act instinctively to save seals." *Natural History* 9:48–48.

———. 2010. "Killer whale predation on penguins in Antarctica." *Polar Biology* 33:1589–94.

———. 2012. "Cooperative hunting behavior, prey selectivity and prey handling by pack ice killer whales (*Orcinus orca*), type B, in Antarctic Peninsula waters." *Marine Mammal Science* 28:16–36.

Pitman, R. L., and P. Ensor. 2003. "Three forms of killer whales (*Orcinus orca*) in Antarctic waters." *Journal of Cetacean Research and Management* 5:131–39.

Poole, J. H., P. L. Tyack, A. S. Stoeger-Horwath, and S. Watwood. 2005. "Animal behaviour: elephants are capable of vocal learning." *Nature* 434:455–56.

Popper, K. R. 2002. *Conjectures and Refutations: The Growth of Scientific Knowledge*. London: Routledge & Kegan.

Powell, A., S. Shennan, and M. G. Thomas. 2009. "Late Pleistocene demography and the appearance of modern human behavior." *Science* 324:1298–1301.

Premack, D., and M. D. Hauser. 2001. "A whale of a tale: calling it culture doesn't help." *Behavioral and Brain Sciences* 24:350–51.

Premack, D., and A. J. Premack. 1996. "Why animals lack pedagogy and some cultures have more of it than others." In *Handbook of Education and Human Development: New Models of Learning, Teaching and Schooling*, edited by D. R. Olson and N. Torrance, 302–23. Oxford: Blackwell Press.

Pryor, K. W. 2001. "Cultural transmission of behavior in animals: how a modern training technology uses spontaneous social imitation in cetaceans and facilitates social imitation in horses and dogs." *Behavioral and Brain Sciences* 24:352.

Psarakos, S., D. L. Herzing, and K. Marten. 2003. "Mixed-species associations between

Pantropical spotted dolphins (*Stenella attenuata*) and Hawaiian spinner dolphins (*Stenella longirostris*) off Oahu, Hawaii." *Aquatic Mammals* 29:390–95.

Putland, D. A., J. A. Nicholls, M. J. Noad, and A. W. Goldizen. 2006. "Imitating the neighbours: vocal dialect matching in a mimic-model system." *Biology Letters* 2:367–70.

Quick, N. J., and V. M. Janik. 2012. "Bottlenose dolphins exchange signature whistles when meeting at sea." *Proceedings of the Royal Society of London*, ser. B 279:2539–45.

Raihani, N. J., and A. R. Ridley. 2008. "Experimental evidence for teaching in wild pied babblers." *Animal Behaviour* 75:3–11.

Rankin, S., and J. Barlow. 2005. "Source of the North Pacific 'boing' sound attributed to minke whales." *Journal of the Acoustical Society of America* 118:3346–51.

Rasmussen, K., D. M. Palacios, J. Calambokidis, M. T. Saborío, L. Dalla Rosa, E. R. Secchi, G. H. Steiger, J. M. Allen, and G. S. Stone. 2007. "Southern Hemisphere humpback whales wintering off Central America: insights from water temperature into the longest mammalian migration." *Biology Letters* 3:302–5.

Read, A. J. 2005. "Bycatch and depredation." In *Marine Mammal Research: Conservation beyond Crisis*, edited by J. E. Reynolds, W. F. Perrin, R. R. Reeves, S. Montgomery, and T. J. Ragen, 5–17. Baltimore: Johns Hopkins University Press.

Read, D. W. 2011. *How Culture Makes Us Human: Primate Social Evolution and the Formation of Human Societies*. Walnut Creek, CA: Left Coast Press.

Reader, S., and D. Biro. 2010. "Experimental identification of social learning in wild animals." *Learning and Behavior* 38:265–83.

Reeves, R. R., W. F. Perrin, B. L. Taylor, C. S. Baker, and M. Mesnick. 2004. *Report of the Workshop on Shortcomings of Cetacean Taxonomy in Relation to Needs of Conservation and Management, April 30–May 2, 2004, La Jolla, California*. NOAA technical memorandum NOAA-TM-NMFS-SWFSC 363. La Jolla, CA: U.S. Department of Commerce, National Oceanic and Atmospheric Administration, National Marine Fisheries Service, Southwest Fisheries Science Center.

Reeves, R. R., T. D. Smith, and E. A. Josephson. 2007. "Near-annihilation of a species: right whaling in the North Atlantic." In *The Urban Whale: North Atlantic Right Whales at the Crossroads*, edited by S. D. Kraus and R. M. Rolland, 39–74. Cambridge, MA: Harvard University Press.

Reiss, D. 1990. "The dolphin: an alien intelligence." In *First Contact: The Search for Extraterrestrial Intelligence*, edited by B. Bova and B. Preiss, 31–39. New York: NAL Books.

Reiss, D., and B. McCowan. 1993. "Spontaneous vocal mimicry and production by bottlenose dolphins (*Tursiops truncatus*): evidence for vocal learning." *Journal of Comparative Psychology* 107:301–12.

Rendell, L. 2012. "Sperm whale communications and culture." *Journal of the American Cetacean Society* 41:21–27.

Rendell, L., R. Boyd, M. Enquist, M. W. Feldman, L. Fogarty, and K. N. Laland. 2011a. "How copying affects the amount, evenness and persistence of cultural knowledge: insights from the social learning strategies tournament." *Philosophical Transactions of the Royal Society of London*, ser. B 366:1118.

Rendell, L., L. Fogarty, W. J. E. Hoppitt, T. J. H. Morgan, M. M. Webster, and K. N. Laland.

2011b. "Cognitive culture: theoretical and empirical insights into social learning strategies." *Trends in Cognitive Sciences* 15:68–76.

Rendell, L., L. Fogarty, and K. N. Laland. 2011. "Runaway cultural niche construction." *Philosophical Transactions of the Royal Society of London*, ser. B 366:823–35.

Rendell, L., S. L. Mesnick, M. L. Dalebout, J. Burtenshaw, and H. Whitehead. 2012. "Can genetic differences explain vocal dialect variation in sperm whales, *Physeter macrocephalus?*" *Behavior Genetics* 42:332–43.

Rendell, L., and H. Whitehead. 2001a. "Cetacean culture: still afloat after the first naval engagement of the culture wars." *Behavioral and Brain Sciences* 24:360–73.

———. 2001b. "Culture in whales and dolphins." *Behavioral and Brain Sciences* 24:309–24.

———. 2003. "Vocal clans in sperm whales (*Physeter macrocephalus*)." *Proceedings of the Royal Society of London*, ser. B 270:225–31.

———. 2004. "Do sperm whales share coda vocalizations? insights into coda usage from acoustic size measurement." *Animal Behaviour* 67:865–74.

———. 2005. "Spatial and temporal variation in sperm whale coda vocalisations: stable usage and local dialects." *Animal Behaviour* 70:191–98.

Reynolds, J. E., III, and J. R. Wilcox. 1986. "Distribution and abundance of the West Indian manatee *Trichechus manatus* around selected Florida power plants following winter cold fronts: 1984–1985." *Biological Conservation* 38:103–13.

Rice, D. W. 2009. "Baleen." In *Encyclopedia of Marine Mammals*, edited by W. F. Perrin, B. Würsig, and J. G. M. Thewissen, 78–80. 2nd ed. San Diego: Academic Press.

Richards, D. G. 1986. "Dolphin vocal mimicry and vocal object labelling." In *Dolphin Cognition and Behaviour: A Comparative Approach*, edited by F. G. Wood, 273–88. Hillsdale, NJ: Lawrence Erlbaum.

Richerson, P. J., and R. Boyd. 2005. *Not by Genes Alone: How Culture Transformed Human Evolution*. Chicago: Chicago University Press.

———. 2013. "Rethinking paleoanthropology: a world queerer than we supposed." In *Evolution of Mind*, edited by G. Hatfield and H. Pittman, 263–302. Philadelphia: University of Pennsylvania Museum of Archaeology and Anthropology. Distributed by University of Pennsylvania Press.

Richerson, P. J., R. Boyd, and J. Henrich. 2010. "Gene-culture coevolution in the age of genomics." *Proceedings of the National Academy of Sciences of the United States of America* 107:8985–92.

Ridgway, S. H., and W. W. L. Au. 1999. "Hearing and echolocation: dolphin." In *Elsevier's Encyclopedia of Neuroscience*, edited by G. Adelman and B. H. Smith, 858–62. Amsterdam: Elsevier Science.

Ridgway, S., D. Carder, M. Jeffries, and M. Todd. 2012. "Spontaneous human speech mimicry by a cetacean." *Current Biology* 22: R860–R861.

Riesch, R., L. G. Barrett-Lennard, G. M. Ellis, J. K. B. Ford, and V. B. Deecke. 2012. "Cultural traditions and the evolution of reproductive isolation: ecological speciation in killer whales?" *Biological Journal of the Linnean Society* 106:1–17.

Riesch, R., J. K. B. Ford, and F. Thomsen. 2006. "Stability and group specificity of stereotyped whistles in resident killer whales, *Orcinus orca*, off British Columbia." *Animal Behaviour* 71:79–91.

Ritchie, E. G., and C. N. Johnson. 2009. "Predator interactions, mesopredator release and biodiversity conservation." *Ecology Letters* 12:982–98.

Ritter, F. 2007. "Behavioral responses of rough-toothed dolphins to a dead newborn calf." *Marine Mammal Science* 23:429–33.

Rizzolatti, G., and L. Craighero. 2004. "The mirror-neuron system." *Annual Review of Neuroscience* 27:169–92.

Robineau, D. 1995. "Upon the so-called symbiosis between the imragen fishermen of Mauritania and the dolphins." *Mammalia* 59:460–63.

Robson, F. D. 1988. *Pictures in the Dolphin Mind*. Auckland: Reed Methuen.

———. 1984. *Strandings: Ways to Save Whales—a Humane Conservationist's Guide*. Johannesburg: Science Press.

Robson, F. D., and P. J. H. van Bree. 1971. "Some remarks on a mass stranding of sperm whales, *Physeter macrocephalus*, near Gisborne, New Zealand, on March 18, 1970." *Zeitschrift für Saugetierkunde* 36:55–60.

Rogers, A. R. 1988. "Does biology constrain culture?" *American Anthropologist* 90:819–31.

Rossbach, K. A., and D. L. Herzing. 1997. "Underwater observations of benthic-feeding bottlenose dolphins (*Tursiops truncatus*) near Grand Bahama Island, Bahamas." *Marine Mammal Science* 13:498–504.

Rothenberg, D. 2010. *Thousand Mile Song: Whale Music in a Sea of Sound*. New York: Basic Books.

Rowlands, M. 2012. "The kindness of beasts." *Aeon*, October 24, http://www.aeonmagazine.com/being-human/mark-rowlands-animal-morality/.

Ryan, S. J. 2006. "The role of culture in conservation planning for small or endangered populations." *Conservation Biology* 20:1321–24.

Sacks, B. N., D. L. Bannasch, B. B. Chomel, and H. B. Ernest. 2008. "Coyotes demonstrate how habitat specialization by individuals of a generalist species can diversify populations in a heterogeneous ecoregion." *Molecular Biology and Evolution* 25:1384–94.

Samuels, A., and P. Tyack. 2000. "Flukeprints: a history of studying cetacean societies." In *Cetacean Societies*, edited by J. Mann, R. C. Connor, P. L. Tyack, and H. Whitehead, 9–44. Chicago: University of Chicago Press.

Sapolsky, R. M., and L. J. Share. 2004. "A pacific culture among wild baboons: its emergence and transmission." *Public Library of Science Biology* 2:534–41.

Sargeant, B. L., and J. Mann. 2009. "From social learning to culture: intrapopulation variation in bottlenose dolphins". In *The Question of Animal Culture*, edited by K. N. Laland and B. G. Galef Jr., 152–73. Cambridge, MA: Harvard University Press.

Sargeant, B. L., J. Mann, P. Berggren, and M. Krützen. 2005. "Specialization and development of beach hunting, a rare foraging behavior, by wild bottlenose dolphins (*Tursiops* sp.)." *Canadian Journal of Zoology* 83:1400–1410.

Sargeant, B. L., A. J. Wirsing, M. R. Heithaus, and J. Mann. 2007. "Can environmental heterogeneity explain individual foraging variation in wild bottlenose dolphins (*Tursiops* sp.)?" *Behavioral Ecology and Sociobiology* 61:679–88.

Saulitis, E. 2013. *Into Great Silence: A Memoir of Discovery and Loss among Vanishing Orcas*. Boston: Beacon Press.

Sayigh, L. S., H. C. Esch, R. S. Wells, and V. M. Janik. 2007. "Facts about signature whistles of bottlenose dolphins, *Tursiops truncatus*." *Animal Behaviour* 74:1631–42.

Sayigh, L. S., P. L. Tyack, R. S. Wells, and M. D. Scott. 1990. "Signature whistles of free-ranging bottlenose dolphins, *Tursiops truncatus:* stability and mother-offspring comparisons." *Behavioural Ecology and Sociobiology* 26:247–60.

Sayigh, L. S., P. L. Tyack, R. S. Wells, M. D. Scott, and A. B. Irvine. 1995. "Sex difference in signature whistle production of free-ranging bottlenose dolphins, *Tursiops truncatus*." *Behavioural Ecology and Sociobiology* 36:171–77.

Schorr, G. S., E. A. Falcone, D. J. Moretti, and R. D. Andrews. 2014. "First long-term behavioral records from Cuvier's beaked whales (*Ziphius cavirostris*) reveal record-breaking dives." *PLoS ONE* e92633.

Schultz, E. A., and R. H. Lavenda. 2009. *Cultural Anthropology: A Perspective on the Human Condition*. 7th ed. New York: Oxford University Press.

Schulz, T. M., H. Whitehead, S. Gero, and L. Rendell. 2008. "Overlapping and matching of codas in vocal interactions between sperm whales: insights into communication function." *Animal Behaviour* 76:1977–88.

———. 2010. "Individual vocal production in a sperm whale (*Physeter macrocephalus*) social unit." *Marine Mammal Science* 27:149–66.

Schusterman, R. J. 1978. "Vocal communication in pinnipeds." In *Behavior of Captive Wild Animals*, edited by H. Markowitz and V. J. Stevens, 247–308. Chicago: Nelson-Hall.

Schusterman, R. J., R. F. Balliet, and S. St. John. 1970. "Vocal displays under water by the gray seal, the harbor seal, and the Steller sea lion." *Psychonomic Science* 18:303–5.

Scott, T. M., and S. S. Sadove. 1997. "Sperm whale, *Physeter macrocephalus*, sightings in the shallow shelf waters off Long Island, New York." *Marine Mammal Science* 13:317–21.

Seppänen, J. T., and J. T. Forsman. 2007. "Interspecific social learning: novel preference can be acquired from a competing species." *Current Biology* 17:1248–52.

Sergeant, D. E. 1982. "Mass strandings of toothed whales (Odontoceti) as a population phenomenon." *Scientific Reports of the Whales Research Institute* 34:1–47.

Sharpe, F. 2001. "Social foraging of the Southeast Alaskan humpback whale." PhD thesis, Simon Fraser University.

Sharpe, F., and L. Dill. 1997. "The behavior of Pacific herring schools in response to artificial humpback whale bubbles." *Canadian Journal of Zoology* 75:725–30.

Sheldon, R. W., A. Prakash, and W. H. Sutcliffe. 1972. "The size distribution of particles in the ocean." *Limnology and Oceanography* 17:327–40.

Sherry, D. F., and B. G. Galef Jr. 1984. "Cultural transmission without imitation: milk bottle opening by birds." *Animal Behaviour* 32:937–38.

Shettleworth, S. J. 1998. *Cognition, Evolution and Behaviour*. New York: Oxford University Press.

Siemann, L. A. 1994. "Mitochondrial DNA sequence variation in North Atlantic long-finned pilot whales, *Globicephala melas*." PhD thesis, Massachusetts Institute of Technology.

Sigler, M. F., C. R. Lunsford, J. M. Straley, and J. B. Liddle. 2008. "Sperm whale depredation of sablefish longline gear in the northeast Pacific Ocean." *Marine Mammal Science* 24:16–27.

Silber, G. K., and D. Fertl. 1995. "Intentional beaching by bottlenose dolphins (*Tursiops truncatus*) in the Colorado River Delta, Mexico." *Aquatic Mammals* 21:183–86.

Silva, M. A., R. Prieto, S. Magalhães, R. Cabecinhas, A. Cruz, J. Gonçalves, and R. Santos. 2003. "Occurrence and distribution of cetaceans in the waters around the Azores (Portugal), Summer and Autumn 1999–2000." *Aquatic Mammals* 29:77–83.

Similä, T., and F. Ugarte. 1993. "Surface and underwater observations of cooperatively feeding killer whales in northern Norway." *Canadian Journal of Zoology* 71:1494–99.

Simmonds, M. P. 1991. "Cetacean mass mortalities and their potential relationship with pollution." In *Symposium: Whales: Biology—Threats—Conservation*, edited by J. J. Symoens, 217–45. Brussels: Royal Academy of Overseas Sciences.

———. 1997. "The meaning of cetacean strandings." *Bulletin de l'Institut Royal des Sciences Naturelles de Belgique Biologie* 67-SUPPL: 29–34.

Simöes-Lopes, P. C., M. E. Fabián, and J. O. Menegheti. 1998. "Dolphin interactions with the mullet artisanal fishing on southern Brazil: a qualitative and quantitative approach." *Revista Brasileira de Zoologia* 15:709–26.

Simon, M., M. B. Hanson, L. Murrey, J. Tougaard, and F. Ugarte. 2009. "From captivity to the wild and back: an attempt to release Keiko the killer whale." *Marine Mammal Science* 25:693–705.

Simon, M., F. Ugarte, M. Wahlberg, and L. A. Miller. 2006. "Icelandic killer whales *Orcinus orca* use a pulsed call suitable for manipulating the schooling behaviour of herring *Clupea harengus*." *Bioacoustics* 16:57–74.

Sjare, B., and I. Stirling. 1996. "The breeding behavior of Atlantic walruses, *Odobenus rosmarus rosmarus*, in the Canadian High Arctic." *Canadian Journal of Zoology* 74:897–911.

Slagsvold, T., and K. L. Wiebe. 2011. "Social learning in birds and its role in shaping a foraging niche." *Philosophical Transactions of the Royal Society of London*, ser. B 366:969–77.

Slater, P. J. B. 2001. "There's CULTURE and 'Culture.'" *Behavioral and Brain Sciences* 24:356–57.

Smith, F. A., A. G. Boyer, J. H. Brown, D. P. Costa, T. Dayan, S. K. M. Ernest, A. R. Evans, et al. 2010. "The evolution of maximum body size of terrestrial mammals." *Science* 330:1216–19.

Smith, J. N., A. W. Goldizen, R. A. Dunlop, and M. J. Noad. 2008. "Songs of male humpback whales, *Megaptera novaeangliae*, are involved in intersexual interactions." *Animal Behaviour* 76:467–77.

Smith, K., and S. Kirby. 2008. "Cultural evolution: implications for understanding the human language faculty and its evolution." *Philosophical Transactions of the Royal Society of London*, ser. B 363:3591–3603.

Smolker, R., and J. W. Pepper. 1999. "Whistle convergence among allied male bottlenose dolphins (Delphinidae, *Tursiops* sp.)." *Ethology* 105:595–617.

Smolker, R. A., A. F. Richards, R. C. Connor, J. Mann, and P. Berggren. 1997. "Sponge-carrying by Indian Ocean bottlenose dolphins: possible tool-use by a delphinid." *Ethology* 103:454–65.

Sober, E. 2005. "Comparative psychology meets evolutionary biology: Morgan's canon and cladistic parsimony." In *Thinking with Animals: New Perspectives on Anthropomorphism*, edited by L. Daston and G. Mitman, 85–99. New York: Columbia University Press.

Sousa-Lima, R., A. P. Paglia, and G. A. B. Da Fonseca. 2002. "Signature information and individual recognition in the isolation calls of Amazonian manatees, *Trichechus inunguis* (Mammalia: Sirenia)." *Animal Behaviour* 63:301–10.

Spector, D. A. 1994. "Definition in biology: the case of 'bird song.'" *Journal of Theoretical Biology* 168:373–81.

Sperber, D. 2006. "Why a deep understanding of cultural evolution is incompatible with shallow psychology." In *Roots of Human Sociality*, edited by N. Enfield and S. Levinson, 431–49. New York: Berg.

Springer, A. M., J. A. Estes, G. B. van Vliet, T. M. Williams, D. F. Doak, E. M. Danner, K. A. Firney, and B. Pfister. 2003. "Sequential megafaunal collapse in the North Pacific Ocean: a legacy of industrial whaling?" *Proceedings of the National Academy of Sciences of the United States of America* 100:12223–28.

Stafford, K. M., C. G. Fox, and D. S. Clark. 1998. "Long-range acoustic detection and localization of blue whale calls in the northeast Pacific Ocean." *Journal of the Acoustical Society of America* 104:3616–25.

Stafford, K. M., S. E. Moore, K. L. Laidre, and M. Heide-Jørgensen. 2008. "Bowhead whale springtime song off West Greenland." *Journal of the Acoustical Society of America* 124:3315–23.

Stearns, S. C., and R. F. Hoekstra. 2000. *Evolution*. Oxford: Oxford University Press.

Steele, J. H. 1985. "A comparison of terrestrial and marine ecological systems." *Nature* 313:355–58.

Sterelny, K. 2009. "Peacekeeping in the culture wars." In *The Question of Animal Culture*, edited by K. N. Laland and B. G. Galef Jr., 288–304. Cambridge, MA: Harvard University Press.

Stevens, J. D., R. Bonfil, N. K. Dulvy, and P. A. Walker. 2000. "The effects of fishing on sharks, rays, and chimaeras (chondrichthyans), and the implications for marine ecosystems." *ICES Journal of Marine Science: Journal du Conseil* 57:476–94.

Stevick, P. T., B. J. McConnell, and P. S. Hammond. 2002. "Patterns of movement." In *Marine Mammal Biology: An Evolutionary Approach*, edited by A. R. Hoelzel, 185–216. Oxford: Blackwell.

Stewart, R. E. A., and F. H. Fay. 2001. "Walrus." In *The New Encyclopedia of Mammals*, edited by D. MacDonald, 174–79. 2nd ed. Oxford: Oxford University Press.

Strager, H. 1995. "Pod-specific call repertoires and compound calls of killer whales (*Orcinus orca* Linnaeus 1758), in the waters off northern Norway." *Canadian Journal of Zoology* 73:1037–47.

Straley, J. 2012. "Sperm whales and fisheries: an Alaskan perspective of a global problem." *Journal of the American Cetacean Society* 41:38–41.

Strimling, P., M. Enquist, and K. Eriksson. 2009. "Repeated learning makes cultural evolution unique." *Proceedings of the National Academy of Sciences of the United States of America* 106:13870–74.

Sullivan, J. J. 2013. "One of us." *Lapham's Quarterly* 6:191–98.

Suzuki, R., J. R. Buck, and P. L. Tyack. 2006. "Information entropy of humpback whale songs." *Journal of the Acoustical Society of America* 119:1849–66.

Svensson, E., J. Abbott, and R. Härdling. 2005. "Female polymorphism, frequency dependence, and rapid evolutionary dynamics in natural populations." *American Naturalist* 165:567–76.

Swartz, S. 1986. "Gray whale migratory, social and breeding behavior." *Reports of the International Whaling Commission (Special Issue)* 8:207–29.

Számadó, S., and E. Szathmáry. 2006. "Selective scenarios for the emergence of natural language." *Trends in Ecology and Evolution* 21:555–61.

Tayler, C. K., and G. S. Saayman. 1973. "Imitative behaviour by Indian Ocean bottlenose dolphins (*Tursiops aduncus*) in captivity." *Behaviour* 44:286–98.

Teaney, D. O. 2004. "The insignificant killer whale: a case study of inherent flaws in the wildlife services' distinct population segment policy and a proposed solution." *Environmental Law* 34:647–1247.

Tebbich, S., M. Taborsky, B. Fessl, and D. Blomqvist. 2001. "Do woodpecker finches acquire tool-use by social learning?" *Proceedings of the Royal Society of London*, ser. B 268:2189–93.

Templeton, C. N., A. A. Ríos-Chelén, E. Quirós-Guerrero, N. I. Mann, and P. J. B. Slater. 2013. "Female happy wrens select songs to cooperate with their mates rather than confront intruders." *Biology Letters* 9:20120863.

Terkel, J. 1996. "Cultural transmission of feeding behavior in the black rat (*Rattus rattus*)." In *Social Learning in Animals: The Roots of Culture*, edited by C. M. Heyes and B. G. Galef Jr., 17–47. San Diego: Academic Press.

Tervo, O. M., S. E. Parks, M. F. Christoffersen, L. A. Miller, and R. M. Kristensen. 2011. "Annual changes in the winter song of bowhead whales (*Balaena mysticetus*) in Disko Bay, Western Greenland." *Marine Mammal Science* 27:E241–E252.

Texier, P., G. Porraz, J. Parkington, J. Rigaud, C. Poggenpoel, C. Miller, C. Tribolo, et al. 2010. "A Howiesons Poort tradition of engraving ostrich eggshell containers dated to 60,000 years ago at Diepkloof Rock Shelter, South Africa." *Proceedings of the National Academy of Sciences of the United States of America* 107:6180–85.

Thewissen, J. 2009. "Archaeocetes, archaic." In *Encyclopedia of Marine Mammals*, edited by W. F. Perrin, B. Würsig and J. G. M. Thewissen, 46–48. 2nd ed. San Diego: Academic Press.

Thinh, V. N., C. Hallam, C. Roos, and K. Hammerschmidt. 2011. "Concordance between vocal and genetic diversity in crested gibbons." *BMC Evolutionary Biology* 11:36.

Thompson, P. O., L. T. Findley, and O. Vidal. 1992. "20-Hz pulses and other vocalizations of fin whales, *Balaenoptera physalus*, in the Gulf of California, Mexico." *Journal of the Acoustical Society of America* 92:3051–57.

Thomsen, F., D. Franck, and J. K. Ford. 2002. "On the communicative significance of whistles in wild killer whales (*Orcinus orca*)." *Naturwissenschaften* 89:404–7.

Thornton, A., and K. McAuliffe. 2006. "Teaching in wild meerkats." *Science* 313:227–29.

Thornton, A., and N. J. Raihani. 2008. "The evolution of teaching." *Animal Behaviour* 75:1823–36.

Thorpe, W. H. 1961. *Bird-Song: The Biology of Vocal Communication and Expression in Birds*. Oxford: Oxford University Press.

Tiedemann, R., and M. Milinkovitch. 1999. "Culture and genetic evolution in whales." *Science* 284:2055a.

Timmermann, A., M. Latif, R. Voss, and A. Grötzner. 1998. "Northern hemispheric interdecadal variability: a coupled air-sea mode." *Journal of Climate* 11:1906–31.

Tint Tun. 2004. "Irrawaddy dolphins in Hsthe-Mandalay segment of the Ayeyawady River and cooperative fishing between Irrawaddy dolphin, *Orcaella brevirostris*, and cast-net fishermen in Myanmar." Report to the Wildlife Conservation Society, Bronx, NY 16195. Available at https://sites.google.com/site/tinttunmm/irrawaddydolphin.

———. 2005. "Castnet fisheries in cooperation with Irrawaddy dolphins (Ayeyawady Dolphins) at Hsthe, Myitkangyi and Myazun Villages, Mandalay Division, in Myanmar." Report to the Wildlife Conservation Society, Bronx, NY 16194. Available at https://sites.google.com/site/tinttunmm/irrawaddydolphin.

Tishkoff, S. A., F. A. Reed, A. Ranciaro, B. F. Voight, C. C. Babbitt, J. S. Silverman, K. Powell, et al. 2007. "Convergent adaptation of human lactase persistence in Africa and Europe." *Nature Genetics* 39:31–40.

Tomasello, M. 1994. "The question of chimpanzee culture." In *Chimpanzee Cultures*, edited by R. W. Wrangham, W. C. McGrew, F. B. M. de Waal, and P. G. Heltne, 301–17. Cambridge, MA: Harvard University Press.

———. 1999a. *The Cultural Origins of Human Cognition*. Cambridge, MA: Harvard University Press.

———. 1999b. "The human adaptation for culture." *Annual Review of Anthropology* 28:509–29.

———. 2009. "The question of chimpanzee culture, plus postscript." In *The Question of Animal Culture*, edited by K. N. Laland and B. G. Galef Jr., 198–221. Cambridge, MA: Harvard University Press.

Tomasello, M., M. Carpenter, and U. Liszkowski. 2007. "A new look at infant pointing." *Child Development* 78:705–22.

Torres, L. G., and A. J. Read. 2009. "Where to catch a fish? the influence of foraging tactics on the ecology of bottlenose dolphins (*Tursiops truncatus*) in Florida Bay, Florida." *Marine Mammal Science* 25:797–815.

Torres, L. G., P. E. Rosel, C. D'Agrosa, and A. J. Read. 2003. "Improving management of overlapping bottlenose dolphin ecotypes through spatial analysis and genetics." *Marine Mammal Science* 19:502–14.

Tost, J., ed. 2008. *Epigenetics*. Norfolk, UK: Caister Academic Press.

Traill, L. W., C. J. A. Bradshaw, and B. W. Brook. 2007. "Minimum viable population size: a meta-analysis of 30 years of published estimates." *Biological Conservation* 139:159–66.

Trivers, R. 1985. *Social Evolution*. Menlo Park, CA: Benjamin/Cummings.

Tyack, P. 1983. “Differential response of humpback whales, *Megaptera novaeangliae*, to playback of song or social sounds.” *Behavioural Ecology and Sociobiology* 13:49–55.

———. 1986. “Whistle repertoires of two bottlenosed dolphins, *Tursiops truncatus*: mimicry of signature whistles?” *Behavioural Ecology and Sociobiology* 18:251–57.

———. 2001. “Cetacean culture: humans of the sea?” *Behavioral and Brain Sciences* 24:358–59.

Tyack, P. L., and E. H. Miller. 2002. “Vocal anatomy, acoustic communication and echolocation.” In *Marine Mammal Biology: An Evolutionary Approach*, edited by A. R. Hoelzel, 142–84. Oxford: Blackwell.

Tyack, P., and H. Whitehead. 1983. “Male competition in large groups of wintering humpback whales.” *Behaviour* 83:132–54.

Tylor, E. B. 1871. *Primitive Culture*. London: Murray.

Tyne, J. A., N. R. Loneragan, A. M. Kopps, S. J. Allen, M. Krützen, and L. Bejder. 2012. “Ecological characteristics contribute to sponge distribution and tool use in bottlenose dolphins *Tursiops* sp.” *Marine Ecology Progress Series* 444:143–53.

Uhen, M. D. 2007. “Evolution of marine mammals: back to the sea after 300 million years.” *Anatomical Record: Advances in Integrative Anatomy and Evolutionary Biology* 290:514–22.

———. 2010. “The origin(s) of whales.” *Annual Review of Earth and Planetary Sciences* 38:189–219.

Urian, K. W., S. Hofmann, R. S. Wells, and A. J. Read. 2009. “Fine-scale population structure of bottlenose dolphins (*Tursiops truncatus*) in Tampa Bay, Florida.” *Marine Mammal Science* 25:619–38.

Valenzuela, L. O., M. Sironi, V. J. Rowntree, and J. Seger. 2009. “Isotopic and genetic evidence for culturally inherited site fidelity to feeding grounds in southern right whales (*Eubalaena australis*).” *Molecular Ecology* 18:782–91.

van Schaik, C. 2006. “Why are some animals so smart?” *Scientific American* 294 (4): 64–71.

van Schaik, C. P., M. Ancrenaz, G. Borgen, B. Galdikas, C. D. Knott, I. Singleton, A. Suzuki, S. S. Utami, and M. Merrill. 2003. “Orangutan cultures and the evolution of material culture.” *Science* 299:102–5.

Vaughn, R. L., M. Degrati, and C. J. McFadden. 2010. “Dusky dolphins foraging in daylight.” In *The Dusky Dolphin: Master Acrobat off Different Shores*, edited by B. G. Würsig and M. Würsig, 115–32. Amsterdam: Elsevier/Academic Press.

Visser, I. N. 1999. “Benthic foraging on stingrays by killer whales (*Orcinus orca*) in New Zealand waters.” *Marine Mammal Science* 15:220–27.

Vygotsky, L. 1978. *Mind in Society*. Cambridge, MA: Harvard University Press.

Warner, R. R. 1988. “Traditionality of mating-site preferences in a coral reef fish.” *Nature* 335:719–21.

Watkins, W. A., M. A. Daher, J. E. George, and D. Rodriguez. 2004. “Twelve years of tracking 52-Hz whale calls from a unique source in the North Pacific.” *Deep Sea Research Part 1: Oceanographic Research Papers* 51:1889–1901.

Watkins, W. A., and W. E. Schevill. 1977. “Sperm whale codas.” *Journal of the Acoustical Society of America* 62:1486–90.

Watkins, W. A., P. Tyack, K. E. Moore, and J. E. Bird. 1987. "The 20-Hz signals of finback whales (*Balaenoptera physalus*)." *Journal of the Acoustical Society of America* 82:1901–12.

Watwood, S. L., P. O. Miller, M. Johnson, P. T. Madsen, and P. L. Tyack. 2006. "Deep-diving foraging behaviour of sperm whales (*Physeter macrocephalus*)." *Journal of Animal Ecology* 75:814–25.

Watwood, S. L., P. L. Tyack, and R. S. Wells. 2004. "Whistle sharing in paired male bottlenose dolphins, *Tursiops truncatus*." *Behavioral Ecology and Sociobiology* 55:531–43.

Weihs, D. 2004. "The hydrodynamics of dolphin drafting." *Journal of Biology* 3:1–16.

Weilgart, L. S. 2007. "The impacts of anthropogenic noise on cetaceans and implications for management." *Canadian Journal of Zoology* 85:1091–1116.

Weilgart, L. S., and H. Whitehead. 1988. "Distinctive vocalizations from mature male sperm whales (*Physeter macrocephalus*)." *Canadian Journal of Zoology* 66:1931–37.

Weinrich, M. T. 1991. "Stable social associations among humpback whales (*Megaptera novaeangliae*) in the Southern Gulf of Maine." *Canadian Journal of Zoology* 69:3012–18.

Weinrich, M. T., H. Rosenbaum, C. S. Baker, A. L. Blackmer, and H. Whitehead. 2006. "The influence of maternal lineages on social affiliations among humpback whales (*Megaptera novaeangliae*) on their feeding grounds in the Southern Gulf of Maine." *Journal of Heredity* 97:226–34.

Weinrich, M. T., M. R. Schilling, and C. R. Belt. 1992. "Evidence for acquisition of a novel feeding behaviour: lobtail feeding in humpback whales, *Megaptera novaeangliae*." *Animal Behaviour* 44:1059–72.

Weller, D. W., B. Würsig, H. Whitehead, J. C. Norris, S. K. Lynn, R. W. Davis, N. Clauss, and P. Brown. 1996. "Observations of an interaction between sperm whales and short-finned pilot whales in the Gulf of Mexico." *Marine Mammal Science* 12:588–94.

Wellings, H. P. 1964. *Shore Whaling at Twofold Bay: Assisted by the Renowned Killer Whales*. Eden, Australia: Magnetic Voice.

Wellman, B. 2001. "Computer networks as social networks." *Science* 293:2031–34.

Wellman, H. M., and J. G. Miller. 2008. "Including deontic reasoning as fundamental to theory of mind." *Human Development* 51:105–35.

Wells, R. S., K. Bassos-Hull, and K. S. Norris. 1998. "Experimental return to the wild of two bottlenose dolphins." *Marine Mammal Science* 14:51–71.

Wells, R. S., M. D. Scott, and A. B. Irvine. 1987. "The social structure of free-ranging bottlenose dolphins." In *Current Mammalogy*, edited by H. H. Genoways, 1:247–305. New York: Plenum Press.

West, M. J., A. P. King, and D. J. White. 2003. "Discovering culture in birds: the role of learning and development." In *Animal Social Complexity: Intelligence, Culture, and Individualized Societies*, edited by F. B. M. de Waal and P. L. Tyack, 470–92. Cambridge, MA: Harvard University Press.

West, S. A., C. El Mouden, and A. Gardner. 2011. "Sixteen common misconceptions about the evolution of cooperation in humans." *Evolution and Human Behavior* 32:231–62.

Wheatcroft, D., and T. D. Price. 2013. "Learning and signal copying facilitate communication among bird species." *Proceedings of the Royal Society of London*, ser. B 280:20123070.

Wheeler, P. E. 1991. "The thermoregulatory advantages of hominid bipedalism in open equatorial environments: the contribution of increased convective heat loss and cutaneous evaporative cooling." *Journal of Human Evolution* 21:107–15.

White, T. I. 2007. *In Defense of Dolphins: The New Moral Frontier*. Malden, MA: Blackwell.

Whitehead, H. 1983. "Structure and stability of humpback whale groups off Newfoundland." *Canadian Journal of Zoology* 61:1391–97.

———. 1996. "Babysitting, dive synchrony, and indications of alloparental care in sperm whales." *Behavioural Ecology and Sociobiology* 38:237–44.

———. 1998. "Cultural selection and genetic diversity in matrilineal whales." *Science* 282:1708–11.

———. 1999. "Variation in the visually observable behavior of groups of Galápagos sperm whales." *Marine Mammal Science* 15:1181–97.

———. 2002. "Estimates of the current global population size and historical trajectory for sperm whales." *Marine Ecology Progress Series* 242:295–304.

———. 2003a. "Society and culture in the deep and open ocean: the sperm whale." In *Animal Social Complexity: Intelligence, Culture and Individualized Societies*, edited by F. B. M. de Waal and P. L. Tyack, 444–64. Cambridge, MA: Harvard University Press.

———. 2003b. *Sperm Whales: Social Evolution in the Ocean*. Chicago: Chicago University Press.

———. 2005. "Genetic diversity in the matrilineal whales: models of cultural hitchhiking and group-specific non-heritable demographic variation." *Marine Mammal Science* 21:58–79.

———. 2007. "Learning, climate and the evolution of cultural capacity." *Journal of Theoretical Biology* 245:341–50.

———. 2009a. "Estimating abundance from one-dimensional passive acoustic surveys." *Journal of Wildlife Management* 73:1000–1009.

———. 2009b. "How might we study culture? a perspective from the ocean." In *The Question of Animal Culture*, edited by K. N. Laland and B. G. Galef Jr., 125–51. Cambridge, MA: Harvard University Press.

———. 2010. "Conserving and managing animals that learn socially and share cultures." *Learning and Behavior* 38:329–36.

———. 2011. "The cultures of whales and dolphins." In *Whales and Dolphins: Cognition, Culture, Conservation and Human Perceptions*, edited by P. Brakes and M. P. Simmonds, 149–65. London: Earthscan.

———, ed. 2012. "Sperm Whale: Whale of Extremes." Special issue, *Whalewatcher*, vol. 41, no. 1.

Whitehead, H., R. Antunes, S. Gero, S. N. P. Wong, D. Engelhaupt, and L. Rendell. 2012. "Multilevel societies of female sperm whales (*Physeter macrocephalus*) in the Atlantic and Pacific: why are they so different?" *International Journal of Primatology* 33:1142–64.

Whitehead, H., and C. Carlson. 1988. "Social behaviour of feeding finback whales off Newfoundland: comparisons with the sympatric humpback whale." *Canadian Journal of Zoology* 66:221.

Whitehead, H., and C. Glass. 1985. "Orcas (killer whales) attack humpback whales." *Journal of Mammalogy* 66:183–85.

Whitehead, H., and D. Lusseau. 2012. "Animal social networks as substrate for cultural behavioural diversity." *Journal of Theoretical Biology* 294:19–28.

Whitehead, H., C. D. MacLeod, and P. Rodhouse. 2003. "Differences in niche breadth among some teuthivorous mesopelagic marine mammals." *Marine Mammal Science* 19:400–406.

Whitehead, H., and J. Mann. 2000. "Female reproductive strategies of cetaceans." In *Cetacean Societies*, edited by J. Mann, R. Connor, P. L. Tyack, and H. Whitehead, 219–46. Chicago: University of Chicago Press.

Whitehead, H., and R. Reeves. 2005. "Killer whales and whaling: the scavenging hypothesis." *Biology Letters* 1:415–18.

Whitehead, H., and L. Rendell. 2004. "Movements, habitat use and feeding success of cultural clans of South Pacific sperm whales." *Journal of Animal Ecology* 73:190–96.

Whitehead, H., L. Rendell, R. W. Osborne, and B. Würsig. 2004. "Culture and conservation of non-humans with reference to whales and dolphins: review and new directions." *Biological Conservation* 120:431–41.

Whitehead, H., and P. Richerson. 2009. "The evolution of conformist social learning can cause population collapse in realistically variable environments." *Evolution and Human Behavior* 30:261–73.

Whitehead, H., P. J. Richerson, and R. Boyd. 2002. "Cultural selection and genetic diversity in humans." *Selection* 3:115–25.

Whiten, A. 2001. "Imitation and cultural transmission in apes and cetaceans." *Behavioral and Brain Sciences* 24:359–60.

———. 2011. "The scope of culture in chimpanzees, humans and ancestral apes." *Philosophical Transactions of the Royal Society of London*, ser. B 366:997–1007.

Whiten, A., J. Goodall, W. C. McGrew, T. Nishida, V. Reynolds, Y. Sugiyama, C. E. G. Tutin, R. W. Wrangham, and C. Boesch. 1999. "Cultures in chimpanzees." *Nature* 399:682–85.

———. 2001. "Charting cultural variation in chimpanzees." *Behaviour* 138:1481–516.

Whiten, A., V. Horner, C. Litchfield, and S. Marshall-Pescini. 2004. "How do apes ape?" *Learning and Behaviour* 32:36–52.

Whiten, A., V. Horner, and de Waal, F. B. M. 2005. "Conformity to cultural norms of tool use in chimpanzees." *Nature* 437:737–40.

Whiten, A., N. McGuigan, S. Marshall-Pescini, and L. M. Hopper. 2009. "Emulation, imitation, over-imitation and the scope of culture for child and chimpanzee." *Philosophical Transactions of the Royal Society of London*, ser. B 364:2417–28.

Whiten, A., A. Spiteri, V. Horner, K. E. Bonnie, S. P. Lambeth, S. Schapiro, and F. B. M. de Waal. 2007. "Transmission of multiple traditions within and between chimpanzee groups." *Current Biology* 17:1038–43.

Wilkinson, A., K. Kuenstner, J. Mueller, and L. Huber. 2010. "Social learning in a non-social reptile (*Geochelone carbonaria*)." *Biology Letters* 6:614–16.

Williams, R., and D. Lusseau. 2006. "A killer whale social network is vulnerable to targeted removals." *Biology Letters* 2:497–500.

Wilson, D. S., and L. A. Dugatkin. 1997. "Group selection and assortative interactions." *American Naturalist* 149:336–51.

Wilson, E. O. 1975. *Sociobiology: The New Synthesis*. Cambridge, MA: Belknap Press.

———. 1979. *On Human Nature*. Cambridge, MA: Harvard University Press.

———. 1994. *Naturalist*. Washington, DC: Island Press.

Wiszniewski, J., S. Corrigan, L. B. Beheregaray, and L. M. Möller. 2012. "Male reproductive success increases with alliance size in Indo-Pacific bottlenose dolphins (*Tursiops aduncus*)." *Journal of Animal Ecology* 81:423–31.

Wiszniewski, J., L. Beheregaray, S. Allen, and L. Möller. 2010. "Environmental and social influences on the genetic structure of bottlenose dolphins (*Tursiops aduncus*) in Southeastern Australia." *Conservation Genetics* 11:1405–19.

Worm, B., and R. A. Myers. 2003. "Meta-analysis of cod-shrimp interactions reveals top-down control in oceanic food webs." *Ecology* 84:162–73.

Worm, B., M. Sandow, A. Oschlies, H. K. Lotze, and R. A. Myers. 2005. "Global patterns of predator diversity in the open oceans." *Science* 309:1365–69.

Würsig, B. 2008. "Intelligence and cognition." In *Encyclopedia of Marine Mammals*, edited by W. F. Perrin, B. Würsig, and J. G. M. Thewissen, 616–23. 2nd ed. San Diego: Academic Press.

Würsig, B., and C. W. Clark. 1993. "Behavior." In *The Bowhead Whale*, edited by J. J. Burns, J. J. Montague, and C. J. Cowles, 157–99. Lawrence, KS: Society for Marine Mammalogy.

Würsig, B., E. M. Dorsey, M. A. Fraker, R. S. Payne, and W. J. Richardson. 1985. "Behavior of bowhead whales, *Balaena mysticetus*, summering in the Beaufort Sea: a description." *Fishery Bulletin* 83:357–77.

Würsig, B., and M. Würsig. 1980. "Behavior and ecology of the dusky dolphin, *Lagenorhynchus obscurus*, in the South Atlantic." *Fishery Bulletin* 77:871–90.

———. "Preface." In *The Dusky Dolphin: Master Acrobat off Different Shores*, edited by B. Würsig and M. Würsig, ix–xiii. Amsterdam: Elsevier/Academic Press.

Xitco, M., J. Gory, and S. Kuczaj. 2001. "Spontaneous pointing by bottlenose dolphins (*Tursiops truncatus*)." *Animal Cognition* 4:115–23.

———. 2004. "Dolphin pointing is linked to the attentional behavior of a receiver." *Animal Cognition* 7:231–38.

Yano, K., and M. E. Dahlheim. 1995. "Killer whale, *Orcinus orca*, depredation on long-line catches of bottomfish in the southeastern Bering Sea and adjacent waters." *Fishery Bulletin* 93:355–72.

Young Harper, J., and B. A. Schulte. 2005. "Social interactions in captive female Florida manatees." *Zoo Biology* 24:135–44.

Yurk, H., L. Barrett-Lennard, J. K. B. Ford, and C. O. Matkin. 2002. "Cultural transmission within maternal lineages: vocal clans in resident killer whales in southern Alaska." *Animal Behaviour* 63:1103–19.

Zachariassen, M. 1993. "Pilot whale catches in the Faroe Islands, 1709–1992." *Reports of the International Whaling Commission (Special Issue)* 14:69–88.

Zappes, C. A., A. Andriolo, P. C. Simoes-Lopes, and A. Di Beneditto Paula Madeira. 2011.

"Human-dolphin (*Tursiops truncatus* Montagu, 1821) cooperative fishery and its influence on cast net fishing activities in Barra de Imbe/Tramandai, Southern Brazil." *Ocean and Coastal Management* 54:427–32.

Zhou, W., D. Sornette, R. A. Hill, and R. I. M. Dunbar. 2005. "Discrete hierarchical organization of social group sizes." *Proceedings of the Royal Society of London*, ser. B 272:439–44.

Ziliak, S. T., and D. N. McCloskey. 2008. *The Cult of Statistical Significance: How the Standard Error Costs Us Jobs, Justice, and Lives*. Ann Arbor: University of Michigan Press.

INDEX

The letter *f* following a page number denotes a figure, the letter *t* following a page number denotes a table, and the letter *n* following a page number denotes a note.